Statistics for Biology and Health

Series Editors
Mitchell Gail
Jonathan M. Samet
B. Singer
Anastasios Tsiatis

More information about this series at http://www.springer.com/series/2848

Richard K. Burdick • David J. LeBlond
Lori B. Pfahler • Jorge Quiroz • Leslie Sidor
Kimberly Vukovinsky • Lanju Zhang

Statistical Applications for Chemistry, Manufacturing and Controls (CMC) in the Pharmaceutical Industry

 Springer

Richard K. Burdick
Elion Labs
Louisville, Colorado, USA

David J. LeBlond
CMC Statistics
Wadsworth, IL, USA

Lori B. Pfahler
Merck & Co., Inc.
Telford, PA, USA

Jorge Quiroz
Merck & Co., Inc.
Kenilworth, NJ, USA

Leslie Sidor
Biogen
Cambridge, Massachusetts, USA

Kimberly Vukovinsky
Pfizer
Old Saybrook, Connecticut, USA

Lanju Zhang
Nonclinical Statistics, Abbvie Inc.
North Chicago, IL, USA

ISSN 1431-8776
Statistics for Biology and Health
ISBN 978-3-319-84338-4
DOI 10.1007/978-3-319-50186-4

ISSN 2197-5671   (electronic)

ISBN 978-3-319-50186-4   (eBook)

Printed on acid-free paper

This Springer imprint is published by Springer Nature
The registered company is Springer International Publishing AG
The registered company address is: Gewerbestrasse 11, 6330 Cham, Switzerland

Contents

Chapter 1
Introduction

Keywords Chemistry, Manufacturing, and Controls (CMC) statistics • Clinical statistics • Pharmaceutical activities • Regulatory guidance • Statistical methods

1.1 Objectives

The motivation for this book came from an American Association of Pharmaceutical Scientists (AAPS) short course on statistical methods applied to Chemistry, Manufacturing, and Controls (CMC) applications presented by four of the authors. One of the course participants asked us for a good reference book, and the only book we could recommend was written over 20 years ago by Chow and Liu (1995). We agreed that a more recent book would serve a need in our industry. This book presents statistical techniques that are critically important to CMC activities.

Statistical methods are presented with a focus on applications unique to the CMC pharmaceutical industry. The target audience consists of statisticians and other scientists who are responsible for performing statistical analyses within a CMC environment. Basic statistical concepts are addressed in Chap. 2 followed by applications to specific topics related to development and manufacturing. The mathematical level assumes an elementary understanding of statistical methods. The ability to use Excel or statistical packages such as Minitab, JMP, or *R* will provide more value to the reader.

Since we began this project, an edited book has been published on the same topic by Zhang (2016). The chapters in Zhang discuss statistical methods for CMC as well as drug discovery and nonclinical development. We believe our book complements Zhang by providing more detailed statistical analyses and examples.

1.2 Regulatory Guidance for CMC Applications

Persons responsible for statistical analyses in CMC applications should be familiar with guidance and regulations that pertain to the pharmaceutical industry. The legality of CMC issues is covered in the Code of Federal Regulations (CFR),

© Springer International Publishing AG 2017
R.K. Burdick et al., *Statistical Applications for Chemistry, Manufacturing and Controls (CMC) in the Pharmaceutical Industry*, Statistics for Biology and Health, DOI 10.1007/978-3-319-50186-4_1

Title 21, Food and Drugs Administration (FDA). Several relevant sections of this code are reported in Table 1.1.

In addition to the CFR, regulatory agencies have produced a number of useful documents that direct the approaches used in statistical analysis. Tables 1.2, 1.3, 1.4, and 1.5 report documents referenced and discussed in this book.

1.3 Use of Statistical Tools in Pharmaceutical Development and Manufacturing

This book focuses on statistical methods used in the development and manufacturting of pharmaceutical products. An excellent description of this area is presented by Peterson et al. (2009). Pharmaceutical products are developed over five parallel activities:

1. Clinical trials,
2. Preclinical assessment,
3. Active pharmaceutical ingredient (API) development,
4. Drug product (DP) formulation, and
5. Analytical method development.

Table 1.1 Important sections of 21 CFR

Source	Title
Code of Federal Regulations, Title 21, Food and Drugs Administration (FDA), Part 210 (21 CFR 210)	Current good manufacturing practice in manufacturing, processing, packing, or holding of drugs
21 CFR 211	Current good manufacturing practice for finished pharmaceuticals
21 CFR 600	Biological products: general
21 CFR 820	Quality system regulations

Table 1.2 Useful regulatory statistical guidance ASTM international

Title	Chapter
E29: Standard practice for using significant digits in test data to determine conformance to specifications	2
E2281: Standard practice for process capability and performance measurement	5
E2475: Standard guide for process understanding related to pharmaceutical manufacture and control	4
E2587: Standard practice for use of control charts in statistical process control	5
E2709: Standard practice for demonstrating capability to comply with an acceptance procedure	7
E2810: Standard practice for demonstrating capability to comply with the test for uniformity of dosage units	7

Table 1.3 Useful regulatory statistical guidance Food and Drug Administration, Center for Drugs Evaluation Research (FDA,CDER)

Title	Chapter
Guidance for industry: immediate release solid oral dosage forms, scale-up and postapproval changes: chemistry, manufacturing and controls, in vitro dissolution testing, and in vivo bioequivalence documentation (1995)	7
Guidance for industry: demonstration of comparability of human biological products, including therapeutic biotechnology-derived products (1996)	9
Guidance for industry: SUPAC-MR modified release solid oral dosage forms, scale-up and postapproval changes: chemistry, manufacturing and controls, in vitro dissolution testing, and in vivo bioequivalence documentation (1997a)	7
Guidance for industry: dissolution testing of immediate release solid oral dosage forms (1997b)	7
Guidance for industry: ANDAs: blend uniformity analysis (1999 withdrawn 2002)	7
Guidance for industry: powder blend and finished dosage units—stratified in-process dosage unit sampling and assessment (October 2003 withdrawn 2013)	7
Guidance for industry: process validation: general principles and practices (2011)	3, 5, 6, 9
Guidance for industry: quality considerations in demonstrating biosimilarity of a therapeutic protein product to a reference product (2015a)	9
Guidance for industry: scientific considerations in demonstrating biosimilarity to a reference product (2015b)	9
Guidance for industry: biosimilars: questions and answers regarding implementation of the biologics price competition and innovation act of 2009 (2015c)	9
Guidance for industry: analytical procedures and methods validation for drugs and biologics (2015d)	6

Figure 1.1 from Peterson et al. displays the timeline for these activities.

While the most common area for statisticians to work is in the clinical area (activities 1 and 2), the focus of this book is on paths 3–5 in Fig. 1.1. Key research questions and statistical methods used to help answer them are shown in Table 1.6.

Statistical quality control methods are applied throughout all activities in Phase IV. These methods are discussed in Chap. 5.

1.4 Differences Between Clinical and CMC Statisticians

To better understand the nature of CMC statisical analysis, it is useful to contrast this work to that of the clinical statistician. The role of a clinical statistician is well established. It is required and integrated into regulations and internal business processes. Often, these predefined roles and responsibilities are outlined in company procedures. Given the key role they play in the clinical drug development process, the clinical statistician is well linked into clinical project teams with strong management support. Among their many responsibilities, clinical statisticians are responsible for statistical design of clinical trials and statistical analysis plans included in protocols which are sent to the FDA for review. These protocols are

Table 1.4 Useful regulatory statistical guidance International Conference on Harmonization (ICH)

Title	Chapter
Q5C stability testing of biotechnological/biological products (1995)	8
Q1B photostability testing of new drug substances and products (1996)	8
Q1C stability testing for new dosage forms (1997)	8
Q6A specifications: test procedures and acceptance criteria for new drug substances and new drug products: chemical substances (1999a)	3, 7, 8
Q6B specifications: test procedures and acceptance criteria for biotechnological/ biological products (1999b)	7, 8
Q7 good manufacturing practice guide for active pharmaceutical ingredients (2000)	4
Q1D bracketing and matrixing designs for stability testing of new drug substances and products (2002)	8
Q1A(R2) stability testing of new drug substances and products (2003a)	8
Q1E evaluation for stability data (2003b)	7, 8
Q3A impurities in new drug substances (2003c)	8
Q3B (revised) impurities in new drug products (2003d)	8
Q5E comparability of biotechnological/biological products subject to changes in their manufacturing process (2004)	2, 9
Q2(R1) validation of analytical procedures: text and methodology (2005a)	6, 8
Q9 quality risk management (2005b)	3–5
Q10 pharmaceutical quality system (2008)	3, 5
Q8(R2) pharmaceutical development (2009)	3, 5
Q11 development and manufacture of drug substances (chemical entities and bio-technological/biological entities) (2012)	3

Table 1.5 Useful regulatory statistical guidance United States US Pharmacopeial (USP)

Title	Chapter
⟨905⟩ Uniformity of dosage units	7
⟨1010⟩ Analytical data—interpretation and treatment	2, 6
⟨1030⟩ Biological assay chapters—overview and glossary	6
⟨1032⟩ Design and development of biological assays	6
⟨1033⟩ Biological assay validation	6
⟨1160⟩ Pharmaceutical calculations in prescription compounding	8
⟨1223⟩ Validation of alternative microbiological methods	6
⟨1224⟩ Transfer of analytical procedures	6
⟨1225⟩ Validation of compendial procedures	6
General notices 3.10: conformance to standards, applicability of standards	7
General notices 7.20: rounding rules	2, 7

very detailed and provide clear articulation of the exact analyses to be followed and the specific endpoints that must be met for clinical success. These protocols typically are based on regulatory requirements. The FDA has a team of statistical reviewers that evaluates the protocols and the definitive pass/fail nature of these

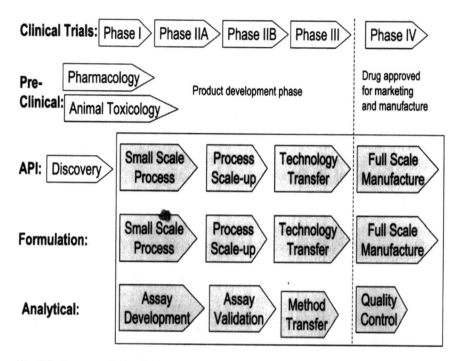

Fig. 1.1 Pharmaceutical activities

Table 1.6 Research questions and statistical methods

Activity	Research questions	Book chapter with application
API development	How can API be scaled to a level required for commercial market?	4
API development	How do we effectively characterize our product and develop a knowledge base for the API?	3, 4
DP formulation	What shelf life limits and specifications need to be established to ensure DP is safe and efficacious?	7, 8
DP formulation	If our process is transferred to another manufacturing site, how do we ensure product safety and efficacy are not impacted?	9
Analytical method development	How do we determine if analytical methods are fit for use?	6
Analytical method development	How do we know if a method will perform the same in two different labs?	6

criteria heightens the criticality of the statistician on the cross-functional team. There is much external guidance that must be followed and the clinical statisticians must follow strict documentation expectations. Data systems are well developed and SAS and R are common data analysis tools. Other software packages are rarely used except for occasional exploratory work. Upon completion of the clinical study, the clinical statistician co-authors reports with clinical colleagues which are included in the regulatory submissions. Given the existing system, the clinical statistician does not spend a significant portion of time training their clinical colleagues to perform their own statistical analyses. Most of their interaction with colleagues involves discussions to help them understand and correctly interpret the statistical analyses completed by the statistician. In order to be the most successful, clinical statisticians must understand the science of the disease so that they can add value to the project team.

Although there are similarities between the roles of the CMC and clinical statisticians, there are differences in both type and degree. CMC statisticians work with scientists to help develop new drug substance and drug product manufacturing processes, develop and validate analytical procedures, improve existing processes and products, and troubleshoot systems when issues arise. Unlike the clinical statistician, the role of the CMC statistician does not have a regulatory requirement and as such, the nature of the role can vary both within and across companies. Factors impacting these differences include the technical and interpersonal skills of the statistician, the nature of management support, and the strength of the partnership with individual collaborators. Small, relatively short duration studies are common as opposed to large clinical trials and access to large data sets created to satisfy Good Manufacturing Practice (GMP) or regulatory requirements are the exception rather than the rule. Unlike the large and well-defined statistical departments in the clinical organization, CMC statisticians often work alone or in very small groups. The CMC statistician may report to management in the area they support, or to the broader clinical organization. Good documentation practices are important for the CMC statistician to adhere to GMP requirements but there are no statistical protocols sent to regulatory agencies for review. In the CMC area, studies are performed and documented internally so as to be available if requested by regulatory agencies. Documentation in these reports must be clear so that analyses can be explained and reproduced when necessary. Given that data sets are often small, data analysis packages such as Minitab and JMP are commonly used to perform calculations. SAS and R are also employed with larger data sets, or when requests are made from regulatory agencies. Because the CMC statistical workforce is relatively small, CMC statisticians spend time teaching their scientific colleagues how to peform their own statistical analyses. This is one reason why statistical packages that do not require written code (e.g., JMP and Minitab) are often selected for analysis. Similar to the clinical statistician, the CMC statistician often contributes to the contents of a regulatory submission. CMC statisticians help write sections describing process and formulation development, stability, justification of specifications, process and product comparability, and analytical method validations. Similar to clinical statisticians, the CMC statistician must understand the science and engineering concepts of their collaborators in order to be successful.

1.5 How to Use This Book

It is possible to gain a working understanding of the methods in this book with no advanced statistical training. In fact, one objective of this book is to make these methods available for scientists who do not possess a degree in statistics. Professional statisticians will also find it helpful to have these methods in a single source for their own use and for training others.

Chapter 2 provides statistical methods that are useful for performing the analyses required to address research questions in the CMC manufacturing environment. We recommend that the reader begin by reading Sects. 2.1–2.5. This provides both a high level view of statistical applications and some specific examples for simple data sets. After reading this material, the reader may complete Chap. 2, or jump to any particular application of interest in Chaps. 3–9. Where needed, Chaps. 3–9 refer back to statistical methods in Chap. 2 where the reader is provided a more thorough understanding of the statistical method. Worked numerical examples are provided in all chapters.

The reader may use any number of statistical packages to help work the examples. Many of the examples can be performed using Excel. Some require user-friendly statistical packages, such as Minitab and JMP. Additionally, we have provided data sets and program codes written in SAS and R at the website for many of the examples. Discussions of the examples will focus on the output rather than specific code used to generate the output.

References

ASTM E29 (2013) Standard practice for using significant digits in test data to determine conformance with specifications. ASTM International, West Conshohocken

ASTM E2281 (2015) Standard practice for process capability and performance measurement. ASTM International, West Conshohocken

ASTM E2475 (2010) Standard guide for process understanding related to pharmaceutical manufacture and control. ASTM International, West Conshohocken

ASTM E2587 (2016) Standard practice for use of control charts in statistical process control. ASTM International, West Conshohocken

ASTM E2709 (2014) Standard practice for demonstrating capability to comply with an acceptance procedure. ASTM International, West Conshohocken

ASTM E2810 (2011) Standard practice for demonstrating capability to comply with the test for uniformity of dosage units. ASTM International, West Conshohocken

Chow S-C, Liu J-P (1995) Statistical design and analysis in pharmaceutical science: Validation, process controls, and stability. Marcel Dekker, Inc., New York

Code of Federal Regulations (CFR) (2003)Title 21, Food and Drugs Administration (FDA), Part 210, http://www.accessdata.fda.gov/scripts/cdrh/cfdocs/cfcfr/CFRSearch.cfm?CFRPart=210. Accessed 21 Sept 2015

Food and Drug Administration. Center for Drugs Evaluation Research (1995) Immediate release solid oral dosage forms, scale-up and postapproval changes: chemistry, manufacturing and controls, in vitro dissolution testing, and in vivo bioequivalence documentation, guidance for industry

Food and Drug Administration. Center for Drugs Evaluation Research (1996) Demonstration of comparability of human biological products, including therapeutic biotechnology-derived products, guidance for industry

Food and Drug Administration. Center for Drugs Evaluation Research (1997a) SUPAC-MR modified release solid oral dosage forms, scale-up and postapproval changes: chemistry, manufacturing and controls, in vitro dissolution testing, and in vivo bioequivalence documentation, guidance for industry

Food and Drug Administration. Center for Drugs Evaluation Research (1997b) Dissolution testing of immediate release solid oral dosage forms, guidance for industry

Food and Drug Administration. Center for Drugs Evaluation Research (1999) withdrawn 2002 ANDAs: blend uniformity analysis, guidance for industry.

Food and Drug Administration. Center for Drugs Evaluation Research (2003) withdrawn 2013 Powder blend and finished dosage units—stratified in-process dosage unit sampling and assessment, guidance for industry

Food and Drug Administration. Center for Drugs Evaluation Research (2011) Process validation: general principles and practices, guidance for industry

Food and Drug Administration. Center for Drugs Evaluation Research (2015a) Quality considerations in demonstrating biosimilarity of a therapeutic protein product to a reference product, guidance for industry

Food and Drug Administration. Center for Drugs Evaluation Research (2015b) Scientific considerations in demonstrating biosimilarity to a reference product, guidance for industry

Food and Drug Administration. Center for Drugs Evaluation Research (2015c) Questions and answers regarding implementation of the biologics price competition and innovation act of 2009, guidance for industry

Food and Drug Administration. Center for Drugs Evaluation Research (2015d) Analytical procedures and methods validation for drugs and biologics, guidance for industry

International Conference on Harmonization (1995) Q5C stability testing of biotechnological/biological products

International Conference on Harmonization (1996) Q1B photostability testing of new drug substances and products

International Conference on Harmonization (1997) Q1C stability testing for new dosage forms

International Conference on Harmonization (1999a) Q6A specifications: test procedures and acceptance criteria for new drug substances and new drug products: chemical substances

International Conference on Harmonization (1999b) Q6B specifications: test procedures and acceptance criteria for biotechnological/biological products

International Conference on Harmonization (2000) Q7 good manufacturing practice guide for active pharmaceutical ingredients

International Conference on Harmonization (2002) Q1D bracketing and matrixing designs for stability testing of new drug substances and products

International Conference on Harmonization (2003a) Q1A (R2) stability testing of new drug substances and products

International Conference on Harmonization (2003b) Q1E evaluation for stability data

International Conference on Harmonization (2003c) Q3A impurities in new drug substances

International Conference on Harmonization (2003d) Q3B (Revised) impurities in new drug products

International Conference on Harmonization (2004) Q5E comparability of biotechnological/biological products subject to changes in their manufacturing process

International Conference on Harmonization (2005a) Q2 (R1) validation of analytical procedures: text and methodology

International Conference on Harmonization (2005b) Q9 quality risk management

International Conference on Harmonization (2008) Q10 pharmaceutical quality system

International Conference on Harmonization (2009) Q8 (R2) pharmaceutical development

International Conference on Harmonization (2012) Q11 development and manufacture of drug substances (chemical entities and biotechnological/biological entities)

Peterson JJ, Snee RD, McAllister PR, Schofield TL, Carella AJ (2009) Statistics in pharmaceutical development and manufacturing. J Qual Technol 41(2):111–147

USP 39-NF 34 (2016a) General chapter <905> uniformity of dosage units. US Pharmacopeial Convention, Rockville

USP 39-NF 34 (2016b) General chapter <1010> analytical data—interpretation and treatment. US Pharmacopeial Convention, Rockville

USP 39-NF 34 (2016c) General chapter <1030> biological assay chapters—overview and glossary. US Pharmacopeial Convention, Rockville

USP 39-NF 34 (2016d) General chapter <1032> design and development of biological assays. US Pharmacopeial Convention, Rockville

USP 39-NF 34 (2016e) General chapter <1033> biological assay validation. US Pharmacopeial Convention, Rockville

USP 39-NF 34 (2016f) General chapter <1160> pharmaceutical calculations in prescription compounding. US Pharmacopeial Convention, Rockville

USP 39-NF 34 (2016g) General chapter <1223> validation of alternative microbiological methods. US Pharmacopeial Convention, Rockville

USP 39-NF 34 (2016h) General chapter <1224> transfer of analytical procedures. US Pharmacopeial Convention, Rockville

USP 39-NF 34 (2016i) General chapter <1225> validation of compendial procedures. US Pharmacopeial Convention, Rockville

USP 39-NF 34 (2016j) General notices 3.10: conformance to standards, applicability of standards. US Pharmacopeial Convention, Rockville

USP 39-NF 34 (2016k) General notices 7.20: rounding rules. US Pharmacopeial Convention, Rockville

Zhang L (ed) (2016) Nonclinical statistics for pharmaceutical and biotechnology industries. Springer, Heidelberg

Chapter 2
Statistical Methods for CMC Applications

Keywords Analysis of variance • Bayesian analysis • Confidence intervals • Data reporting • Data rounding • Data transformations • Dependent measures • Equivalence testing • Hypothesis testing • Interaction effects • LOQ values • Mixed models • Multiple regression • Nonlinear models • Non-normal data • Prediction intervals • Quadratic effects • Regression analysis • Residual analysis • Statistical consulting • Statistical intervals • Tolerance intervals • Visualization of data

2.1 Introduction

In this chapter, we provide statistical methods that are useful in CMC applications. Our goal is to provide a description of these methods without delving deeply into the theoretical aspects. References are provided for the reader who desires a more in depth understanding of the material.

2.2 Statistical Analysis in a CMC Environment

As described in Chap. 1, CMC statisticians work directly with individual subject matter experts (SME) from other areas of science. In this section, we describe a typical example of this interaction from the viewpoint of the statistician. Here "statistician" is defined as the person who is responsible for performing the required statistical analyses. This need not be a person with a terminal degree in statistics. The term "client" is used to represent the SME requiring help in performing the statistical analysis. The statistical analysis process consists of the following four steps:

1. Initial client meeting.
2. Planning of statistical analysis.
3. Data analysis.
4. Communication of results to the client.

Each of these steps is now described in the context of an example.

© Springer International Publishing AG 2017
R.K. Burdick et al., *Statistical Applications for Chemistry, Manufacturing and Controls (CMC) in the Pharmaceutical Industry*, Statistics for Biology and Health, DOI 10.1007/978-3-319-50186-4_2

2.2.1 Initial Client Meeting

The initial meeting with a client is where everything begins. The goal for the statistician is to listen to the client and learn about the problem. The statistician should listen for at least 80% of the conversation, and ask clarifying questions for the other 20%. This is not the time to overwhelm the client with statistical jargon and tales of the sophisticated power of a pet statistical procedure. One of the benefits of being a statistician is that you have the opportunity to work with clients in a variety of fields. So take advantage of these sessions and learn something new. You will have an opportunity later in the process to share your expertise in statistics.

An example of a typical initial meeting is shared below. The statistician, Tom, has met the client, Noel, in Noel's office. Noel is a product quality leader who is responsible for all quality issues related to a particular product.

Tom: Good morning, Noel. How have you been doing?

Noel: Fine. Just trying to keep up with all the work.

Tom: Tell me about it! (Note: There is always the obligatory greeting that suggests both parties are the hardest working people in the company.) What can I help you with today?

Noel: As you know, we are transferring our manufacturing process to Ireland, and are in the middle of the comparability phase of the transfer. We need to demonstrate to regulatory agencies that once the process is operating in Ireland, it will be manufacturing product of a similar quality to our present process with no risks to patient safety or product efficacy.

Tom: Right. I have helped on these types of projects in the past. What are some of the details?

Noel: We have produced eight lots in Ireland and want to compare the lot release values for some of our quality attributes with those we have collected from our process here in the USA. The two processes don't have to be exactly the same, but as I have already mentioned, we cannot compromise patient safety or product efficacy. One thing we do to ensure our present process continues to provide safe and efficacious product is to demonstrate that the average purity is no less than 93%. Anything less than 93% would be sufficiently different from our present process that I would be concerned.

Tom: Thanks. So if our process in Ireland has a mean of 93% or greater, we believe the new process is operating as expected. Right?

Noel: Exactly.

Tom: What data do we have available?

Noel: Purity is measured using the reversed-phase high performance liquid chromatography (RP-HPLC) main peak. I have these values for each of the eight lots produced at the Ireland plant. Do you think eight values is a large enough data set to draw any meaningful conclusion?

Tom: Well, eight lots is better than seven lots but not as good as nine lots. We can perform calculations with the available eight lots, but the uncertainty may be too large to provide meaningful results. You will have to make that determination. If there is too much uncertainty to be useful, you will need more data.

Noel: Right. We would like to know now if there are any apparent issues at this point, so let's see what the eight lots tell us.

Tom: Sounds good.

Noel: I will get the data to you as soon as possible. I know you like the data in a certain format. Can you remind me how to prepare it for you?

Tom: Thanks, Noel. It lessens the opportunity for errors if you get me the data in the proper format. I would like the data placed in two columns of an Excel spreadsheet. The first column will be the lot number and the second column will be the purity value. The first row of the spreadsheet will have the label for each column and then there will be eight rows of data. Make sure to report the recorded values for your measurements.

Noel: That doesn't sound too hard. I will get the data verified and then place the spreadsheet in the company information system for you to access.

Tom: That sound's great. What kind of timeframe do we have?

Noel: It should take me about a week to get everything verified, approved, and into the system. If you could do your magic within a week after receiving the data, it would give us time to assess the results and move on to the next step.

Tom: That sounds workable. Please send me a note when the data are ready. Talk to you later.

Noel: Thanks, Tom. I look forward to talking to you again in a couple of weeks.

2.2.2 Planning of Statistical Analysis

After meeting with the client, it is time to formulate a strategy for answering the research question. To do this, it is often necessary to make a statistical inference. Statistical inference concerns the ability to answer a question based on a collected data set. The research question concerns a collection of items called the population or process output. For purposes of our discussion, a population is a finite collection of items, such as all vials within a manufactured lot of drug product. A process is a series of actions or operations used to convert a set of inputs into a set of outputs. The process output over a fixed interval of time constitutes a population of items. When thinking of a process, we often want to make a statement about future items that will result if the process operates in a manner consistent with the observed data. In CMC applications, we are often interested in process outputs where one set of outputs is created with an existing process and the other set of output is created with a new or improved process. The planning of a statistical analysis should consider three components:

1. Statement of the study objective.
2. Data acquisition.
3. Selection of a statistical tool.

2.2.2.1 Statement of the Study Objective

One result of the initial client meeting is an understanding of the research question. In order to develop an effective statistical strategy, it is necessary for the statistician and client to agree on a clear research objective. In working with the client, it is sometimes helpful to ask them how they intend to use the results of the statistical analysis. For example, ask the client what action she would take if presented with a certain outcome of the statistical analysis. Similarly, ask the client what action he would take based on an alternative outcome. If such questions cannot be answered, then the research question is not well-defined, and more discussion is needed to clarify the study objective. The research question in our present example is "Does the Ireland manufacturing process operate at an acceptable capability?" One piece of information that will be used to help answer this question is the process mean for purity. Once the research question has been clearly defined, the next step is to determine the most appropriate data to answer the question.

2.2.2.2 Data Acquisition

The following questions are worth consideration in any discussion of data acquisition.

1. What is the population or process of interest, and how can we ensure the collected data are representative of this group?
   Require an exact definition of the population/process including the time period of interest. In the present example, Noel wants to know something about long-run behavior of a process for which eight items presently exist. If it is planned to run the same process in the future, then the eight available lots are likely representative of future output. The assessment of whether a data set is representative is *not* a statistical question. It requires the judgment of an expert in the field of application (i.e., Noel in our example).
2. Do the data already exist (observational), or will we have to create it experimentally?
   Although the answer to this question is often based on convenience or timeliness, it is important to understand the different types of inference that can be made with each type of data. Generally speaking, inferences of a causal nature require experimental data. That is, if one wishes to provide evidence that changes in factor X cause change in factor Y, then an experimental data set is required in order to properly isolate the relationship of Y and X and protect it from other factors. Although observational data cannot directly demonstrate causality, it does provide a description of how variables relate to each other in the "real world." To better explain in the context of our example, the lot release data is observational since it was collected as part of the manufacturing process. It does not require a separate set of experimental studies to generate the data. In the analysis that follows, we are able to estimate the mean purity of the Ireland

process. However, if it is lower than desired, we have not learned anything to help us determine how to increase the mean and improve the process. To do this, we would likely need a set of experimental studies where inputs of the process are systematically changed to determine the impact on the mean purity. Data generated in this manner is an example of experimental data. In the CMC world, experimental data is generally required in process development, whereas observational data are used in quality operations.

3. What sampling method is used to collect the data?

It is necessary that the data used for analysis be collected using a valid statistical sampling procedure. There are many different types of sampling procedures, and the best approach in any given situation will depend on the structure and availability of the data. Natural groupings of population items often impact the manner in which a statistical sample is selected. Consider the following example. As part of a method validation study, four plates are prepared, each one containing three aliquots. Measurements made on the three aliquots in the same plate are typically more similar than measurements of aliquots in different plates. For this reason, the statistical analysis must account for this relationship. Had the 12 measurements been collected using 12 plates with one aliquot each, a different statistical analysis would be performed. The larger physical units (plate in this example) are called experimental units, and the smaller units (aliquots) are called observational units. Observational units are what are actually measured. In many situations, the experimental unit is the observational unit. As noted, it is always important to identify any grouping of observational units in order to perform the correct statistical analysis.

4. How are the variables defined and measured?

A variable provides information of interest about each individual item in a population. Variables can provide a number (quantitative variable) or a category (qualitative variable). It is important that all variable definitions be included in the statistical report. In our present example, the variable assigns a number that represents lot purity as measured by the RP-HPLC main peak. We might label this variable "Purity." Each variable in a study must be measured for each sampled item.

When one measures the value of a continuous variable, the measured value is never *exactly* equal to the true value. The difference between the measured value and the true value is called measurement error. We say that a measurement procedure is unbiased if there is no systematic tendency for the procedure to underestimate or overestimate the true value. For example, if the weight of 100 people were measured using an unbiased scale, we would expect the reported weight to be greater than the true value for some people and less than the true value for others. On average, the error in measurement would be essentially zero. However, if the scale is not calibrated properly, it might consistently yield readings that are lower (or higher) than the true weights. In such a case, the scale is said to be biased.

In addition to having a small amount of bias, a good measuring device is precise. Precision concerns the reliability of a measurement procedure. If one measures the same item several times, it is hoped that the measurements are more or less the same each time and do not fluctuate wildly. Measurement procedures that exhibit little variability when measuring the same item are said to be precise. One of the reasons that analytical procedure validations are performed is to ensure that the bias and the precision are acceptable. Chapter 6 concerns the validation of analytical procedures.

Another question to ask concerning quantitative variables is whether the numerical values are recorded to enough decimal places to allow a meaningful analysis. It is recommended to always perform calculations with recorded data, rather than data that have been reported to a lesser degree of accuracy. More on this topic is provided in Sect. 2.3.

5. What sample size is required to provide a useful analysis?

This is probably the most frequent question asked of a statistician. The statistician can never answer this question in isolation. It always requires interaction between the client and the statistician. The thing to remember is that as sample size increases, then uncertainty in the statistical inference will decrease. At some point, one must balance the cost and time of selecting additional sample items against the risks associated with an uncertain decision. In many CMC applications, sample sizes are small because a single item, such as a production lot, is produced infrequently or is extremely expensive. Valid statistical analyses can be performed with relatively small data sets, but it is important to report the level of uncertainty in order to fairly represent the usefulness of the conclusions.

2.2.2.3   Selection of a Statistical Tool

Once the research question is well defined and the relevant data have been identified, it is time to select a statistical procedure to help answer the research question. For the Ireland manufacturing problem discussed earlier, it is necessary to estimate the mean of the new process using the data set of eight values. We will use the sample mean of eight production runs to provide a point estimate of the true process mean. A point estimate is useful, but we need to recognize that it is based on only eight process lots. Thus, a point estimate has uncertainty associated with it. One quantifies this uncertainty by computing a statistical interval that provides a range of possible values. The various types of statistical intervals are introduced in Sect. 2.5. For the present example, a confidence interval is the appropriate statistical interval. A confidence interval contains the true unknown value of the purity mean with an associated level of confidence (e.g., 95%). The confidence level of 95% describes the ability of the confidence interval to correctly capture the true value of the purity mean. More discussion is provided on the interpretation of the confidence coefficient in Sect. 2.5.1.

In our example, if all values in the confidence interval exceed 93%, we will conclude that the new process has attained the desired mean. Now that our strategy has been determined, we wait for the data.

2.2.3   Data Analysis

Data analysis consists of both graphical representation of the data and numerical description. Data analysis consists of the following operations:

1. Obtain the data.
2. Plot the data.
3. Estimate the unknown quantities of interest.
4. Quantify uncertainty in point estimates using a statistical interval.

We demonstrate this process within the context of the previous example.

2.2.3.1   Obtain the Data

The big day arrives and Tom receives an email that the data are ready for analysis. The data are presented in Table 2.1.

True to his word, Noel provided the data in a format that is amenable for performing the statistical analysis. Each column in the table represents a variable. The name for each variable is found in the first row. For example, the first column in Table 2.1 reports values for the variable "Lot." The variable "Lot" is a qualitative variable because the values assigned are categories. The variable "Purity" in the second column is quantitative or numerical. Since the unit of measurement for purity is a percentage, it can assume any numerical value between 0 and 100%, inclusive. Each lot is represented by a single purity value as reported by the lab.

2.2.3.2   Plot the Data

Lynne Hare is a well-known statistician who established a rule that should be followed in every data analysis. The rule is termed the "Always, always, always--without exception rule" (AAAWER). The AAAWER is simple—Plot the data. Some years after creation of the AAAWER, he added the corollary—"and look at it." It is an old adage that a picture is worth a thousand words, but it is really true in

Table 2.1 Purity measures (%) from the Ireland process

Lot	Purity (%)
A	94.20
B	92.68
C	94.47
D	94.14
E	95.17
F	94.47
G	94.14
H	95.17

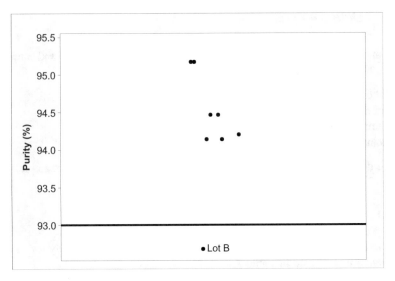

Fig. 2.1 Plot of purity data (%)

data analysis. Most learnings are obtained by simply plotting the data (and looking at the plot). We begin the analysis of the purity data by constructing a plot. The plot is shown in Fig. 2.1.

Recall that the objective of the study is to determine if the process mean in Ireland is at least 93%. Looking at the plot, all of the purity values except one exceed 93%, so it is clear that the mean of the eight values exceeds 93%.

Note that the purity values are all different. It is useful when looking at data to ask yourself the question, "Why aren't these values all the same?" There are two primary reasons why these eight values are not all equal.

1. There is expected variation from lot-to-lot for any manufacturing process.
2. There is measurement error in the data.

As in this example, most variables encountered in CMC applications are impacted by variation from both the manufacturing process and the analytical (measurement) method.

Another aspect to note about the plot is that Lot B is somewhat removed from the other seven lots. Points that are inconsistent with the rest of the data are called outliers. Outliers may arise due to coding errors, or perhaps due to some unique event that occurred during the manufacturing of a particular lot. The SME should be consulted immediately to discuss possible reasons for any detected outlier. Sections 2.4 and 2.12.2 provide more discussion on the statistical definition and identification of outliers.

Table 2.2 Definition of parameters

Parameter symbol	Interpretation
μ (Lower case Greek letter "mu")	Process mean at Ireland site
σ^2 (Lower case Greek letter "sigma" squared	Process variance at Ireland site
σ (Lower case Greek letter "sigma")	Process standard deviation at Ireland site

2.2.3.3   Estimate the Unknown Quantities of Interest

As discussed earlier, one objective of a statistical analysis is to make an inference (conclusion) about an unknown characteristic of a population or process. Both populations and processes are described by summary measures called parameters. The parameter of interest in the example is the process mean. The values of the eight completed lots represent a sample of all process outcomes. One can compute the mean for the eight lots in Table 2.1, but since this is only a sample, one cannot know with certainty the exact value of the true process mean.

It is a convention in the statistical literature to represent an unknown parameter value with a Greek letter. Table 2.2 shows Greek notation used to represent three parameters in the example. Think of the Greek symbol as you would an acronym. It is simply a shorthand notation to replace all the words in column 2 of Table 2.2.

Let the measured value of the variable "Purity" for lot i be represented as Y_i. Since there are eight lots, the index i goes from 1 to 8, inclusive. The sample of eight measured values is represented as $Y_1, Y_2, \ldots, Y_8$. The sample mean is computed by adding all eight values and dividing by the number of values (8). Representing as a formula we have

$$\bar{Y} = \frac{\sum_{i=1}^{n} Y_i}{n} = \frac{94.20 + \cdots + 95.17}{8} = 94.305\% \qquad (2.1)$$

where n is the sample size. Summary measures computed from a sample such as $\bar{Y}$ are referred to as statistics. A point estimator is a function of one or more statistics used to provide a "best guess" of the true (unknown) parameter value. For example, the sample mean $\bar{Y}$ is the point estimator of the true and unknown process mean, μ. The realized value of $\bar{Y}$, 94.305, is the point estimate for μ.

A point estimator for the process standard deviation, σ, is the sample standard deviation

$$S = \sqrt{\frac{\sum_{i=1}^{n}(Y_i - \bar{Y})^2}{n-1}} = \sqrt{\frac{(94.20 - 94.305)^2 + \cdots + (95.17 - 94.305)^2}{8-1}}$$
$$= 0.780\%. \qquad (2.2)$$

The square of the standard deviation is called the variance, and S^2 is the point estimator for σ^2.

The point estimate for the process mean, 94.305 %, exceeds the required value of 93%. However, we must ask the question, "How close to the true value of μ is the estimate?"

2.2.3.4 Quantify the Uncertainty in the Point Estimates Using a Statistical Interval

Point estimates are useful, but they do not tell us anything about the uncertainty associated with the estimates. Recall we are sampling only eight lots from the process, and we wish to make an inference concerning future lots. The best way to quantify the uncertainty in this example is to compute a statistical confidence interval. For this example, the following formula provides a 95% lower bound on the true process mean (see Eq. (2.8) later in this chapter).

$$L = \bar{Y} - 1.895 \times \frac{S}{\sqrt{n}}. \tag{2.3}$$

In our example,

$$L = 94.305 - 1.895 \times \frac{0.780}{\sqrt{8}} = 93.78\%. \tag{2.4}$$

The 95% lower bound exceeds the 93% criterion, providing statistical evidence that the process mean will exceed 93% if the process continues to operate in the future as it has in the past.

2.2.4 Communication of Results to Client

The data analysis is complete and now it is time for another visit with the client. Unlike the first visit where the statistician took the role of listener, the statistician now assumes the role of teacher.

Tom: Good morning, Noel. How have you been doing?
Noel: Great. Good to see you again, Tom. What can you tell me about the analysis?
Tom: As you will recall, we wanted to see if the new process has a mean greater than or equal to 93%. I looked at the sample of eight lots you sent me and computed an estimate of the new process mean. My conclusion is that the new process is expected to have a mean no less than 93.78% with 95% confidence. Since 93.78% exceeds your requirement of 93%, we can expect the new process to deliver the desired level of quality.

Noel: That sounds great. As we discussed before, I am a little concerned that I only have eight lots in the study. Can we really make this conclusion with such a small sample size?

Tom: Yes, as long as you are comfortable with the 95% confidence level. If you desire a greater level of confidence, such as 99%, I could compute a new bound, and this value would be less than 93.78%. However, such a high level of confidence with such a small data set may make our bound somewhat non-informative. The point estimate based on the eight lots is 94.305%. If we were to wait for more lots to increase sample size, we would lessen the difference between the point estimate and the lower bound.

Noel: No. I am still comfortable with the 95% confidence level since this is typically the level expected by the regulatory agencies. Did you see anything that looked strange in the data?

Tom: Maybe. Here is a plot of the data. (Tom shows Noel Fig. 2.1). As you can see, Lot B is somewhat removed from the other values. It might be worth checking if the value was properly recorded, or if there was something unique about that production run. It is not an extreme aberration, but it never hurts to check. I monitored the reasonableness of the assumptions required to use the formula for the confidence bound, and there is no reason to believe the results are not appropriately derived.

Noel: That's all good news. So do you think I can report that our initial analysis suggests that our process mean is in alignment with expectations?

Tom: Yes, I think that is reasonable. We should be good as long as the process continues to perform in the future as it has for these first eight runs.

Noel: Great. Thanks, Tom. I appreciate your time.

Tom: Thank you, Noel. I always appreciate the opportunity to work with you. Have a good day.

In this section we introduced the steps needed to perform a statistical analysis. We now consider elements of this process in more detail.

2.3 Data Rounding and Reporting of Results

If you ever want to bring a meeting to a screeching halt, bring up the topic of rounding and reporting results of a statistical analysis. It seems everyone has their own opinion on how this should be done. Believe it or not, there are some best practices that should be followed. The process of measuring and reporting data is discussed in three regulatory documents:

1. Volume III, Sect. 4.3, of the Office of Regulatory Affairs (ORA) Laboratory Manual.
2. ASTM E29: Standard Practice for Using Significant Digits in Test Data to Determine Conformance with Specifications.
3. USP General Notices 7.20: Rounding Rules.

Table 2.3 Terminology

Term	Definition
Recorded value	The value output by a measuring system (e.g., analytical method)
Effectively unrounded value	A value which has a sufficient number of decimal places so there is negligible impact on subsequent statistical analyses. The number of decimal places should never be less than the meaningful part of the recorded value consistent with the precision of the measuring device
Reported value	A rounded version of an effectively unrounded value or a rounded version of a statistical computation based on effectively unrounded values

The objective of these documents is to provide an approach for reporting numbers that avoids false perception of numerical precision, while at the same time, prevents the loss of information for subsequent statistical analyses. We now present an example to demonstrate best practices based on terminology presented in Table 2.3.

Consider a laboratory that measures the percentage of total area under the main peak using a reverse-phase HPLC method where peaks are detected by UV absorbance in milli-absorbance units (mAU) over a course of time (seconds). The software package used with the equipment reports the percentage to five decimal places. Assume this value for a given test sample is 97.84652%. This is the recorded value as defined in Table 2.3.

Although the statistical software reports the value to five decimal places, it is important to know how many of these digits for this single test result provide meaningful information concerning the true value of the measured sample. ASTM E29 (2013) is useful for this purpose. The number of informative digits is related to the precision of the measuring device. Here, precision is defined as the standard deviation created when measuring the same item many times under exactly the same condition (e.g., with the same operator and equipment). This is commonly referred to as the repeatability standard deviation. ASTM E29 provides a very simple rule for determining the number of informative digits: Report results to the place of the first significant digit in the standard deviation if that digit is 2 or greater. If the digit is 1, report it an additional place. Table 2.4 provides some examples.

Based on a validation of the analytical method in our example, it is known that the repeatability standard deviation is 0.023%. Since the first significant digit is in the second decimal place, and the value exceeds 1, values after the second decimal place provide no meaningful information for this single test result.

When the laboratory provides values to a client, they may or may not know how the client intends to use the data. Thus, they often provide values with at least one more decimal place than the number of informative decimal places. Truncation rather than rounding should be used at this stage in order to avoid double rounding by the client. The value provided to the client is defined as the effectively unrounded value in Table 2.3. In our example, the last informative decimal is in the second position. Thus, the lab might truncate the recorded value 97.84652%

Table 2.4 Determining the number of informative digits

Repeatability standard deviation	First significant digit position	Number of informative decimal places
0.0056	5 is in third decimal position	3
0.0016	1 is in third decimal position	4
0.216	2 is in first decimal position	1
0.167	1 is in first decimal position	2
1.363	1 is in the unit position	1
2.951	2 is in the unit position	0

after the third decimal place and report the effectively unrounded value 97.846%. The additional decimal place would usually be sufficient to prevent any negative impact on future statistical analyses. If the laboratory knows that the value is to be compared to a specification limit, Borman and Chatfield (2015) provide an alternative method and recommend that the effectively unrounded value be represented with at least two decimal places more than the number of decimal places in the reported value to be compared to the specification (usually the reported value and specification are given to the same precision). When no estimate of the repeatability standard deviation is available, then rules for retention of significant digits are described in Sect. 7.4.1 of ASTM E29.

Any subsequent statistical analyses by the client should use effectively unrounded data values. One should perform all calculations with the effectively unrounded values and round to a reported value at the end of all calculations. Borman and Chatfield provide examples of the consequences that occur by not following this practice. These consequences include making decisions not consistent with the data. Rules for determining retention of significant digits during computations are provided in Sect. 7.4.1 of ASTM E29.

Once a calculation is completed, or if an effectively unrounded value is used with no subsequent calculations, rounding is often allowed before comparing the value to a limit. For example, USP General Notices 7.20 (2016b) states that "the observed or calculated values shall be rounded off to the number of decimal places that is in agreement with the limit expression." In our example, suppose the effectively unrounded value of 97.846% is to be compared to a release specification limit of no less than 97.0%. Since the specification limit has only one decimal, the value 97.846% must be rounded to one decimal. The resulting value after rounding is defined as the reported value in Table 2.3. Burgess (2013) provides a decision tree that describes best practices for rounding. To explain this process, consider the effectively unrounded value of 97.846% that needs to be rounded to one decimal. A graphical representation is shown in Fig. 2.2 where X is the digit to be rounded, Y is the digit to be dropped, and Z represents all digits to the right of Y.

Best practice for rounding uses the following three rules:

1. If Y is less than 5, X remains unchanged.
2. If Y is greater than 5, then X increases by 1.

Fig. 2.2 Representation of effectively unrounded value for rounding

3. If Y is equal to 5, and Z has all 0s or is absent, X increases by 1 if it is odd and is left unchanged if it is even. If Z contains at least one non-zero element, then X increases by 1.

The three rules above are described in the ORA Laboratory Manual and in ASTM E29. However, rule 3 does not appear in USP 7.20. In this guidance, if Y is equal to 5, X always increases by 1 regardless of Z. As noted by Burgess, ignoring rule 3 will bias data upward. Burgess also notes that the Excel rounding function ignores rule 3, but provides Excel statements that incorporate the rule. Following rule 3 often avoids annoying problems. For example, suppose we are reporting percentage of area for two peaks with an HPLC method, and the sum of the two peaks is by definition 100%. Suppose two measurements that must be rounded to one decimal are 10.05 and 89.95%. If rule 3 is ignored, the rounded values are 10.1 and 90.0% which no longer sum to 100%. In contrast, by using rule 3, the rounded percentages are 10.0 and 90.0% which do add to 100%.

In our example, the effectively unrounded value of 97.846% must be rounded to one decimal. Since $Y = 4$, X remains unchanged by rule 1, and the reported value to compare to the specification is 97.8%. Since this value meets the specification of no less than 97.0%, the lot satisfies the specification limit.

2.4 Use of Tables and Graphs

Early in any statistical analysis it is necessary to summarize and visually display individual values. The purpose of this process is to quickly communicate the data to others. Each table or graph should be able to answer a relevant research question for the study of interest.

The first decision to make is whether to use a table or a graph. Tables should be considered when

- The reader needs exact numbers.
- There are many summary statistics comparing several data sets.
- Detail is important to your conclusions.
- The audience is predominantly numerical thinkers.
- You need to make large amounts of information accessible to the reader.
- Visual organization is important.
- You cannot readily compose a sentence to convey the same information.

Graphs are more useful when

- Shapes, trends, relationships, or unusual values are of interest.
- The distribution is lopsided, there is more than one peak, or there is anything else unusual in the data.
- Exact values are of less interest.
- You want to compare data sets.
- There are more than 25 observations.

Some guidelines for producing useful tables are the following:

- Like elements should read down in columns, not across in rows.
- Related data should be grouped spatially.
- Lines should not be used to create cells. Lines should be used only to relate and divide information.
- Align the decimal point in columns of numbers.
- Use bold and italics only occasionally for emphasis.
- Use lower case for all body text.
- Times and Univers are good all-around font choices.
- Fonts should be at least 10 point.
- Standard conditions should not be repeated.
- The table should stand alone, without the text.
- Remove non-significant digits.

When graphs are more appropriate, selection of the correct graph is essential. Some basic graphs and their use are listed in Table 2.5.

Table 2.6 presents a small data set to be presented in a graphical manner.

The response in this example is concentration measured as a percentage. Each lot is measured six times at six different time periods. The variable Lot is qualitative

Table 2.5 Basic graphs

Type of variable	Type of graph	Format of horizontal and vertical axes
One qualitative	Bar chart	Count, percentage, or proportion on one axis. Qualitative categories on the other axis.
Two qualitative	Side-by-side bar chart	Count, percentage, or proportion on one axis. Qualitative categories of one variable on the other axis. Other qualitative variable categories are side-by-side.
One quantitative	Boxplot	Numerical values on one axis. No unit of measure on the other axis.
	Histogram	Numerical values on one axis. Other axis reports count or density.
One quantitative and one qualitative	Side-by-side boxplots	Numerical values of quantitative variable on one axis. Other axis reports categories of qualitative factor
Two quantitative	Scatterplot	Numerical values of quantitative variables on each axis. If a response–predictor relationship exists, response is on the verical axis and the predictor on the horizontal axis.

Table 2.6 Example data for graphing

Lot	Concentration (%)	Time (months)
A	102.1	0
A	101.4	6
A	101	12
A	101.1	18
A	100.8	24
A	99.6	30
B	100	0
B	100	6
B	100.2	12
B	98.8	18
B	99.8	24
B	99	30
C	97.6	0
C	98.3	6
C	98.1	12
C	97.1	18
C	96.5	24
C	96	30

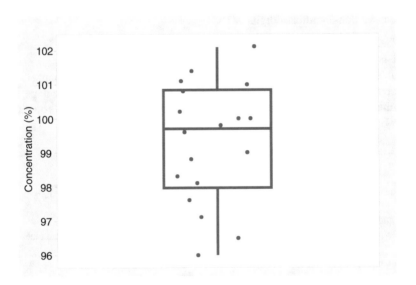

Fig. 2.3 Boxplot with all concentration values

and Concentration and Time are quantitative. Based on Table 2.5, it might be of interest to examine a boxplot of all concentrations in a single boxplot as well as side-by-side boxplots across the qualitative variable Lot. These graphs are shown in Figs. 2.3 and 2.4, respectively.

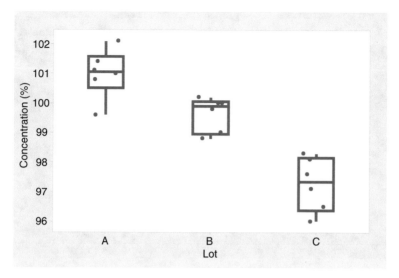

Fig. 2.4 Boxplots of concentration by lot

The three horizontal lines from bottom to top in a boxplot represent the first quartile (Q_1), the median, and the third quartile (Q_3). The first quartile exceeds 25% of the data values, and the third quartile exceeds 75% of the data values. The vertical lines are referred to as whiskers. The upper whisker extends to the highest data value within the upper inner fence, $Q_3 + 1.5(Q_3 - Q_1)$. The lower whisker exends to the lowest value within the lower inner fence, $Q_1 - 1.5(Q_3 - Q_1)$. Values below the lower inner fence or above the upper inner fence are described as outliers. Values between the upper inner fence and $Q_3 + 3(Q_3 - Q_1)$ are sometimes called mild outliers, and those that exceed $Q_3 + 3(Q_3 - Q_1)$ are called extreme outliers. Similarly, values between the lower inner fence and $Q_1 - 3(Q_3 - Q_1)$ are mild outliers, and those less than $Q_1 - 3(Q_3 - Q_1)$ are extreme outliers. Figure 2.4 suggests the concentration level differs among the three lots. Quantification of these differences relative to expected process and measurement error is needed to derive any conclusion.

Finally, the most important research question for this example is whether the lots degrade at similar rates. This is the so-called stability research question (Chap. 8 provides more on stability models). Table 2.5 suggests a scatterplot is a useful graph for this purpose. A scatterplot is provided in Fig. 2.5 with straight lines fit through the data to help visualize the slopes.

Figure 2.5 suggests that the slopes are reasonably similar. As with the previous figure, quantitative assessment of the visual difference relative to identifiable sources of variation is needed to draw a conclusion. Since the horizontal axis in Fig. 2.5 measures time, it is sometimes referred to as a time series plot.

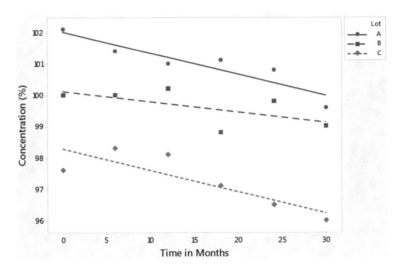

Fig. 2.5 Scatterplot of concentration on time by lot

Other recommendations for producing effective graphs include the following:

- Use large, bold, prominent plotting symbols.
- If symbols overlay one another, use open symbols and jitter the data on any scale that is not quantitative.
- Don't let lines obscure the data points.
- Connect the points with lines if there is a natural order such as time sequence. This helps the reader track the movement of the data.
- Enclose the data in a rectangle. Make sure all the data points are inside and not touching the rectangle.
- Place tick marks outside the rectangle on all four sides, from 3 to 10 tick marks per side.
- If data labels are needed inside the rectangle, make the plotting symbol more prominent than the label by using small letters for the label.
- The axes (especially the y-axis) do not need to include zero.
- Avoid the use of 3D when the third dimension has no information. Bar charts and 3D don't go together.
- Try to avoid pie charts. If absolutely necessary to use one, do not use exploding or 3D pie slices and label the proportion on each pie slice.

There are many excellent references on this topic. Two favorites are Tufte (1983) and Cleveland (1985).

2.5 Statistical Intervals

As noted earlier, it is important in any statistical analysis to quantify the amount of uncertainty in point estimates. This is best done using a statistical interval. There are three types of statistical intervals that are useful for this purpose:

1. Confidence intervals,
2. Prediction intervals, and
3. Tolerance intervals.

An excellent resource for learning more about each type of interval is the book by Hahn and Meeker (1991). The book by Krishnamoorthy and Mathew (2009) is an excellent resource for tolerance intervals, and the books by Burdick et al. (2005) and Burdick and Graybill (1992) provide more information for confidence intervals for functions of variances.

2.5.1 Confidence Intervals

Confidence intervals are used to quantify the uncertainty associated with the estimation process. For demonstration, consider the mean parameter μ described in Table 2.2. A $100(1 - \alpha)\%$ two-sided confidence interval for μ is a random interval with a lower bound L and an upper bound U. These bounds are functions of the sample values such that $\Pr[L \leq \mu \leq U] = 100 \times (1 - \alpha)\%$. The term $100 \times (1 - \alpha)\%$ is called the confidence coefficient, and is selected prior to data collection. Typical values for the confidence coefficient are 90, 95, and 99% (i.e., α is equal to $0.10, 0.05,$ and $0.01,$ respectively). The confidence coefficient describes the quality of the statistical process. In particular, the confidence coefficient defines the success rate in capturing the true value of μ. Figure 2.6 provides a visual interpretation of the confidence coefficient.

Assume the purpose of a study is to estimate the true mean concentration (μ) of a population of 5000 vials based on a sample of 50 vials. Further assume the true value of μ is 0.500 mg/mL. The computed 95% two-sided confidence intervals for 100 samples of size 50 selected from this population are shown as the vertical lines in Fig. 2.6. Note that a different interval is computed for each sample due to the randomness of the sampling process. There are 95 vertical lines that include the parameter value 0.500 mg/mL and five vertical dashed lines that do not include the true value of 0.500 mg/mL. The percentage of intervals that correctly capture the true value of 0.500 mg/mL is the confidence coefficient. Of course in practice, we select only a single sample. Our confidence that the single interval we obtain contains the true value is equal to the percentage of all such intervals that contain the true value. This percentage is the confidence coefficient. If one increases the confidence coefficient, the confidence intervals will widen, causing a greater percentage of the intervals to contain the true value. However, as the confidence

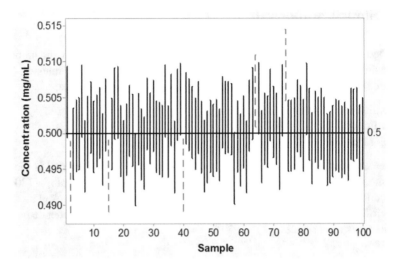

Fig. 2.6 Results of 100 computed confidence intervals

intervals widen, the information provided is less useful. Thus, there must always be some compromise between the assurance the true value is captured and the utility of the interval as defined by the length.

A confidence interval that satisfies $\Pr[L \leq \mu \leq U] = 100 \times (1 - \alpha)\%$ is called an exact two-sided confidence interval. Often exact intervals do not exist, and $\Pr[L \leq \mu \leq U]$ is only approximately equal to $100 \times (1 - \alpha)\%$. An approximate interval that has a realized confidence coefficient greater than the stated level (i.e., $\Pr[L \leq \mu \leq U] > 100 \times (1 - \alpha)\%$) is called a conservative interval. An approximate interval that has a realized confidence coefficient less than the stated level (i.e., $\Pr[L \leq \mu \leq U] < 100 \times (1 - \alpha)\%$) is called a liberal interval. In general, conservative intervals are preferred when only approximate intervals are available. Liberal intervals can be recommended if the confidence coefficient is not too much less than the desired level.

If a parameter value is of interest only when it is "too high" or "too low", then one-sided confidence bounds are computed. The exact one-sided $100(1 - \alpha)\%$ lower confidence bound satisfies the statement $\Pr[L \leq \mu] = 100 \times (1 - \alpha)\%$ and the exact one-sided $100 \times (1 - \alpha)\%$ upper bound satisfies $\Pr[\mu \leq U] = 100 \times (1 - \alpha)\%$.

2.5.2 Prediction Intervals

A $100 \times (1 - \alpha)\%$ two-sided prediction interval for a single new observation is a random interval with a lower bound L and an upper bound U, that will contain the next randomly selected observation from a process or population with stated confidence level $100 \times (1 - \alpha)\%$. As with confidence intervals, the coefficient 100

$\times (1 - \alpha)\%$ refers to the percentage of computed prediction intervals that actually include the next new value.

2.5.3 Tolerance Intervals

A $100 \times (1 - \alpha)\%$ two-sided tolerance interval is a random interval with a lower bound L and an upper bound U, that contains at least a specified proportion (P) of future process outcomes with stated level of confidence $100 \times (1 - \alpha)\%$. Tolerance intervals are used in a variety of CMC applications. Most notably, tolerance intervals are used in setting specifications and assessment of pharmaceutical quality as described in Dong et al. (2015).

2.5.4 Individual Versus Mean

Prediction intervals and tolerance intervals each describe the behavior of individual values, whereas a confidence interval describes something about a *summary measure* of individual values. It is important to understand this distinction in order to properly interpret statistical intervals. The distinction between an inference concerning an individual value and one concerning a summary measure is an important concept in the analysis of CMC data. Understanding whether an inference concerns a single lot in a manufacturing process or a summary measure of all past lots is necessary to make an informed decision. We provide several applications related to this distinction throughout the book.

2.5.5 Formula Notation

Throughout the book we provide formulas for computing statistical intervals that can be computed in a spreadsheet program. These formulas employ percentile values of familiar statistical distributions. Table 2.7 presents these distributions and the notation used to represent them in formulas. In all cases the subscript α represents the area to the *left* of the percentile.

Table 2.7 Percentile notation

Distribution	Notation (α is area to left)
Standard normal	Z_α
T	$t_{\alpha:df}$
Chi-squared	$\chi^2_{\alpha:df}$
F	$F_{\alpha:df_1,df_2}$

Table 2.8 Percentile examples

Described percentile	Percentile value	Excel command
$Z_{0.05}$	-1.644	=NORMINV(0.05,0,1)
$Z_{0.90}$	1.281	=NORMINV(0.9,0,1)
$t_{0.95:6}$	1.943	=TINV(0.1,6)
$t_{0.10:10}$	-1.372	=-TINV(0.2,10)
$\chi^2_{0.05:10}$	3.940	=CHIINV(0.95,10)
$\chi^2_{0.95:10}$	18.307	=CHIINV(0.05,10)
$F_{0.05:3,6}$	0.111	=FINV(0.95,3,6)
$F_{0.95:6,3}$	8.940	=FINV(0.05,6,3)

Table 2.8 provides some examples that illustrate the notation, and reports functions of Excel used to compute them. The TINV function in Excel reports percentiles for computing two-sided intervals. To obtain t-values consistent with our notation, use the following rule:

1. If α (the area to the left) exceeds 0.50, then use the value $2 \times (1 - \alpha)$ as the reported probability in TINV, and
2. If α is less than 0.50, then use the value $2 \times \alpha$ as the reported probability and attach a negative sign to the result.

In Excel, the functions CHIINV and FINV require you to report the area to the right, whereas NORMINV requires you to report the area to the left. All values in Table 2.8 are truncated at three decimal places.

2.6 Intervals for One Population (Independent Measurements)

We consider two statistical models for analyzing data collected from one population or process. In this section, all measurements are independent. In Sect. 2.7, measurements are dependent.

In the Irish manufacturing example discussed earlier, a lot release value for purity was measured for each of eight manufactured lots. A very simple statistical model representing the measured purity value for a single lot is

Measured Purity = True average purity of manufacturing process
$$+ \text{ Random value due to process and measurement variation}$$

$$(2.5)$$

We now replace the words with symbols and write

$$Y = \mu + E \tag{2.6}$$

where Y is the measured purity, μ is the true average purity for the process, and E is the random value that changes from lot-to-lot and from measurement to measurement within a lot. Both μ and E are unknown—the only thing we can observe is Y. The term μ is said to be fixed, because it never changes as long as the process does not change. On the other hand, the term E is random because it changes from measurement to measurement. All random variables such as E have an associated probability distribution. It is assumed here that E has a normal probability distribution with mean zero and an unknown variance σ^2. We will discuss the importance of these assumptions concerning E in Sect. 2.12.2.

Note that if the process were perfectly consistent, and our analytical method could always make an exact measurement, then $E = 0$ and every lot would have a value of μ. Because this is not a realistic expectation for any manufacturing process or analytical method, it is necessary to incorporate E in all statistical models.

You will note the symbols used in model (2.6) vary. A random variable such as E is represented with an upper case Latin letter whereas a fixed quantity such as μ is represented with a lower case Greek letter. We use this convention throughout the book. (Note that since Y is a function of the random variable E, it is also a random variable.)

Finally, we modify model (2.6) by adding subscripts in order to represent a collection of n values instead of a single value. The final model that we use for the rest of this section is

$$Y_i = \mu + E_i \quad i = 1, 2, \ldots, n. \tag{2.7}$$

An important assumption concerning E_i not mentioned previously is that the value associated with the ith lot is independent (not impacted) by the values assigned to other lots. Thus, we describe model (2.7) as the one population model with independent measurements. As we will see in Sect. 2.7, items are sometimes grouped in such a manner that the value of one item is influenced by the values of items in its group. In this case, the measured Y_i values are dependent rather than independent. It is always important to know if data are independent or not, so that the correct analysis can be performed.

Table 2.9 reports summary formulas used to describe the collection of values described in (2.7). Since the data represent a sample of process values, the summary measures are called the sample statistics. These sample statistics are used to compute statistical intervals to describe the sampled process.

Table 2.10 reports the parameters, their point estimators, and computed estimates using the example data in Table 2.1. As described in Sect. 2.2, parameters are unknown summary measures of the process or population of interest. Sample statistics are used to estimate parameters.

Table 2.9 Sample statistics
for one population with
independent measurements

Statistic	Symbol	Formula
Sample mean	$\bar{Y}$	$\displaystyle\sum_{i=1}^{n} Y_i \over n$ $\sum_{i=1}^{n} Y_i \over n$
Sample variance	S^2	$\displaystyle\sum_{i=1}^{n} (Y_i - \bar{Y})^2 \over n-1$ $\sum_{i=1}^{n}(Y_i-\bar{Y})^2 \over n-1$
Sample standard deviation	S	$\sqrt{S^2}$

Table 2.10 One population
mean with independent
measurements example

Parameters	Point estimator	Computed estimate Table 2.1
μ	$\bar{Y}$	94.305
σ^2	S^2	0.608
σ	S	0.780

2.6.1 Confidence Interval for Mean

The bounds for a $100(1-\alpha)\%$ two-sided confidence interval for μ are

$$L = \bar{Y} - t_{1-\alpha/2:n-1}\sqrt{\frac{S^2}{n}}$$
$$U = \bar{Y} + t_{1-\alpha/2:n-1}\sqrt{\frac{S^2}{n}}$$

(2.8)

Assume in our lot release example that we desire a 95% two-sided confidence interval for μ. Thus, $100(1-\alpha)\% = 95\%$ and so $1-\alpha = 0.95$, $\alpha = 0.05$, and $\alpha/2 = 0.025$. With $n = 8$, then $t_{1-0.025:8-1} = t_{0.975:7} = 2.364$. Using the computed point estimates in Table 2.10 the computed 95% confidence interval on μ is

$$L = 94.305 - 2.364 \times \sqrt{\frac{0.608}{8}} = 93.65\%$$
$$U = 94.305 + 2.364 \times \sqrt{\frac{0.608}{8}} = 94.96\%.$$

(2.9)

With 95% confidence, we can state the true value of μ is between 93.65 and 94.96%.

2.6.2 Confidence Interval for Variance

The bounds for a $100(1 - \alpha)\%$ two-sided confidence interval on σ^2 are

$$
\begin{aligned}
L &= \frac{(n-1)S^2}{\chi^2_{1-\alpha/2:n-1}} \\
U &= \frac{(n-1)S^2}{\chi^2_{\alpha/2:n-1}}.
\end{aligned}
\tag{2.10}
$$

The computed 95% interval using the statistics in Table 2.10 is

$$
\begin{aligned}
L &= \frac{(8-1)0.608}{16.013} = 0.27 \\
U &= \frac{(8-1)0.608}{1.690} = 2.52
\end{aligned}
\tag{2.11}
$$

2.6.3 Confidence Interval on the Standard Deviation

Based on these calculations, the 95% confidence interval on the standard deviation σ is $L = \sqrt{0.27} = 0.52\%$ and $U = \sqrt{2.52} = 1.59\%$.

2.6.4 Confidence Interval on the Percent Relative Standard Deviation

In terms of the population parameters, the percent relative standard deviation (% RSD) is expressed as

$$
\%\text{RSD} = \frac{\sigma}{\mu} \times 100\%.
\tag{2.12}
$$

In words, it is the standard deviation expressed as a percentage relative to the average. It has also been referred to as the coefficient of variation (CV) in the statistical literature. Analytical laboratories often use %RSD to describe the performance of an analytical method. Since it is a relative measure, it is useful for comparisons across dissimilar magnitudes or dissimilar units of measure. However, as noted by Torbeck (2010), there are situations where it is not useful. For example, it is not informative for measurement scales such as pH where zero has no physical meaning. It can be misleading if the original unit of measure is a percentage such as percent recovery or percent of main peak. Care must also be taken when data have very small averages. In such a situation, a minor shift in the standard deviation can cause a very big shift in the %RSD.

A naïve method for computing a confidence interval on %RSD is to compute the interval in (2.10), take the square root, and then divide both bounds by $\bar{Y}$. The problem is this does not properly account for the variability in $\bar{Y}$ as an estimator of μ. This results in an interval that is too short. Vangel (1996) offers two alternative intervals to mitigate this problem. However, we have found the naïve interval works well for practical purposes. In our example, the two-sided 95% naïve interval is

$$
L = \sqrt{\frac{(8-1)0.608}{16.013}} \times \frac{1}{94.305} \times 100\% = 0.55\%
$$

$$
U = \sqrt{\frac{(8-1)0.608}{1.690}} \times \frac{1}{94.305} \times 100\% = 1.68\%.
$$

(2.13)

To compare, the two-sided modified McKay method recommended by Vangel is from 0.547 to 1.683%. There is no practical difference between these two intervals.

2.6.5 Confidence Interval for Proportion Out of Specification

In many applications, it is desired to estimate the probability that a reportable value measured on a quality attribute will exceed a predefined specification limit. Such a measurement is said to be out of specification (OOS). Exact confidence intervals can be obtained where there is a single specification limit, and a good approximation is available when there are both a lower and upper specification limit. Computation of these intervals requires the ability to determine percentiles of a non-central t-distribution. To begin, consider a process where a reportable value is within specification if it is no less than a lower specification limit (LSL). That is, the reportable value, Y, is in specification if $LSL \leq Y$. The bounds for an exact 100$(1-\alpha)$% confidence interval on the probability of an OOS are

$$
L = 1 - \Phi\left(\frac{\lambda_L}{\sqrt{n}}\right)
$$

$$
U = 1 - \Phi\left(\frac{\lambda_U}{\sqrt{n}}\right)
$$

where λ_L and λ_U are determined such that

$$
\Pr[t_{\lambda_L:n-1} \leq \sqrt{n} \times K_{LSL}] = \frac{\alpha}{2}
$$

$$
\Pr[t_{\lambda_U:n-1} \leq \sqrt{n} \times K_{LSL}] = 1 - \frac{\alpha}{2},
$$

$$
K_{LSL} = \left(\frac{\bar{Y} - LSL}{S}\right)
$$

(2.14)

$t_{\lambda_L:n-1}$ is a non-central t-variate with non-centrality parameter λ_L and degrees of freedom $n - 1$, $t_{\lambda_U:n-1}$ is a non-central t-variate with non-centrality parameter λ_U and degrees for freedom $n - 1$, and $\Phi(\bullet)$ is the area in a standard normal curve to the left of $(\bullet)$. Generally, only an upper bound is desired, but for completeness formulas for both bounds are provided below.

Assume LSL $= 88\%$. For a 95% two-sided interval, solve the following equations for λ_L and λ_U

$$\Pr\left[t_{\lambda_L:n-1} \leq \sqrt{8}\left(\frac{94.305 - 88}{0.780}\right)\right] = \frac{0.05}{2}$$

$$\Pr[t_{\lambda_L:n-1} \leq 22.868] = 0.025 \qquad (2.15)$$

$$\Pr[t_{\lambda_U:n-1} \leq 22.868] = 0.975.$$

Unfortunately, Excel does not have non-central t-distribution percentiles in its base package to solve these equations. SAS provides the values for the non-centrality parameters using the function "tnonct" and Minitab calculates cumulative probability for the non-central t-distribution so that it can be used to solve for the two non-centrality parameters iteratively. For our example, $\lambda_L = 34.72$ and $\lambda_U = 11.04$. Thus the computed bounds are

$$L = 1 - \Phi\left(\frac{\lambda_L}{\sqrt{n}}\right) = 1 - \Phi\left(\frac{34.72}{\sqrt{8}}\right) = 1 - \Phi(12.28) = 0$$

$$U = 1 - \Phi\left(\frac{\lambda_U}{\sqrt{n}}\right) = 1 - \Phi\left(\frac{11.04}{\sqrt{8}}\right) = 1 - \Phi(3.90) = 0.000047. \qquad (2.16)$$

As seems obvious from the plot in Fig. 2.1, the likelihood of an OOS is essentially non-existent if the process remains in control. If only an USL is provided, one simply replaces K_{LSL} with $K_{USL} = \left(\frac{USL - \bar{Y}}{S}\right)$ in (2.14).

The calculations become a bit more complex when there are two specification limits. The following algorithm reported by Mee (1988) is used to compute a 100 $(1 - \alpha)\%$ two-sided confidence interval:

1. Compute K_{LSL} and K_{USL}. If either of these values exceeds $(n - 1)/\sqrt{n}$, then set $K^* = \min(K_{LSL}, K_{USL})$ and go to step 4. Otherwise, compute the maximum likelihood point estimator (MLE) of the proportion outside the specification limits using the formula

$$\hat{\pi} = \Phi\left(-\sqrt{\frac{n}{n-1}} \times K_{LSL}\right) + \Phi\left(-\sqrt{\frac{n}{n-1}} \times K_{USL}\right). \qquad (2.17)$$

2. Determine the $100\hat{\pi}\%$ percentile of a beta distribution with parameters $(n - 2)$ $/2$ and $(n - 2)/2$ using the inverse cumulative distribution function (Excel, Minitab, and SAS all have this function). Denote this value as b.

3. Compute

$$K^* = \frac{(1 - 2 \times b) \times (n - 1)}{\sqrt{n}}.$$ (2.18)

4. Solve the equations

$$\Pr\left[t_{\lambda_L:n-1} \le \sqrt{n} \times K^*\right] = \frac{\alpha}{2} \text{ to get } \lambda_L, \text{ and}$$

$$\Pr\left[t_{\lambda_U:n-1} \le \sqrt{n} \times K^*\right] = 1 - \frac{\alpha}{2}, \text{ to get } \lambda_U.$$ (2.19)

Apply (2.16) to obtain L and U.

This approximation is improved by using the minimum variance unbiased estimator in place of the MLE as discussed by Mee. However, it is a bit more complicated to compute, and for demonstration, we have selected the MLE.

Consider a simple example with $\bar{Y} = 10, S = 2, n = 20, LSL = 8$, and $USL = 13$. The computations following the steps above for a 90% confidence interval are

1. $K_{LSL} = (10 - 8)/2 = 1$, $K_{USL} = (13 - 10)/2 = 1.5$. Since both of these values are less than $(20 - 1)/\sqrt{20} = 4.25$, compute

$$\hat{\pi} = \Phi\left(-\sqrt{\frac{n}{n-1}} \times K_{LSL}\right) + \Phi\left(-\sqrt{\frac{n}{n-1}} \times K_{USL}\right)$$

$$= \Phi\left(-\sqrt{\frac{20}{19}} \times 1\right) + \Phi\left(-\sqrt{\frac{20}{19}} \times 1.5\right)$$ (2.20)

$$= \Phi(-1.026) + \Phi(-1.539)$$

$$= 0.152 + 0.062 = 0.214.$$

2. $b = 0.406$ using the Excel function "=betainv(0.214, (20-2)/20,(20-2)/20)".
3. $K^* = \frac{(1-2\times b)\times(n-1)}{\sqrt{n}} = \frac{(1-2\times 0.406)\times(20-1)}{\sqrt{20}} = 0.796$.
4. $\lambda_L = 5.786, \lambda_U = 1.267, L = 0.098$, and $U = 0.388$.

2.6.6 Prediction Interval for the Next Observed Process Value

The bounds for a $100(1 - \alpha)\%$ two-sided prediction interval for the next observation from a process are

$$L = \bar{Y} - t_{1-\alpha/2:n-1}\sqrt{\left(1 + \frac{1}{n}\right) \times S^2}$$

$$U = \bar{Y} + t_{1-\alpha/2:n-1}\sqrt{\left(1 + \frac{1}{n}\right) \times S^2}.$$

$$(2.21)$$

The computed 95% prediction interval using the statistics in Table 2.10 is

$$L = 94.305 - 2.364\sqrt{\left(1 + \frac{1}{8}\right) \times 0.608} = 92.35\%$$

$$U = 94.305 + 2.364\sqrt{\left(1 + \frac{1}{8}\right) \times 0.608} = 96.26\%.$$

$$(2.22)$$

By replacing $t_{1-\alpha/2:n-1}$ in Eq. (2.21) with $t_{1-\alpha/(2m):n-1}$ the prediction interval contains *all* of the next m future process values for the desired level of confidence. Once m becomes sufficiently large, it is more informative to compute the tolerance interval described in the next section.

2.6.7 Tolerance Interval for all Future Process Values

A two-sided tolerance interval is an interval expected to contain at least $100P$ % of a population with $100(1 - \alpha)\%$ confidence. For example, if $P = 0.99$ and $\alpha = 0.05$, one can state with 95% confidence that 99% of the population will fall in the computed interval. For model (2.7), a two-sided tolerance interval is written as

$$L = \bar{Y} - K\sqrt{S^2}$$
$$U = \bar{Y} + K\sqrt{S^2}$$

$$(2.23)$$

where $\bar{Y}$ and S^2 are as defined in Table 2.9 and K is a constant obtained from either statistical software or a table of values (see, e.g., tables in Appendix B of Krishnamoorthy and Mathew or tables in Appendix A of Hahn and Meeker).

The exact value for K from Table A.10b of Hahn and Meeker for a two-sided 95% tolerance interval that contains 99% of the process values with $n = 8$ is $K = 4.889$. Thus, using the statistics in Table 2.10 with this value of K

$$L = 94.305 - 4.889\sqrt{0.608} = 90.49\%$$
$$U = 94.305 + 4.889\sqrt{0.608} = 98.12\%$$

$$(2.24)$$

Thus, with 95% confidence we expect the range from 90.49 to 98.12% will include at least 99% of all future purity values.

There are several simple approximations for K in Eq. (2.23) that can be used if you don't have exact values. One good approximation proposed by Howe (1969) is

$$K = \sqrt{\frac{\left(1 + \frac{1}{n_e}\right) Z^2_{(1+P)/2} \times \nu}{\chi^2_{\alpha:\nu}}} \tag{2.25}$$

where $\chi^2_{\alpha:\nu}$ is the chi-squared percentile with ν degrees of freedom and area α to the left, and $Z^2_{(1+P)/2}$ is a standard normal percentile with area $(1 + P)/2$ to the left. The term n_e represents the number of observations used to estimate the mean (sometimes called the effective sample size) and $\nu = n - 1$ in this example. The approximate value of K is

$$K = \sqrt{\frac{\left(1 + \frac{1}{n_e}\right) Z^2_{(1+P)/2} \times \nu}{\chi^2_{\alpha:\nu}}}$$

$$K = \sqrt{\frac{\left(1 + \frac{1}{8}\right) 2.576^2 \times 7}{2.167}} = 4.910 \tag{2.26}$$

which closely matches the exact value of 4.889.

Most typically as in this example, n_e is equal to the sample size n, and ν is equal to $n - 1$. However, this need not always be the case. In Chap. 9, an example is given where a tolerance interval is computed where the center of the interval is the sample mean of one data set of size n_1, and the standard deviation is computed as the pooled estimate of this data set and a second data set of size n_2. In this situation, $n_e = n_1$ and $\nu = n_1 + n_2 - 2$.

In some cases, only one-sided bounds are necessary. Typically, one desires either a $100(1 - \alpha)\%$ lower tolerance bound to be exceeded by at least $100P\%$ of the population, or a $100(1 - \alpha)\%$ upper tolerance bound to exceed at least $100P\%$ of the population where $P > 0.5$. Exact tolerance bounds can be computed using a non-central t-distribution. In particular, a $100(1 - \alpha)\%$ lower tolerance bound to be exceeded by at least $100P\%$ of the population is

$$L = \bar{Y} - K_1\sqrt{S^2}$$

$$K_1 = \frac{t_{1-\alpha:n-1,ncp}}{\sqrt{n}} \tag{2.27}$$

$$ncp = Z_P \times \sqrt{n}$$

where $t_{1-\alpha:n-1,ncp}$ is the percentile of a non-central t-distribution with area $1 - \alpha$ to the left, degrees of freedom $n - 1$ with non-centrality parameter ncp, and Z_P is a standard normal percentile with area P to the left. A $100(1 - \alpha)\%$ upper tolerance bound to exceed at least $100P\%$ of the population is

$$U = \bar{Y} + K_1 \sqrt{S^2}.$$ (2.28)

Tabled values of K_1 are available in the same sources referenced for the two-sided interval or determined with statistical software that provides the non-central t-distribution.

Consider a 99% upper tolerance bound that will exceed at least 90% of the population. Using (2.28) with $\alpha = 0.01$ and $P = 0.90$,

$$ncp = Z_{.9} \times \sqrt{8} = 1.28 \times \sqrt{8} = 3.62$$

$$K_1 = \frac{t_{1-\alpha:n-1,ncp}}{\sqrt{n}} = \frac{9.88}{\sqrt{8}} = 3.49$$ (2.29)

$$U = 94.305 + 3.50\sqrt{0.608} = 97.03\%.$$

The veracity of computed tolerance and prediction intervals are very much dependent on the assumption that the data arise from a normal population or process. Methods to validate this assumption are presented later in this chapter. If the normal population assumption is not reasonable, one can compute a nonparametric tolerance interval. The references listed above as well as the article by Gitlow and Awad (2013) offer tables and examples for these intervals. Nonparametric tolerance intervals require more data than normal-based intervals to provide useful intervals. In our present example where $n = 8$, the interval covered by the minimum to the maximum values, i.e., the interval from 92.68 to 95.17%, has a confidence of only 18.7% of including at least 90% of the entire population. To increase the confidence to 95% requires a sample size of at least $n = 46$. That is, if one selects a sample of size 46 with 95% confidence, the range from the minimum to the maximum sample value includes at least 90% of the population (see Table A.17 of Hahn and Meeker).

2.6.8 Comparison of Statistical Intervals

Figure 2.7 shows a plot of the sample data with the confidence interval for the mean, the prediction interval, and the tolerance interval.

All intervals are computed with 95% confidence and the tolerance interval contains 99% of future values. The tolerance interval is represented by the solid lines, the prediction interval is the line with long dashes, and the confidence interval for the mean is the line with short dashes. As expected, the tolerance interval is widest because there is greater uncertainty about the behavior of all future individual values than there is for the next individual value (the prediction interval) or the average of all values (the confidence interval). The fact that some of the individual values fall outside the confidence interval for the mean is not surprising. Remember that the confidence interval is expected to include the average 95% of the time, but it will not include all the individual values that make up the average.

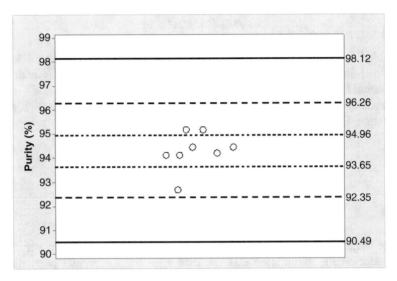

Fig. 2.7 Plot of lot release data with statistical intervals

2.6.9   Data Sets with LOQ Values

Measured values are sometimes subject to a limit of quantitation (LOQ). The LOQ is the lowest amount of analyte in a sample that can be determined with acceptable precision and accuracy under the stated experimental conditions. Values that are less than LOQ are typically reported as "LOQ" with no numerical value. The measured values are said to be left-censored. One strategy for computing descriptive statistics in this situation is to replace the label "LOQ" with the numerical value of LOQ and compute the desired statistics. This strategy is called the "Substitution" method.

Unfortunately, the substitution method will overestimate the mean (because one is increasing all values less than LOQ to LOQ) and underestimate the standard deviation (because a shorter range of the data is artificially created).

There is a large body of research that considers estimation of means and standard deviations with normally distributed left-censored data. One very simple approach that works well was suggested by Persson and Rootzen (1977). This is an adjusted maximum likelihood approach that does not require an iterative estimation process. Thus, it can be easily implemented in Excel.

Let n represent the total sample size including observations marked as "LOQ," and k represent the number of observations greater than LOQ. Thus, the number of "LOQ" values is $n - k$. The sample mean $(\bar{Y})$ and sample standard deviation (S) for the Persson and Rootzen approach are computed using the sample mean and sample standard deviation computed from the k values that exceed LOQ, denoted $\bar{Y}_1$ and S_1, respectively. In particular, the estimators are

$$Y = Y_1 - r \times c \qquad\qquad\qquad Y = Y_1 - r \times c$$

$$S = \sqrt{\left(\frac{k-1}{k}\right) \times S_1^2 - r \times (\lambda - r) \times c^2} \quad S = \sqrt{\left(\frac{k-1}{k}\right) \times S_1^2 - r \times (\lambda - r) \times c^2}$$

$$(2.30)$$

where

$$c = \frac{1}{2}\left[\lambda(\bar{Y}_1 - LOQ) + \sqrt{(4 + \lambda^2) \times (\bar{Y}_1 - LOQ)^2 + 4\left(1 - \frac{1}{k}\right)S_1^2}\right]$$

$$(2.31)$$

$$r = \left(\frac{n}{k}\right) \times \left(\frac{1}{\sqrt{2\pi}}\right)\exp\left(-\frac{\lambda^2}{2}\right)$$

and λ is the percentile of a standard normal distribution with area $(1 - k/n)$ to the left. Once the mean and standard deviation are estimated in this manner, they can be used to compute statistical intervals.

Table 2.11 reports a sample of $n = 48$ nucleic acid measurements (ng/mg protein) taken from the first column of a purification process. There are $k = 41$ values that exceed the LOQ of 0.007 ng/mg.

Table 2.12 shows the computed quantities defined in Eqs. (2.30) and (2.31).

As shown in Fig. 2.8, the data in Table 2.11 are extremely skewed to the right. Thus, it is not appropriate to compute a normal-based prediction or tolerance interval with these data.

However, one can still compute a confidence interval on the mean because the sample size is large enough for the central limit theorem to apply. Thus, a 95% confidence interval on the mean using Eq. (2.8) is

Table 2.11 Example data

Obs	Value	Obs	Value	Obs	Value	Obs	Value
1	0.043	13	0.010	25	0.015	37	LOQ
2	0.033	14	0.012	26	0.013	38	0.011
3	0.032	15	0.013	27	0.019	39	LOQ
4	0.035	16	0.011	28	0.019	40	0.095
5	0.034	17	0.014	29	0.017	41	0.023
6	0.037	18	0.012	30	0.019	42	0.015
7	0.040	19	0.012	31	0.017	43	0.017
8	0.034	20	0.011	32	LOQ	44	0.020
9	0.012	21	0.026	33	LOQ	45	LOQ
10	0.014	22	0.011	34	0.017	46	0.025
11	0.016	23	0.009	35	LOQ	47	0.033
12	0.014	24	0.010	36	LOQ	48	0.024

Table 2.12 Computed
statistics for example problem

LOQ	0.007
n	48
k	41
$\bar{Y}_1$	0.0218
S_1	0.0151
λ	-1.0545
c	0.0146
r	0.2679
$\bar{Y}$	0.0179
S	0.0173

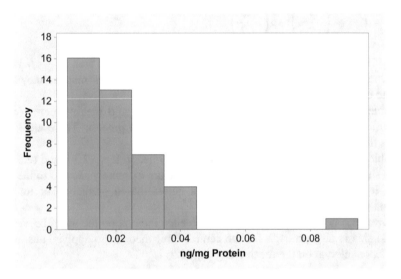

Fig. 2.8 Example data (ng/mg protein)

$$L = 0.018 - 2.01 \times \frac{0.017}{\sqrt{48}} = 0.013 \text{ ng/mg protein}$$

$$U = 0.018 + 2.01 \times \frac{0.017}{\sqrt{48}} = 0.023 \text{ ng/mg protein}$$

(2.32)

2.6.10   Non-Normal Data

As noted in the previous example, unless that data are well represented with a normal probability model, the statistical interval formulas presented to this point are not appropriate. This is especially true for prediction and tolerance intervals, but

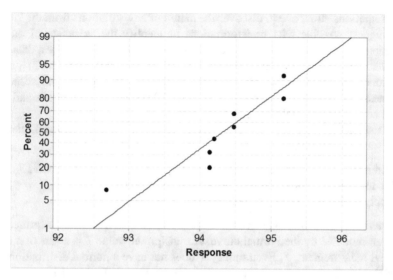

Fig. 2.9 Normal quantile plot of purity data

less so for the confidence interval on the mean. A confidence interval for the mean is not as sensitive to the assumption of normality because it is based on the sampling distribution of the sample mean. For the sample mean, the central limit theorem ensures that the sampling distribution tends toward a normal probability distribution regardless of the underlying distribution of individual observations. So with relatively large samples, Eq. (2.8) works well even if the underlying distribution is not normal. However, prediction intervals and tolerance intervals may not perform well if the population is non-normal. The validity of these intervals depends on the distribution of the individual values which are not impacted by the central limit theorem. More discussion on this topic is found in Sect 4.10 of Hahn and Meeker.

A useful plot for determining whether data are reasonably represented with a normal distribution is the normal quantile plot (also called normal probability plot). Figure 2.9 shows a normal quantile plot of the purity values in Table 2.1. The horizontal axis reports the value of the measurement and the vertical axis reports the percentage of values in the data set that are less than or equal to the measured value. (There is a continuity adjustment made to these values.) The vertical axis is scaled so that if the data can be described with the normal probability model, the measurements will fall in a straight line that can be covered with a "fat pencil."

It seems reasonable to suggest the measurements in Fig. 2.9 fall in a straight line, so use of the previously computed statistical intervals for this data set is appropriate.

Statistical tests can be used to test the null hypothesis that data are normal. However, we prefer graphical representations over statistical tests. The results of a statistical test are dependent on sample size—large samples often lead to the conclusion that data are not normal, and small samples are often too small to reject the null hypothesis of normality in even extreme situations. Additionally, graphical representations allow one to discover the nature of any variations from the "ideal," and suggest possible data transformations or identify the existence of outliers. Additional discussion on this issue and other useful graphs to display a data set is provided by Vukovinsky and Pfahler (2014). Boylan and Cho (2012) provide additional examples of how to interpret normal quantile plots with respect to skewness, kurtosis, and variability. Vining (2010) provides calculations and additional advice on the application of these plots.

When data are not well represented with a normal probability model, there are three ways to proceed.

1. Determine a transformation that will make the data fit a normal model.
2. Model the data with a more appropriate distribution.
3. Apply a distribution-free or nonparametric approach.

The first approach is to determine a transformation so that the transformed data are well modeled by the normal curve. A transformed value T is a function of the original measurement, Y. Even though Y does not have a normal distribution, it is possible that T can be represented by the normal model. Table 2.13 reports some common transformations and conditions for Y that might yield a successful transformation.

Power transformations, the most common of which are Box-Cox transformations, are also useful tools. These transformations are of the form

$$T = \begin{cases} \dfrac{Y^\lambda - 1}{\lambda} & \lambda \neq 0 \\ \ln(Y) & \lambda = 0 \end{cases} \tag{2.33}$$

where λ is selected to best transform the data set to normality. Information on Box-Cox transformations is provided in Sect. 6.5.2 of the NIST/SEMATECH e-Handbook of Statistical Methods (2016).

Table 2.13 Useful transformations for normality

Transformation name	Transformation	Back transformation	Conditions for Y
Log	$T = \ln(Y)$	$Y = e^T$	Useful when Y is skewed to the right with values ranging over several orders of magnitude. If Y has a log-normal distribution, then T will have a normal distribution
Arcsine	$T = 2 \times \sin^{-1}(\sqrt{Y})$	$Y = \left[\sin\left(\frac{T}{2}\right)\right]^2$	Useful if Y is a proportion, i.e., $0 \leq Y \leq 1$. Works well for proportions that are "pushed up" against a finite bound such as 0 or 1
Logit	$T = \ln\left(\frac{Y}{1-Y}\right)$	$Y = \frac{e^T}{1+e^T}$	Works in situations similar to the arcsine transformation where $0 \leq Y \leq 1$

Table 2.14 Counts of sub-visible particles of size 25 μm

Number of particles in vial	Number of vials
0	4
1	21
2	11
3	2

After completing an analysis with transformed values (T), results are typically "back-transformed" to the original scale of measurement (Y). For example, if a lower tolerance bound, L, is computed using Eq. (2.23) with transformed data $T = \ln(Y)$, then the reported lower tolerance bound in the original units is e^L.

It has been our experience that back-transformed prediction and tolerance bounds are often too wide to be useful when the original distribution is extremely skewed and the sample size is small. From a practical standpoint, this is what makes a transformation strategy difficult to employ in many CMC applications. Additionally, consumers of your analysis may not like the fact you are "changing the data," even though the resulting analysis is perfectly fine from a statistical standpoint. Units of measure on transformed data can also present problems in interpretation.

As an alternative to data transformation, one can compute statistical intervals with a more appropriate probability distribution. With the advancement of statistical software, this approach is well within the capabilities of most investigators.

This approach is demonstrated by computing a tolerance interval for counts of sub-visible particles (SbVP) found in a vial or syringe. Historically, variables that are measured as counts in a fixed unit of space or time are modeled with the Poisson distribution, and this model is a good representation of SbVP counts. Table 2.14 summarizes counts of sub-visible particles of size greater than 25 μm for a sample of 38 vials taken from a manufacturing process. For example, of the 38 vials in the sample, four had no particles, and 21 had one particle each.

Krishnamoorthy et al. (2011) propose one-sided and two-sided tolerance intervals on the mean λ of the Poisson distribution, based on either the exact or score-based confidence interval on λ. To compute a two-sided $100(1 - \alpha)\%$ tolerance interval that contains $100P\%$ of a population, start with a $100(1 - 2\alpha)\%$ confidence interval on the mean of the Poisson distribution, λ. One recommended confidence interval on λ with confidence coefficient $100(1 - 2\alpha)\%$ is

$$
L = \frac{\chi^2_{\alpha:2m}}{2n}
$$
$$
U = \frac{\chi^2_{1-\alpha:2m+2}}{2n}
$$
(2.34)

where n is the sample size and m is the sum of all n values (counts). The upper tolerance bound is now the $\frac{1+P}{2}$ percentile of a Poisson distribution with mean U from Eq. (2.34), and the lower tolerance bound is the $\frac{1-P}{2}$ percentile of a Poisson distribution with mean L from Eq. (2.34).

This calculation can be performed using the function "Quantile" in SAS, the inverse cumulative function in the "Calc" menu of Minitab, or the "Poisson" function in Excel. Using the data in Table 2.14, $n = 38$ and $m = 0 \times 4 + 1 \times 21 + 2 \times 11 + 3 \times 2 = 49$. To construct a 95 % two-sided tolerance interval that contains 99.5 % of future values, $\alpha = 0.025$ and $P = 0.995$. The two-sided confidence interval on λ from Eq. (2.34) is

$$L = \frac{\chi^2_{\alpha:2m}}{2n}$$

$$L = \frac{\chi^2_{0.025:98}}{2 \times 38} = \frac{72.5}{2 \times 38} = 0.954$$

$$U = \frac{\chi^2_{1-\alpha:2m+2}}{2n} \tag{2.35}$$

$$U = \frac{\chi^2_{0.975:100}}{2 \times 38} = \frac{129.56}{2 \times 38} = 1.705.$$

To compute the upper bound on the tolerance interval, use the Excel function "Poisson" to find the smallest integer such that probability exceeds $\frac{1+P}{2} = \frac{1+0.995}{2} = 0.9975$ with mean$=U = 1.705$. This yields "Poisson(6,1.705,1)=0.9981". So the upper tolerance bound is 6 particles. The lower tolerance bound is 0 since "Poisson (0,0.954,1)=0.3852" exceeds $\frac{1-P}{2} = 0.0025$, and the lower bound can be no less than zero.

It is instructive to note that if these data had been used to compute a two-sided tolerance interval using the normal-based formula in Eq. (2.23), $L = -1.292$, and $U = 3.871$. As can be seen from this result, the normal-based upper bound is too small because it does not account for the skewness of the distribution. The same approach can be applied for a binomial distribution as described in Krishnamoorthy et al. A generalized linear model (see, e.g., Myers et al. 2002) offers a more general approach for constructing statistical intervals under various distributional probability models.

Nonparametric methods can also be considered for relatively large data sets. Chapter 5 of Hahn and Meeker (1991) provides distribution-free formulas for several statistical intervals.

2.7   Intervals for One Population (Dependent Measurements)

Measurements in a laboratory are often collected with observational units nested within experimental units. A common example is an analytical method where aliquots (observational units) are contained within plates (experimental units). An

Table 2.15 Measurements on drug product from laboratory

Plate	Aliquot	Purity (%)
1	1	96.672
1	2	96.606
2	3	96.793
2	4	96.883
3	5	96.253
3	6	96.298
4	7	96.074
4	8	96.075
5	9	96.098
5	10	96.071
6	11	96.870
6	12	96.755

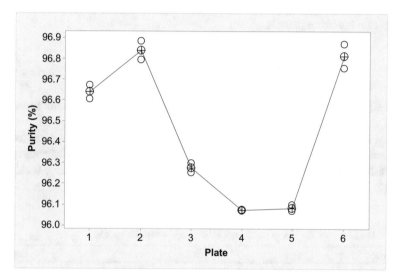

Fig. 2.10 Plot of drug product measurements

example of a typical data set is shown in Table 2.15 where the reading from a single aliquot is a purity measure expressed as a percentage.

The data are plotted in Fig. 2.10. The circles with the crosses represent the averages for each plate.

Measurements have been taken from 12 aliquots and there are two aliquots on each plate. Because values of aliquots on the same plate are more similar than values of aliquots from different plates, the measurements within the same plate are dependent (correlated). This situation is demonstrated in Fig. 2.10 where it is seen that the means from plate to plate vary more than the values within any given plate. The statistical model used to represent these data is

$$Y_{ij} = \mu + A_i + E_{ij}$$
$$i = 1, \ldots, a \text{ (plates)}; \quad j = 1, \ldots, r \text{ (aliquots per plate)};$$

(2.36)

where Y_{ij} is the measured value for aliquot j on plate i, a is the number of plates, r is the number of aliquots on each plate, μ is the true average, A_i is a random error contributed by plate i, and E_{ij} is a random error specific to aliquot j from plate i.

For the design in Table 2.15, $a = 6$ plates and $r = 2$ aliquots per plate. The total sample size is $n = a \times r = 12$. Unlike model (2.7) in Sect. 2.6 where there is a single random variable, the model in (2.36) has two random variables, A_i and E_{ij}. As a matter of terminology, when there is more than one random error term in the model, the error term on the far right of the equation (E_{ij}) is referred to as the residual error. The two random variables are assumed to be independent, and each one has a normal distribution with mean zero. The variance associated with the normal population of the A_i is σ_A^2 and the variance associated with the normal population of the E_{ij} is σ_E^2. Based on these assumptions, the correlation between two aliquots on the same plate is

$$\rho = \frac{\sigma_A^2}{\sigma_A^2 + \sigma_E^2}.$$

(2.37)

Thus, if $\sigma_A^2 = 0$, measurements on the same plate are independent and model (2.36) simplifies to the independent model (2.7). Note that ρ can also be described as the proportion of the total error in Y_{ij} that is attributed to the plate-to-plate variation. Table 2.16 reports statistics and Table 2.17 reports the parameters, their point estimators, and computed estimates for the model in (2.36) using the data in Table 2.15.

Table 2.16 Statistics for one population with dependent measurements

Statistic	Symbol	Formula	Estimate
Sample mean	$\bar{Y}$	$\dfrac{\sum\limits_{i=1}^{a} \sum\limits_{j=1}^{r} Y_{ij}}{ar}$	96.454
Among group (plate) mean square	S_A^2	$\dfrac{r \sum\limits_{i=1}^{a} (\bar{Y}_i - \bar{Y})^2}{a-1}$ where $\bar{Y}_i = \dfrac{\sum\limits_{j=1}^{r} Y_{ij}}{r}$	0.249
Within group mean square	S_E^2	$\dfrac{\sum\limits_{i=1}^{a} \sum\limits_{j=1}^{r} (Y_{ij} - \bar{Y}_i)^2}{a(r-1)}$	0.00237

Table 2.17 One population mean with dependent measurements example

Parameter	Point estimator	Estimate
μ	$\bar{Y}$	96.454
σ_A^2	$\dfrac{S_A^2 - S_E^2}{r}$	0.12
σ_E^2	S_E^2	0.00237
ρ	$\dfrac{S_A^2 - S_E^2}{S_A^2 + (r-1)S_E^2}$	0.98

Statistical intervals are now presented for the parameters in Table 2.17. When ρ is relatively high (say greater than 0.75), interval lengths are primarily determined by the value of a, so one should make a as large as possible, and r as small as possible ($r = 2$).

2.7.1   Confidence Interval for Mean

The bounds for a $100(1 - \alpha)\%$ two-sided confidence interval on μ are

$$L = \bar{Y} - t_{1-\alpha/2:a-1}\sqrt{\frac{S_A^2}{ar}}$$

$$U = \bar{Y} + t_{1-\alpha/2:a-1}\sqrt{\frac{S_A^2}{ar}} \qquad (2.38)$$

A $100(1 - \alpha)\% = 95\%$ two-sided interval on μ has $\alpha/2 = 0.025$ and $t_{1-0.025,6-1} = t_{0.975,5} = 2.570$. Using the statistics from Table 2.16 the computed 95% confidence interval is

$$L = 96.454 - 2.570\sqrt{\frac{0.249}{12}} = 96.1\%$$

$$U = 96.454 + 2.570\sqrt{\frac{0.249}{12}} = 96.8\%. \qquad (2.39)$$

2.7.2   Confidence Intervals for Individual Variances, the Sum of the Variances, and the Ratio

There are two variances of interest in model (2.36), σ_A^2 and σ_E^2. Exact confidence intervals can be computed for both σ_E^2 and ρ. However, only an approximate interval can be computed for σ_A^2. The bounds for the exact $100(1 - \alpha)\%$ two-sided confidence interval on the variance σ_E^2 are

$$L = \frac{a(r-1)S_E^2}{\chi^2_{1-\alpha/2:a(r-1)}}$$

$$U = \frac{a(r-1)S_E^2}{\chi^2_{\alpha/2:a(r-1)}}.$$

(2.40)

The 95% confidence interval on σ_E^2 using the computed statistics in Table 2.16 is

$$L = \frac{6 \times 0.00237}{14.449} = 0.00098$$

$$U = \frac{6 \times 0.00237}{1.237} = 0.011.$$

(2.41)

The 95% confidence interval on the standard deviation σ_E is obtained by taking the square root of L and U in Eq. (2.41). This interval is from 0.031% to 0.11%.

The exact $100(1-\alpha)\%$ two-sided confidence interval on ρ is based on the F-distribution and given by the formula

$$L = \frac{L_1}{1+L_1}$$

$$U = \frac{U_1}{1+U_1}$$

$$L_1 = \frac{S_A^2}{rF_{1-\alpha/2:a-1,a(r-1)}S_E^2} - \frac{1}{r}$$

$$U_1 = \frac{S_A^2}{rF_{\alpha/2:a-1,a(r-1)}S_E^2} - \frac{1}{r}.$$

(2.42)

The 95% confidence interval on ρ using the computed statistics in Table 2.16 is

$$L_1 = \frac{0.249}{2 \times 5.987 \times 0.00237} - \frac{1}{2} = 8.28$$

$$U_1 = \frac{0.249}{2 \times 0.143 \times 0.00237} - \frac{1}{2} = 366$$

$$L = \frac{L_1}{1+L_1} = 0.89$$

$$U = \frac{U_1}{1+U_1} = 1.00.$$

(2.43)

As depicted in Fig. 2.10, the correlation among plates explains most of the variation in the measured values. Thus, selecting a value for r greater than two will not do much to shorten the statistical intervals.

The parameter σ_A^2 is estimated using a difference of two mean squares, S_A^2 and S_E^2. Burdick et al. (pages 16–17) provide two methods for constructing approximate confidence intervals for σ_A^2 under model (2.36). The lengths of the two intervals are comparable and both generally maintain the stated confidence level. The more simple $100(1 - \alpha)\%$ two-sided confidence interval on σ_A^2 is

$$L = \frac{(a - 1)\left(S_A^2 - S_E^2 \times F_{1-\alpha/2:a-1,a(r-1)}\right)}{r \times \chi_{1-\alpha/2:a-1}^2}$$

$$U = \frac{(a - 1)\left(S_A^2 - S_E^2 \times F_{\alpha/2:a-1,a(r-1)}\right)}{r \times \chi_{\alpha/2:a-1}^2}.$$

(2.44)

The lower bound will be negative if $S_A^2/S_E^2 < F_{1-\alpha/2:a-1,a(r-1)}$, and should be increased to zero. Some software packages provide confidence bounds for σ_A^2 based on large sample likelihood approximations. In general, Eq. (2.44) is preferred to these likelihood approximations, because the likelihood equations do not generally maintain the stated confidence level (see, e.g., Yu and Burdick 1995). The 95% confidence interval for σ_A^2 in this example is

$$L = \frac{(5)(0.249 - 0.00237 \times 5.987)}{2 \times 12.832} = 0.05$$

$$U = \frac{(5)(0.249 - 0.00237 \times 0.143)}{2 \times 0.831} = 0.75.$$

(2.45)

Finally, there is generally interest in the total variance, $\sigma_{Total}^2 = \sigma_A^2 + \sigma_E^2$. As will be discussed in Chap. 6, the intermediate precision of an analytical method can be represented in this manner. Intermediate precision is the sum of all components within a laboratory that contribute to the variability of the analytical method. Such factors include analyst, day, and equipment. The point estimator for σ_{Total}^2 is

$$S_{Total}^2 = \frac{S_A^2}{r} + \frac{(r - 1)S_E^2}{r}.$$

(2.46)

Nijhuis and Van den Heuvel (2007) recommend the following approximate confidence interval for σ_{Total}^2. This interval is based on the modified large sample (MLS) method developed by Graybill and Wang (1980). This $100(1 - \alpha)\%$ two-sided confidence interval on σ_{Total}^2 is

$$L = S_{Total}^2 - \sqrt{\left(\frac{G_1 S_A^2}{r}\right)^2 + \left(\frac{(r-1)G_2 S_E^2}{r}\right)^2}$$

$$U = S_{Total}^2 + \sqrt{\left(\frac{H_1 S_A^2}{r}\right)^2 + \left(\frac{(r-1)H_2 S_E^2}{r}\right)^2}$$

$$G_1 = 1 - \frac{a-1}{\chi_{1-\alpha/2:a-1}^2}$$

$$G_2 = 1 - \frac{a(r-1)}{\chi_{1-\alpha/2:a(r-1)}^2} \tag{2.47}$$

$$H_1 = \frac{a-1}{\chi_{\alpha/2:a-1}^2} - 1$$

$$H_2 = \frac{a(r-1)}{\chi_{\alpha/2:a(r-1)}^2} - 1.$$

To complete our example, the point estimator for σ_{Total}^2 is

$$S_{Total}^2 = \frac{0.249}{2} + \frac{(2-1)0.00237}{2} = 0.1257. \tag{2.48}$$

The computed 95% confidence interval on σ_{Total}^2 is

$$G_1 = 1 - \frac{5}{12.832} = 0.610$$

$$G_2 = 1 - \frac{6(2-1)}{17.53455} = 0.585$$

$$H_1 = \frac{5}{0.831} - 1 = 5.105$$

$$H_2 = \frac{6(2-1)}{1.237} - 1 = 3.849$$

$$L = 0.1257 - \sqrt{\left(\frac{0.610 \times 0.249}{2}\right)^2 + \left(\frac{(2-1) \times 0.585 \times 0.00237}{2}\right)^2} = 0.05$$

$$U = 0.1257 + \sqrt{\left(\frac{5.105 \times 0.249}{2}\right)^2 + \left(\frac{(2-1) \times 3.849 \times 0.00237}{2}\right)^2} = 0.75.$$

$$\tag{2.49}$$

2.7.3 Prediction Interval for the Next Observed Process Value

An approximate formula for computing a prediction interval for model (2.36) is obtained by replacing S^2 in (2.21) with S^2_{Total} and the degrees of freedom $(n-1)$ with the value m defined by the Satterthwaite approximation. (The Satterthwaite approximation is discussed in Sect. 2.12.7 of this book.)

The resulting $100(1-\alpha)\%$ two-sided prediction interval is

$$L = \bar{Y} - t_{1-\alpha/2:m} \sqrt{\left(1 + \frac{1}{a \times r}\right) \times S^2_{Total}}$$

$$U = \bar{Y} + t_{1-\alpha/2:m} \sqrt{\left(1 + \frac{1}{a \times r}\right) \times S^2_{Total}} \tag{2.50}$$

$$m = \frac{S^4_{Total}}{\dfrac{S^4_A}{r^2 \times (a-1)} + \dfrac{(r-1) \times S^4_E}{r^2 \times a}}.$$

If using software that does not allow non-integer values for the t-distribution (e.g., Excel), then round m to the nearest integer. Interval (2.50) provides a range that will include the next observed observation, Y_{ij}, with the given level of confidence. Using data from the example problem, the computed 95% two-sided prediction interval for the individual value Y_{ij} is

$$L = 96.454 - 2.57 \sqrt{\left(1 + \frac{1}{12}\right) \times 0.1257} = 95.5\%$$

$$U = 96.454 + 2.57 \sqrt{\left(1 + \frac{1}{12}\right) \times 0.1257} = 97.4\% \tag{2.51}$$

$$m = \frac{0.1257^2}{\dfrac{0.249^2}{(6-1)2^2} + \dfrac{0.00237(2-1)}{6 \times 2^2}} = 5.10 = 5 \text{ (rounded)}$$

Mee (1984) provides an alternative interval that also uses the Satterthwaite approximation. The two intervals are virtually identical for practical applications.

2.7.4   Tolerance Interval for All Future Process Values

The same substitution used for the prediction interval can be used to construct a tolerance interval from the interval in (2.23). In particular, the $100(1 - \alpha)\%$ two-sided tolerance interval that includes $100\,P\%$ of all future observations, Y_{ij}, is

$$
\begin{aligned}
L &= \bar{Y} - K\sqrt{S^2_{Total}} \\
U &= \bar{Y} + K\sqrt{S^2_{Total}} \\
K &= \sqrt{\frac{\left(1 + \dfrac{1}{a \times r}\right)Z^2_{(1+P)/2} \times m}{\chi^2_{\alpha:m}}}
\end{aligned}
\tag{2.52}
$$

where m is defined in (2.50) and S^2_{Total} in (2.46). For the present example, the 95% tolerance interval that includes 90% of future values is

$$
\begin{aligned}
L &= 96.454 - 3.577\sqrt{0.1257} = 95.2\% \\
U &= 96.454 + 3.577\sqrt{0.1257} = 97.7\% \\
K &= \sqrt{\frac{\left(1 + \dfrac{1}{12}\right)1.645^2 \times 5}{1.146}} = 3.577.
\end{aligned}
\tag{2.53}
$$

Hoffman and Kringle (2005) and Quiroz et al. (2016) provide alternative intervals that generally better match the stated confidence level. For this example, the tolerance interval recommended by Hoffman and Kringle is from 95.1 to 97.8%.

Figure 2.11 presents a plot of the prediction and tolerance intervals computed above. The prediction interval is the solid line and the tolerance interval the dashed line. Because most of the variation in the data is due to plate, more plates would need to be collected to reduce the width of the intervals.

2.7.5   Modifications for Unbalanced Designs

The design in this example included an equal number of aliquots (r) for each plate. Such a design is described as balanced because there is an equal number of observational units within each experimental unit. In many situations, the number of observational units is not equal for all experimental units. When this occurs, it is necessary to modify the formulas in this section. A simple set of modifications are based on an approach recommended by Thomas and Hultquist (1978). To use this approach, replace S^2_A with S^2_{AU} and r with r_H where

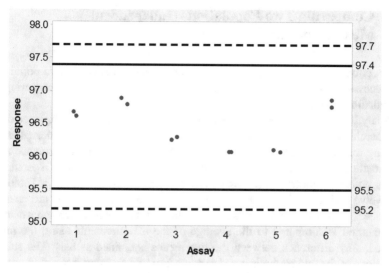

Fig. 2.11 Tolerance and prediction intervals for data in Table 2.15

$$S_{AU}^2 = \frac{r_H \sum_{i=1}^{a} (\bar{Y}_{iU} - \bar{Y}_U)^2}{a-1}$$

$$\bar{Y}_{iU} = \frac{\sum_{j=1}^{r_i} Y_{ij}}{r_i}$$

$$\bar{Y}_U = \frac{\sum_{i=1}^{a} \bar{Y}_{iU}}{a}$$ (2.54)

$$r_H = \frac{a}{\sum_{i=1}^{a} \frac{1}{r_i}}$$

and r_i is the number of observations in the i th experimental unit (plate). The term r_H represents the harmonic mean of the r_i values. Additionally, the term $a(r-1)$ in the definitions of G_2 and H_2 is changed to $\sum_{i=1}^{a} r_i - a$. This modification will provide good approximations in virtually all situations encountered in practice. Burdick et al. (2005) provide worked examples for this approach.

2.8   Comparing Two Populations (Independent Measurements)

This section considers comparisons of the means and variances of two populations or processes. To start with an example, consider the process of comparing two analytical methods.

Analytical methods used to measure quality attributes are often replaced during commercialization with improved methods. Before a new method replaces an existing method, a bridging study is conducted to ensure the performance of the new method will not create any bias that could lead to false out-of-specification measurements. Data from a representative bridging study are shown in Table 2.18.

Twelve vials are selected from a lot of drug product and six vials are randomly assigned to each analytical method. The average of the vial measurements using the new method is compared to the average of the measurements using the present method. Any difference between the two means is defined as bias. The standard deviations of the two methods are also compared to determine if the methods have comparable precision. The statistical model used to represent the data is

$$Y_{ij} = \mu_i + E_{ij} \quad i = 1, 2 \text{ (methods)}; \quad j = 1, \ldots, n_i \text{ (vials per method)}; \quad (2.55)$$

where Y_{ij} is the measured value for vial j using analytical method i, n_i is the number of vials measured using the ith method ($n_1 = n_2 = 6$ in this example), μ_1 and μ_2 are the true averages for the present method and new method, respectively, and E_{ij} is the residual error associated with vial j from method i. It is assumed that the E_{ij} are sampled from a normal population with mean 0 and variance σ_1^2 for the present method and variance σ_2^2 for the new method.

In this example there are two primary contributors to the residual error term: (1) the variability from vial to vial (the manufacturing variability) and (2) the measurement error. Since there is only one measurement per vial, these two errors

Table 2.18 Bridging study with independent measurements	Method	Vial	Protein concentration (mg/mL)
	Present	1	0.426
	Present	2	0.456
	Present	3	0.454
	Present	4	0.444
	Present	5	0.456
	Present	6	0.440
	New	7	0.449
	New	8	0.476
	New	9	0.467
	New	10	0.452
	New	11	0.473
	New	12	0.461

Table 2.19 Statistics for two populations with independent measurements

Statistic	Symbol	Formula
Sample mean group 1	$\bar{Y}_1$	$\dfrac{\sum_{j=1}^{n_1} Y_{1j}}{n_1}$
Sample mean group 2	$\bar{Y}_2$	$\dfrac{\sum_{j=1}^{n_2} Y_{2j}}{n_2}$
Sample variance group 1	S_1^2	$\dfrac{\sum_{j=1}^{n_1}(Y_{1j}-\bar{Y}_1)^2}{n_1-1}$
Sample variance group 2	S_2^2	$\dfrac{\sum_{j=1}^{n_2}(Y_{2j}-\bar{Y}_2)^2}{n_2-1}$

Table 2.20 Two population means with independent measures example

Parameter	Point estimator	Computed estimate Table 2.18 (mg/mL)
μ_1	$\bar{Y}_1$	0.446
μ_2	$\bar{Y}_2$	0.463
σ_1^2	S_1^2	0.0001408
σ_2^2	S_2^2	0.0001212

cannot be separated. In the present example, it is assumed the majority of the variation is due to measurement error since we expect the manufactured drug product to be uniform (homogeneous) from vial to vial within a lot.

Table 2.19 reports the statistics needed to perform the required calculations where group 1 is "Present" and group 2 is "New."

Table 2.20 reports the point estimators using the data in Table 2.18.

2.8.1 Confidence Interval for Difference in Means

There are two versions for the confidence interval on the difference in two means. The appropriate interval depends on the relationship between the two variances, σ_1^2 and σ_2^2. If one assumes the variances are equal, then the appropriate $100(1-\alpha)\%$ two-sided confidence interval on $\mu_1 - \mu_2$ is

$$L = \bar{Y}_1 - \bar{Y}_2 - t_{1-\alpha/2:n_1+n_2-2}\sqrt{S_P^2\left(\frac{1}{n_1}+\frac{1}{n_2}\right)}$$

$$U = \bar{Y}_1 - \bar{Y}_2 + t_{1-\alpha/2:n_1+n_2-2}\sqrt{S_P^2\left(\frac{1}{n_1}+\frac{1}{n_2}\right)} \tag{2.56}$$

$$S_P^2 = \frac{(n_1-1)S_1^2+(n_2-1)S_2^2}{n_1+n_2-2}.$$

Equation (2.56) is referred to as the pooled confidence interval.

To illustrate Eq. (2.56) using the estimates in Table 2.20, a $100(1 - \alpha)\% = 95\%$ two-sided interval on $\mu_1 - \mu_2$ has $\alpha/2 = 0.025$ and $t_{1-0.025:12-2} = t_{0.975:10} = 2.228$. The computed 95% confidence interval is

$$L = 0.446 - 0.463 - 2.228\sqrt{0.000131\left(\frac{1}{6} + \frac{1}{6}\right)} = -0.032 \text{ mg/mL}$$

$$U = 0.446 - 0.463 + 2.228\sqrt{0.000131\left(\frac{1}{6} + \frac{1}{6}\right)} = -0.002 \text{ mg/mL} \qquad (2.57)$$

$$S_P^2 = \frac{(6-1)(0.0001408) + (6-1)(0.0001212)}{6+6-2} = 0.000131$$

If, on the other hand, evidence suggests the variances are not equal, it is more appropriate to use the formula

$$L = \bar{Y}_1 - \bar{Y}_2 - t_{1-\alpha/2:df}\sqrt{\frac{S_1^2}{n_1} + \frac{S_2^2}{n_2}}$$

$$U = \bar{Y}_1 - \bar{Y}_2 + t_{1-\alpha/2:df}\sqrt{\frac{S_1^2}{n_1} + \frac{S_2^2}{n_2}} \qquad (2.58)$$

$$df = \frac{\left(\frac{S_1^2}{n_1} + \frac{S_2^2}{n_2}\right)^2}{\frac{S_1^4}{n_1^2(n_1 - 1)} + \frac{S_2^4}{n_2^2(n_2 - 1)}}.$$

The computed interval for the example problem using Eq. (2.58) is

$$L = 0.446 - 0.463 - 2.228\sqrt{\frac{0.0001408}{6} + \frac{0.0001212}{6}} = -0.032 \text{ mg/mL}$$

$$U = 0.446 - 0.463 + 2.228\sqrt{\frac{0.0001408}{6} + \frac{0.0001212}{6}} = -0.002 \text{ mg/mL}$$

$$df = \frac{\left(\frac{0.0001408}{6} + \frac{0.0001212}{6}\right)^2}{\frac{0.0001408^2}{6^2(6-1)} + \frac{0.0001212^2}{6^2(6-1)}} = 9.94 = 10 \text{ (rounded)}.$$

$$(2.59)$$

Comparison with the pooled confidence interval demonstrates that the intervals are the same in this example. This is generally not the case. It occurs in this example because there are equal observations in each group and S_1^2 and S_2^2 are very close to being equal.

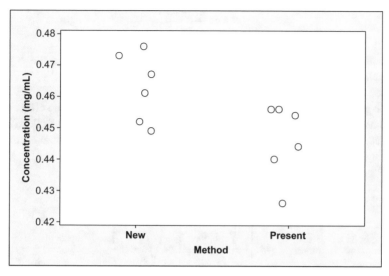

Fig. 2.12 Plot of data in Table 2.18

Whether variances should be considered equal or not can be determined in two ways. First, the data can be plotted after grouping by population. A visual assessment of this plot often provides sufficient guidance for making a decision. Statistical tests can be used to test the null hypothesis that two variances are equal. One such test can be performed using the confidence interval on the ratio of variances provided in Sect. 2.8.3. However, the probability of rejecting the null hypothesis of equal variances depends greatly on the total sample size, $n_1 + n_2$. If the sample size is relatively small, then one will generally not reject the assumed model of equal variance. Conversely, if the sample size is large, one will likely reject the assumption, even in cases where the equality assumption is reasonable. Thus, we believe the pooling decision is better based on a plot of the data combined with subject matter expert knowledge of the specific problem. Figure 2.12 displays the plotted data for the data in Table 2.18.

Since the spreads of the data are similar, it is decided to use the pooled interval in Eq. (2.56) to compute the desired confidence interval.

If there is interest in obtaining individual confidence intervals on the mean for each group, the confidence interval in Eq. (2.8) can be used for this purpose.

2.8.2 Confidence Interval for the Effect Size

It is often useful to express a difference of means relative to the standard deviation of population 1. As will be discussed in Chap. 9, such a measure is useful in the analysis of biosimilarity data. This measure is called the effect size and is defined as

$$\lambda = \frac{\mu_1 - \mu_2}{\sigma_1}. \tag{2.60}$$

Confidence intervals on λ can be computed whether variances are equal or unequal. Each situation is described below.

Case 1: $\sigma_1 = \sigma_2$.

The general approach for constructing a confidence interval on the effect size is based on the inversion confidence interval principle as discussed by Kelley (2007). For Case 1, the maximum likelihood estimator (with degree of freedom correction) for λ is

$$\hat{\lambda} = \frac{\bar{Y}_1 - \bar{Y}_2}{S_P}$$

$$= \left[\frac{\bar{Y}_1 - \bar{Y}_2}{S_P \sqrt{\frac{1}{n_1} + \frac{1}{n_2}}} \right] \times \sqrt{\frac{1}{n_1} + \frac{1}{n_2}} \tag{2.61}$$

$$= t_{calc} \times \sqrt{\frac{1}{n_1} + \frac{1}{n_2}}.$$

The statistic t_{calc} is the familiar test statistic used to test equality of means for two independent groups. It has a non-central t-distribution with non-centrality parameter

$$ncp = \frac{\lambda}{\sqrt{\frac{1}{n_1} + \frac{1}{n_2}}}. \tag{2.62}$$

In order to form a confidence interval on λ based on $\hat{\lambda}$, it is necessary to first form a confidence interval on ncp. Once a two-sided confidence interval is obtained for ncp, each bound is multiplied by $\sqrt{\frac{1}{n_1} + \frac{1}{n_2}}$ to obtain a confidence interval on λ. To demonstrate, the following SAS code is used to compute the confidence interval on λ. The function "tnonct" returns the ncp that yields the specified probability (e.g., 0.95 or 0.05) for the observed value of t_{calc}.

```
twosideconf=0.90;
tcalc=meandiff/sqrt(pooledvar*(1/n1+1/n2));
lbeffect= tnonct(tcalc,n1+n2-2,(1+twosideconf)/2)*sqrt(1/n1+1/n2);
ubeffect= tnonct(tcalc,n1+n2-2, (1-twosideconf)/2)*sqrt(1/n1+1/n2);
```

A more detailed explanation and R code is provided by Kelley. Note that this procedure provides an exact confidence interval on λ.

Case 2: $\sigma_1 \neq \sigma_2$.

This situation is a bit more problematic as only information for the first sample is used to estimate the denominator of λ. A useful approach for this situation is to employ a generalized confidence interval (GCI) as described in Appendix B of Burdick et al. (2005). Hannig et al. (2006) have shown that under very mild conditions, these intervals provide correct frequentist coverage. That is, for practical purposes, they can be considered as "exact" confidence intervals.

For the present application, a GCI can be computed using the following steps:

1. Compute the sample means $\bar{X}_1$ and $\bar{X}_2$ and the sample variances S_1^2 and S_2^2 for the sample data sets of size n_1 and n_2, respectively.
2. Simulate N values of the effect size:

$$\lambda_{sim} = \frac{\bar{Y}_1 - \bar{Y}_2 - Z \times \sqrt{\dfrac{(n_1 - 1) \times S_1^2}{n_1 \times W_1} + \dfrac{(n_2 - 1) \times S_2^2}{n_2 \times W_2}}}{\sqrt{\dfrac{(n_2 - 1) \times S_2^2}{W_2}}} \tag{2.63}$$

where W_1 is a chi-squared random variable with $n_1 - 1$ degrees of freedom, W_2 is a chi-squared random variable with $n_2 - 1$ degrees of freedom, and Z is a standard normal random variable with mean 0 and variance 1. A value of $N \geq 10,000$ is recommended.
3. Order the N simulated λ_{sim} values obtained in Step (2) from least to greatest.
4. Define the lower bound for a two-sided $100(1 - \alpha)\%$ confidence interval as the value in position $N \times (\alpha/2)$ of the ordered data set in Step (3). Define the upper bound as the value in position $N \times (1 - \alpha/2)$ of this same ordered set. For example, if $N = 10,000$ the lower bound of a 95% two-sided confidence interval is the value in position $10,000 \times 0.025 = 250$ and the upper bound is the value in position $10,000 \times 0.975 = 9,750$.

Note that these steps can be computed with any software package that contains sorting and simulation functions including Excel.

To demonstrate, the data in Table 2.20 are used to compute a 95% confidence interval on λ assuming equal variances. Here

$$\hat{\lambda} = \frac{\bar{Y}_1 - \bar{Y}_2}{S_P} = \frac{0.446 - 0.463}{\sqrt{0.000131}} = -1.485$$

$$t_{calc} = \frac{\hat{\lambda}}{\sqrt{\dfrac{1}{n_1} + \dfrac{1}{n_2}}} = \frac{-1.485}{\sqrt{\dfrac{1}{6} + \dfrac{1}{6}}} = -2.573 \tag{2.64}$$

and the resulting 95% two-sided confidence interval on ncp is from -4.778 to -0.271. Converting this interval to the 95% confidence interval on the effect size

W1 uniform	W1	W2 uniform	W2	Z	Mean1	Mean2	SD1	SD2	n1	n2	λ sim
0.361	5.476	0.425	4.924	-0.133	0.446	0.463	0.012	0.011	6	6	-1.455
0.300	6.066	0.474	4.543	0.620	0.446	0.463	0.012	0.011	6	6	-1.818
0.677	3.151	0.566	3.885	-0.045	0.446	0.463	0.012	0.011	6	6	-1.333
0.379	5.310	0.419	4.971	-1.203	0.446	0.463	0.012	0.011	6	6	-0.830
0.020	13.390	0.436	4.834	1.985	0.446	0.463	0.012	0.011	6	6	-2.484
0.610	3.586	0.262	6.486	-0.247	0.446	0.463	0.012	0.011	6	6	-1.581
0.529	4.146	0.836	2.093	0.592	0.446	0.463	0.012	0.011	6	6	-1.304
0.522	4.195	0.424	4.932	0.156	0.446	0.463	0.012	0.011	6	6	-1.632
0.352	5.557	0.839	2.070	-0.379	0.446	0.463	0.012	0.011	6	6	-0.808
0.777	2.499	0.217	7.052	-0.164	0.446	0.463	0.012	0.011	6	6	-1.695

Fig. 2.13 Example excel worksheet

Table 2.21 Summary of 10,000 values of λ_{sim}

Mean	-1.49
Median	-1.42
Standard deviation	0.87
Minimum	-13.05
Maximum	5.74
Count	10000.00
Largest (250)	-0.01
Smallest (250)	-3.38

$$L = \sqrt{\frac{1}{n_1} + \frac{1}{n_2}} \times -4.778 = \sqrt{\frac{1}{6} + \frac{1}{6}} \times -4.778 = -2.76$$

$$U = \sqrt{\frac{1}{n_1} + \frac{1}{n_2}} \times -0.271 = \sqrt{\frac{1}{6} + \frac{1}{6}} \times -0.271 = -0.16.$$

(2.65)

Note the effect size has no units of measure.

The algorithm described for Case 2 is used to compute the confidence interval on the effect size using an unequal variance assumption. Figure 2.13 presents 10 rows of an Excel sheet that demonstrates the required calculation by simulating the random chi-squared values using the uniform distribution and the Excel function CHIINV to obtain W_1 and W_2.

Table 2.21 presents a summary for 10,000 iterations that produces the 95% confidence interval from -3.38 to -0.01.

2.8.3 Confidence Interval for the Ratio of Two Variances

The best way to compare two variances is to examine their ratio. This is because the difference of two variances is reported in squared units of the measurement scale and has no practical interpretation. Conversely, a ratio of two variances has no units and expresses a magnitude of difference (e.g., the variance of group 1 is twice as great as the variance of group 2). If two variances are equal, the ratio is equal to one. The $100(1 - \alpha)\%$ two-sided confidence interval on the ratio σ_1^2/σ_2^2 is

$$L = \frac{S_1^2}{S_2^2 \times F_{1-\alpha/2:n_1-1,n_2-1}}$$

$$U = \frac{S_1^2}{S_2^2 \times F_{\alpha/2:n_1-1,n_2-1}}. \tag{2.66}$$

Using the numerical values in Table 2.18 with $\alpha = 0.05$, $F_{0.975:5,5} = 7.146$, $F_{0.025:5,5} = 0.140$, and

$$L = \frac{0.0001408}{0.0001212 \times 7.146} = 0.16$$

$$U = \frac{0.0001408}{0.0001212 \times 0.140} = 8.3 \tag{2.67}$$

Equation (2.10) can be used to construct confidence intervals on the individual variances.

2.9   Confidence Interval for Difference of Means (Dependent Measurements)

Suppose the data in Table 2.18 were collected in a different manner. In particular, instead of assigning six different vials to each analytical method, a single set of six vials was selected at random from the manufacturing process. Each vial is split into half, and each half is measured using a different analytical method. This new design is called a paired design because each experimental unit (vial) is measured with both treatments. Table 2.22 reports the same numerical values as Table 2.18, but is organized in a manner consistent with the data collection process. As will be demonstrated, the resulting confidence interval on the mean difference between the two methods will differ from those computed in the previous section even though the numerical values in the two examples are the same. This will demonstrate the importance of knowing how data are collected in order to perform the most appropriate statistical analysis.

Table 2.22 Bridging study with paired design

Vial	Concentration (mg/mL) Present method	Concentration (mg/mL) New method	Concentration (mg/mL) Present-new
1	0.426	0.449	−0.023
2	0.456	0.476	−0.020
3	0.454	0.467	−0.013
4	0.444	0.452	−0.008
5	0.456	0.473	−0.017
6	0.440	0.461	−0.021

The fact that each vial is measured using both methods creates a dependent relationship for the measured values on the same vial.

The two population model with dependent measurements is written as

$$Y_{ij} = \mu_i + A_j + E_{ij} \quad i = 1, 2 \text{ (methods)}; \quad j = 1, \ldots, n \text{ (vials)}; \qquad (2.68)$$

where Y_{ij} is the measured value for vial j with analytical method i, n is the total number of vials ($n = 6$ in this example), μ_1 and μ_2 are the true average measurements for the present method and new method, respectively, A_j is random manufacturing error that creates vial-to-vial differences, and E_{ij} is the residual error associated with vial j from method i. It is assumed that A_j are sampled from a normal population of vial effects with mean 0 and variance σ_A^2. The E_{ij} are sampled from a normal population with mean 0 and variance σ_1^2 for the present method and variance σ_2^2 from the new method. The covariance between two measurements within the same vial is σ_A^2.

It is traditional in a paired design to model the difference between the two measured values for each vial. That is, the statistical model considers the difference

$$D_j = Y_{1j} - Y_{2j} \quad j = 1, \ldots, n \text{ (vials)}. \qquad (2.69)$$

By substituting the definitions for Y_{1j} and Y_{2j} from Eq. (2.68) into Eq. (2.69) it is shown that the vial-to-vial variation is removed from the model. That is,

$$
\begin{aligned}
D_j &= Y_{1j} - Y_{2j} \\
&= \mu_1 + A_j + E_{1j} - \left(\mu_2 + A_j + E_{2j} \right) \\
&= \mu_1 - \mu_2 + E_{1j} - E_{2j} \\
&= \mu_D + E_j^*
\end{aligned}
\qquad (2.70)
$$

where $\mu_D = \mu_1 - \mu_2$ is the difference between the two method means and $E_j^* = E_{1j} - E_{2j}$ is a random normal variable with mean 0 and variance $\sigma_D^2 = \sigma_1^2 + \sigma_2^2$.

Note that model (2.70) is now in the same form as the independent model (2.7). Thus, we can use Eq. (2.8) with the D_j values to construct a $100(1 - \alpha)\%$ two-sided confidence interval on μ_D. This interval is

$$L = \bar{D} - t_{1-\alpha/2:n-1} \sqrt{\frac{S_D^2}{n}}$$

$$U = \bar{D} + t_{1-\alpha/2:n-1} \sqrt{\frac{S_D^2}{n}}$$

$$\bar{D} = \frac{\sum\limits_{j=1}^{n} D_j}{n} \qquad (2.71)$$

$$S_D^2 = \frac{\sum\limits_{j=1}^{n} \left(D_j - \bar{D} \right)^2}{n - 1}.$$

The D_j values for the example problem are shown in the last column of Table 2.22. The computed 95% confidence interval on the difference in means $\mu_1 - \mu_2$ is

$$L = -0.017 - 2.571\sqrt{\frac{0.0000316}{6}} = -0.023 \text{ mg/mL}$$

$$U = -0.017 + 2.571\sqrt{\frac{0.0000316}{6}} = -0.011 \text{ mg/mL}$$

(2.72)

where $\bar{D} = -0.017$ and $S_D^2 = 0.0000316$.

Recall the interval for the independent design in Table 2.18 was $L = -0.032$ mg/mL to $U = -0.002$ mg/mL. Note that although the average difference is the same for both intervals, -0.017, the interval based on the paired design is tighter. This occurs because elimination of the vial-to-vial variance reduces the uncertainty in the estimated difference and leads to a tighter confidence interval. This example demonstrates that the manner in which data are collected is an important consideration when selecting an appropriate statistical analysis, and that efficiencies are gained by proper experimental design.

2.10 Basics of Hypothesis Testing

A hypothesis test is a procedure used to determine the amount of evidence a sample provides for concluding that a population parameter is in an interval specified by an investigator. In this section, we review the basics of hypothesis testing.

2.10.1 Statement of Hypotheses

To conduct a hypothesis test, an investigator makes two complementary statements about an unknown parameter. To demonstrate, we use the Greek letter θ (theta) to represent an unknown parameter such as μ (a process mean) or σ^2 (a process variance). An investigator makes two complementary statements about θ. The first statement is called the null hypothesis and is denoted as H_0. The second statement is called the alternative (or research) hypothesis and is denoted as H_a.

In general, an investigator wants to determine the amount of evidence a sample provides to support a claim or conjecture. This claim is chosen as H_a, the alternative hypothesis. If the sample evidence is sufficient to convince an investigator that the null hypothesis H_0 is false (and hence the alternative hypothesis H_a is true), the result of the test is stated as "Reject H_0". If the sample evidence is not sufficient to convince the investigator that the null hypothesis H_0 is false, the result of the test is stated as "Do not reject H_0". The statement "Do not reject H_0" does *not* mean the

Table 2.23 Hypothesis sets of interest

Row	H_0	H_a	Sample must prove...
1	$\theta \leq \theta_0$	$\theta > \theta_0$	θ exceeds θ_0
2	$\theta \geq \theta_0$	$\theta < \theta_0$	θ is less than θ_0
3	$\theta = \theta_0$	$|\theta - \theta_0| > 0$	θ is different from θ_0
4	$|\theta - \theta_0| \geq K$	$|\theta - \theta_0| < K$	θ differs from θ_0 by less than K

sample provides evidence that H_0 is true. It simply means that the sample evidence is not sufficient for the investigator to reject H_0 and claim H_a is true. Four sets of hypotheses are shown in Table 2.23.

The hypothesis in row 4 of Table 2.23 is referred to as the hypothesis of equivalence.

2.10.2 Testing Errors and Power

Since sample data are used to make a conclusion, one might ask the question, "What is the probability of getting sample data that lead to an erroneous conclusion?" That is, what is the probability that the sample data will indicate H_0 should be rejected when H_0 is true? Or, what is the probability that the sample data will indicate that H_0 should not be rejected when H_0 is false?

The following two mistakes (errors) are possible when conducting a hypothesis test:

1. The mistake of rejecting H_0 when H_0 is true. The probability of making this mistake is denoted by the Greek letter α (alpha). It is referred to as a type I error.
2. The mistake of not rejecting H_0 when H_0 is false. The probability of making this mistake is denoted by the Greek letter β (beta). It is referred to as a type II error. The difference $1 - \beta$ is called the power of the test. The power of the test depends on the true value of θ.

The probabilities of making either of the two mistakes are schematically exhibited in Table 2.24.

When conducting a hypothesis test, it is desirable to choose small values for both α and β so that the probability of making a mistake is small. Typically, the value of α is a commonly accepted value such as 0.05. Once α is fixed, it is necessary to determine the required sample size to obtain the desired β. The process of determining the required sample size to ensure both α and β are at acceptable levels is called a power calculation. Power calculations are applied throughout the book as the need arises.

Table 2.24 Probabilities of errors in a statistical test

	If H_0 is true	If H_0 is false
Reject H_0	Probability of a mistake is α	Correct decision
Do not reject H_0	Correct decision	Probability of a mistake is β

2.10.3 Using Confidence Intervals to Conduct Statistical Tests

The simplest manner to conduct a hypothesis test is to compute a confidence interval on the parameter and then compare it to the set of hypotheses. An alternative testing procedure employs a p-value. We prefer using a confidence interval because it encourages one to also consider the practical importance of any differences that might exist.

The rule for using a confidence interval to test a statistical hypothesis is quite simple. If every value contained in the confidence interval is consistent with the alternative hypothesis, then reject H_0 and claim H_a is true. In all other cases, fail to reject H_0. The two-sided confidence interval should have a confidence coefficient of $100(1 - 2\alpha)\%$ for testing the hypotheses in rows 1, 2, and 4 of Table 2.23. The two-sided confidence coefficient for the hypothesis set in row 3 is $100(1 - \alpha)\%$. This convention will provide a type I error rate of α for all four sets of hypotheses.

Recall the earlier example in this chapter where it was desired to demonstrate that the mean purity of a manufacturing process is greater than 93%. In the context of this example, the data must support the condition that the process mean is greater than 93%. This requires testing the set of hypotheses in row 1 of Table 2.23 with $\theta_0 = 93\%$. These hypotheses are

$$
\begin{aligned}
H_0 &: \text{The process mean } \leq 93\% \\
H_a &: \text{The process mean } > 93\%
\end{aligned}
\tag{2.73}
$$

The two-sided 90% confidence interval on the process mean is from $L = 93.78$ % to $U = 94.83\%$. The decision using this confidence interval is to reject H_0 since all values in the confidence interval exceed 93%. Note that since the two-sided confidence coefficient is 90%, then $100(1 - 2\alpha)\% = 90\%$ and $\alpha = 0.05$.

2.10.4 Using p-Values to Conduct a Statistical Test

Another approach for testing a statistical hypothesis that is equivalent to the confidence interval approach is based on a p-value (probability value). The p-value is the smallest value of α for which H_0 can be rejected. The p-value measures the amount of sample evidence in favor of rejecting H_0, with smaller values providing stronger evidence. If a computed p-value is less than the pre-selected value of α, then reject H_0 and claim H_a is true. Otherwise, fail to reject H_0. The p-value approach will provide the same test results as the confidence interval procedure described previously. However, we believe the confidence interval provides a more scientifically enlightened assessment of the data.

2.11   Equivalence Testing

ICH Q5E (2004) states the goal of a comparability study is to determine the potential adverse impact of manufacturing changes on product quality attributes. Pre- and post-change products do not need to have identical quality attributes, but do need to be highly similar with scientific justification that any observed differences will not impact safety or efficacy. One statistical approach for demonstrating similarity is equivalence testing.

The hypothesis set in row 4 of Table 2.23 can be used to demonstrate the equivalence of two process means, μ_1 and μ_2. In particular, the hypotheses are stated as

$$H_0 : |\mu_1 - \mu_2| \geq \text{EAC}$$
$$H_a : |\mu_1 - \mu_2| < \text{EAC}$$

(2.74)

where EAC is referred to as the Equivalence Acceptance Criterion. To test the hypotheses in Eq. (2.74), one constructs a two-sided $100(1 - 2\alpha)\%$ confidence interval on the difference $\mu_1 - \mu_2$. The null hypothesis H_0 in (2.74) is rejected and equivalence is demonstrated if the entire confidence interval from L to U falls in the range from $-\text{EAC}$ to $+\text{EAC}$. As noted by Berger and Hsu (1996), use of an equal-tailed two-sided $100(1 - 2\alpha)\%$ confidence interval provides a test size (type I error rate) of α. This procedure is referred to as the two one-sided statistical test procedure (TOST).

It is important to remember that failure to reject H_0 *does not* imply that the two processes are not equivalent. Figure 2.14 presents three possible outcomes for an equivalence test of means.

Scenario A is the situation where the confidence interval is entirely contained in the range from $-\text{EAC}$ to $+\text{EAC}$. The conclusion here is to reject H_0 and claim the two process means are equivalent. In both Scenarios B and C, at least some of the values in the confidence interval fall outside the range from $-\text{EAC}$ to $+\text{EAC}$. Scenario C provides an inconclusive result, since as noted earlier, failure to reject H_0 does not imply that H_0 is true. In contrast, since all values in the confidence interval exceed EAC in Scenario B, one can reject H_0 in the following set of hypotheses:

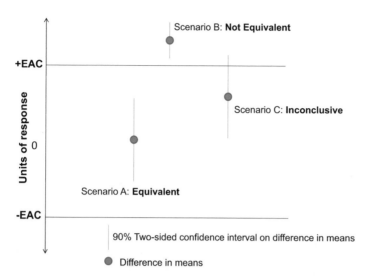

Fig. 2.14   Outcomes of an equivalence test

$$H_0 : |\mu_1 - \mu_2| \leq \text{EAC}$$
$$H_a : |\mu_1 - \mu_2| > \text{EAC}$$
(2.75)

That is, the data provide evidence that the two means *are not* equivalent.

When employing a test of equivalence, the EAC should represent an amount viewed to have no practical importance. Establishment of an EAC is sometimes based on existing information of the pre-change condition to ensure the same level of quality (see, e.g., Appendix E of USP <1010>). Another approach is to base the criterion on the process capability of the pre-change process. Such an approach is described by Limentani et al. (2005).

Consider the bridging study presented in Table 2.22. Prior to conducting the study, it was determined that the two methods would be considered equivalent if the true difference in mean concentration was no greater than 0.03 mg/mL. The hypotheses for this equivalence test are stated as

$$H_0 : |\mu_1 - \mu_2| \geq 0.03$$
$$H_a : |\mu_1 - \mu_2| < 0.03$$
(2.76)

The computed two-sided 90% confidence interval on the difference in means using Eq. (2.71) is

$$L = -0.017 - 2.015\sqrt{\frac{0.0000316}{6}} = -0.022 \text{ mg/mL}$$

$$U = -0.017 + 2.015\sqrt{\frac{0.0000316}{6}} = -0.012 \text{ mg/mL}$$
(2.77)

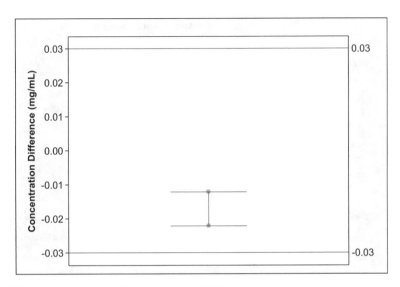

Fig. 2.15 Comparison of confidence interval to EAC

where $\bar{D} = -0.017$ and $S_D^2 = 0.0000316$. Figure 2.15 shows a plot of the confidence interval relative to the range $-$EAC to $+$EAC. Since all values in the confidence interval fall within the range from -0.03 to $+0.03$, the test demonstrates that the two methods are equivalent.

As a final reminder, after the EAC has been established, it is important to apply a power calculation to determine appropriate sample sizes. This is necessary to provide an acceptable probability of passing the test when means are equivalent. Section 9.5.1 later in this book demonstrates such a power calculation.

2.12 Regression Analysis

To this point, we have been concerned with estimation of unknown population and process parameters. Another useful statistical application is prediction. Consider a manufactured lot consisting of individual vials of drug product. Let the measurement Y represent the concentration as a percentage of label. It is known that the value of Y will decrease over time. Every six months, a single vial is selected at random from the lot and the concentration is measured. The time periods when values are collected are 0, 6, 12, 18, 24, and 30 months. These time periods represent values assigned by the variable "Time" which is denoted with the letter X. The variable X is called a predictor variable. The variable "Concentration" denoted by Y is called the response variable. Regression analysis is used to predict Y for a given value of X. Table 2.25 provides an example data set. A scatterplot of the data is provided in Fig. 2.16.

Table 2.25 Concentration
(%) over time

Vial	Concentration (%) (Y)	Time in months (X)
1	102.1	0
2	101.4	6
3	101.0	12
4	101.1	18
5	100.8	24
6	99.6	30

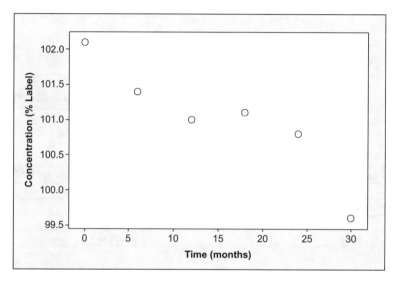

Fig. 2.16 Scatterplot of data in Table 2.25

2.12.1   Linear Regression with One Predictor Variable

In this section we demonstrate how regression analysis is used to obtain prediction
functions when there is a single predictor variable, X.

In many cases it is reasonable to assume that Y is a linear function of X. In such
cases, the form of the statistical model that relates Y and X is

$$Y_i = \beta_0 + \beta_1 X_i + E_i \quad i = 1, \ldots, n; \tag{2.78}$$

where Y_i is the measured response for item i, X_i is the known value of the predictor
variable associated with item i, β_0 is the y-intercept of the linear relationship, β_1 is
the slope of the linear relationship, and E_i is the residual error for item i. The error
term E_i is assumed to have a normal distribution with mean 0 and variance σ^2. Note
that model (2.78) is essentially the same as model (2.7) with $\beta_0 + \beta_1 X_i$ replacing μ.
For this reason, it is helpful to think of the term $\beta_0 + \beta_1 X_i$ as the *average* of Y when

Table 2.26 Statistics for simple linear regression calculations

Statistic	Symbol	Formula
Sample mean for Y	$\bar{Y}$	$\dfrac{\sum\limits_{i=1}^{n} Y_i}{n} = \dfrac{606}{6} = 101$
Sample mean for X	$\bar{X}$	$\dfrac{\sum\limits_{i=1}^{n} X_i}{n} = \dfrac{90}{6} = 15$
Sample slope	b_1	$\dfrac{\sum\limits_{i=1}^{n} (Y_i - \bar{Y})(X_i - \bar{X})}{\sum\limits_{i=1}^{n} (X_i - \bar{X})^2} = \dfrac{-42.6}{630} = -0.0676$
Sample y-intercept	b_0	$\bar{Y} - b_1\bar{X} = 101 - (-0.0676)15 = 102.014$
Sample variance	S^2	$\dfrac{\sum\limits_{i=1}^{n} (Y_i - b_0 - b_1 X_i)^2}{n-2} = \dfrac{0.499429}{4} = 0.1249$

Table 2.27 Simple linear regression example

Parameter	Point estimator	Computed estimate for Table 2.25
β_1	b_1	−0.0676
β_0	b_0	102.014
σ^2	S^2	0.1249
σ	S	0.3534

the predictor variable is equal to a particular value, say X_P. More formally, $\beta_0 + \beta_1 X_P$ is referred to as the *conditional mean* of Y given the predictor $X = X_P$.

Table 2.26 reports the sample statistics needed to compute a regression function for model (2.78). The computed values in Table 2.26 are for the example data in Table 2.25. Table 2.27 reports the parameters, their point estimators, and computed estimates.

In the example, it is desired to predict the value of Y when the value of X is equal to the product expiry, $X_P = 30$ months. The formula for predicting Y given X_P is

$$\hat{Y}_P = b_0 + b_1 X_P. \tag{2.79}$$

Plugging in the values from Table 2.27 to predict the mean of all Y values when $X_P = 30$ months yields

$$\hat{Y}_P = 102.014 - 0.0676(30) = 99.99\%. \tag{2.80}$$

The estimated value $\hat{Y}_P = 99.99\%$ provides the answer to two different questions:

1. What is the best guess for the *average* concentration of *all* vials in the lot at 30 months?
2. What is the best guess for the concentration of an *individual* vial selected at random from the lot at 30 months?

Thus, Eq. (2.79) can be used to draw an inference for both individual values and the average of all individual values. Recall the discussion in Sect. 2.5.4 in which it was noted that uncertainties associated with point estimates vary depending on whether the inference concerns an *individual* or an *average* of individuals. There is always less uncertainty associated with an average than there is for an individual. In our example, there is less uncertainty about the average of all the vials at 30 months than there is for the value of a single vial selected at random at 30 months. We again use interval estimates to quantify the uncertainty associated with both individuals and averages. In particular, formulas are now provided for

1. a confidence interval on the average of all vials at X_P months,
2. a prediction interval for the value of a single vial selected at random from all vials at X_P months, and
3. a tolerance interval that contains $100P\%$ of all vial values from the lot at X_P months.

To begin, the $100(1 - \alpha)\%$ two-sided confidence interval on the mean of all vials at X_P months is

$$L = \hat{Y}_P - t_{1-\alpha/2:n-2} \times S \times d$$
$$U = \hat{Y}_P + t_{1-\alpha/2:n-2} \times S \times d$$
$$d = \sqrt{\frac{1}{n} + \frac{(X_P - \bar{X})^2}{\sum\limits_{i=1}^{n}(X_i - \bar{X})^2}}. \tag{2.81}$$

The product $S \times d$ is called the estimated standard error of the fitted value.

The $100(1 - \alpha)\%$ two-sided prediction interval for a single vial selected at random at X_P months is

$$L = \hat{Y}_P - t_{1-\alpha/2:n-2} \times S \times \sqrt{1 + d^2}$$
$$U = \hat{Y}_P + t_{1-\alpha/2:n-2} \times S \times \sqrt{1 + d^2}. \tag{2.82}$$

An exact $100(1 - \alpha)\%$ one-sided tolerance interval is based on a non-central t-distribution. A one-sided interval is used when interest is only in the value that is greater than $100P\%$ of the population, or the value that is less than $100P\%$ of the population. The $100(1 - \alpha)\%$ one-sided tolerance bound that is less than $100P\%$ of all vials at X_P months is

$$L = \hat{Y}_P - t_{1-\alpha:n-2}(Z_P/d) \times S \times d \tag{2.83}$$

where $t_{1-\alpha:n-2}(Z_P/d)$ is the percentile from a non-central t-distribution with area $1 - \alpha$ to the left, $n - 2$ degrees of freedom, non-centrality parameter (Z_P/d), and Z_P is the standard normal percentile that has area P to the left (e.g., $Z_{0.95} = 1.645$). Similarly, an upper $100(1 - \alpha)\%$ one-sided tolerance bound that is greater than $100P\%$ of all vials at X_P months is

$$U = \hat{Y}_P + t_{1-\alpha:n-2}(Z_P/d) \times S \times d. \tag{2.84}$$

An exact $100(1 - \alpha)\%$ two-sided tolerance interval that contains at least $100P\%$ of all vial values from the lot at X_P months was derived by Eberhardt et al. (1989). It is computationally intensive, and a more simple approximation based on (2.83) and (2.84) is

$$\begin{aligned} L &= \hat{Y}_P - t_{1-\alpha/2:n-2}(Z_P/d) \times S \times d \\ U &= \hat{Y}_P + t_{1-\alpha/2:n-2}(Z_P/d) \times S \times d. \end{aligned} \tag{2.85}$$

Krishnamoorthy and Mathew note that interval (2.85) is an excellent approximation when $d^2 \geq 0.30$. In situations where $d^2 < 0.30$, they recommend a very simple approximation due to Wallis

$$\begin{aligned} L &= \hat{Y}_P - \sqrt{\frac{(n-2) \times \chi^2_{P:1}(d^2)}{\chi^2_{\alpha:n-2}}} \times S \\ U &= \hat{Y}_P + \sqrt{\frac{(n-2) \times \chi^2_{P:1}(d^2)}{\chi^2_{\alpha:n-2}}} \times S \end{aligned} \tag{2.86}$$

where $\chi^2_{P:1}(d^2)$ is the percentile from a non-central χ^2-distribution with area P to the left, one degree of freedom, and non-centrality parameter d^2. Krishnamoorthy and Mathew also provide a slightly more complex approximation, but based on their reported simulation results, Eq. (2.86) seems more than adequate for practical applications when $d^2 < 0.30$. If you are using software that does not provide percentiles for the non-central chi-squared distribution (such as Excel), Krishnamoorthy and Mathew present another approximation that can be employed.

All three intervals are now computed for the stability example with $X_P = 30$ months. We first compute the value of d as

$$d = \sqrt{\frac{1}{n} + \frac{(X_P - \bar{X})^2}{\sum\limits_{i=1}^{n}(X_i - \bar{X})^2}}$$

$$d = \sqrt{\frac{1}{6} + \frac{(30 - 15)^2}{630}} \, 0.724.$$

(2.87)

The 95% two-sided confidence interval on the mean of all vials at $X_P = 30$ months shown in Eq. (2.81) is

$$L = \hat{Y}_P - t_{1-\alpha/2:n-2} \times S \times d$$
$$L = 99.99 - 2.78 \times 0.3534 \times 0.724 = 99.3\%$$
$$U = \hat{Y}_P + t_{1-\alpha/2:n-2} \times S \times d$$
$$U = 99.99 + 2.78 \times 0.3534 \times 0.724 = 100.7\%.$$

(2.88)

The 95% two-sided prediction interval for a single vial selected at random at $X_P = 30$ months shown in Eq. (2.82) is

$$L = \hat{Y}_P - t_{1-\alpha/2:n-2} \times S \times \sqrt{1+d^2}$$
$$L = 99.99 - 2.78 \times 0.3534 \times \sqrt{1 + 0.724^2} = 98.8\%$$
$$U = \hat{Y}_P + t_{1-\alpha/2:n-2} \times S \times \sqrt{1+d^2}$$
$$L = 99.99 + 2.78 \times 0.3534 \times \sqrt{1 + 0.724^2} = 101.2\%.$$

(2.89)

Since $d^2 > 0.30$, Eq. (2.85) is suitable for computing a two-sided tolerance interval. This approximate 95% two-sided tolerance bound that contains at least 90% of the population of individual values at $X_P = 30$ months is

$$L = \hat{Y}_P - t_{1-\alpha/2:n-2}(Z_P/d) \times S \times d$$
$$L = 99.99 - 6.57 \times 0.3534 \times 0.724 = 98.3\%$$
$$U = \hat{Y}_P + t_{1-\alpha/2:n-2}(Z_P/d) \times S \times d$$
$$U = 99.99 + 6.57 \times 0.3534 \times 0.724 = 101.7\%.$$

(2.90)

The two-sided 95% tolerance interval that contains 90% of the population of individual values at $X_P = 30$ months based on the Wallis approximation in Eq. (2.86) is

Table 2.28 Summary of statistical intervals

Focus of inference at $X_P = 30$ months	Statistical interval	Results (%)
Average of all vials	Confidence interval	L = 99.3 to U = 100.7
A single vial selected at random	Prediction interval	L = 98.8 to U = 101.2
All individual vials	Tolerance interval	L = 98.3 to 101.7

$$L = \hat{Y}_P - \sqrt{\frac{(n-2) \times \chi^2_{P:1}\left(d^2\right)}{\chi^2_{\alpha:n-2}}} \times S$$

$$L = 99.99 - \sqrt{\frac{(6-2) \times 4.09}{0.711}} \times 0.3534 = 98.3\%$$

$$U = \hat{Y}_P + \sqrt{\frac{(n-2) \times \chi^2_{P:1}\left(d^2\right)}{\chi^2_{\alpha:n-2}}} \times S \tag{2.91}$$

$$U = 99.99 + \sqrt{\frac{(6-2) \times 4.09}{0.711}} \times 0.3534 = 101.7\%$$

Since $d^2 \geq 0.30$, the two tolerance intervals provide the same results to the reported decimal accuracy. Table 2.28 reports the three computed statistical intervals for the example.

As shown in Table 2.28, the uncertainty is greatest (i.e., the interval is widest) for the tolerance interval, and the uncertainty is least for the confidence interval on the mean. This is consistent with our previous discussion concerning the relationships of the three intervals.

Figure 2.17 presents a plot of the data with the fitted regression line and the three statistical intervals described in this section. Note that all three intervals have the least amount of uncertainty in the middle of the plot, and the uncertainty increases as the value of X_P moves away from the center of the X values.

In addition to making predictions, there is interest in determining if there is evidence of a linear relationship between Y and X. This is most easily accomplished by computing a confidence interval on the slope of the regression, β_1. If $\beta_1 = 0$, then model (2.78) is no longer a function of X, and Y and X are not linearly related. Thus, X is not a useful predictor of Y when $\beta_1 = 0$. Statistical evidence of a linear relationship between Y and X is provided by testing the set of hypotheses $H_0 : \beta_1 = 0$ against $H_a : |\beta_1| > 0$. As described in Sect. 2.10.3, these hypotheses can be tested with test size α by computing a $100(1 - \alpha)\%$ two-sided confidence interval on β_1. If the interval does not contain the null value of 0, then there is statistical evidence of a linear relationship between Y and X.

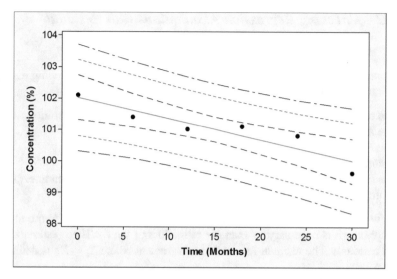

Fig. 2.17 Statistical intervals for stability regression example

A $100(1 - \alpha)\%$ two-sided confidence interval for the slope β_1 in model (2.78) is

$$L = b_1 - t_{1-\alpha/2:n-2} \times k$$
$$U = b_1 + t_{1-\alpha/2:n-2} \times k$$
$$k = \frac{S}{\sqrt{\sum\limits_{i=1}^{n}(X_i - \bar{X})^2}}. \tag{2.92}$$

The term k is referred to as the estimated standard error of the regression coefficient.

The 95% confidence interval for the example problem is

$$L = -0.06762 - 2.78 \times \frac{0.3534}{\sqrt{630}} = -0.107$$
$$U = -0.06762 + 2.78 \times \frac{0.3534}{\sqrt{630}} = -0.029. \tag{2.93}$$

Since both bounds in the interval are negative, the value 0 is not included, and one concludes there is statistical evidence at the $\alpha = 0.05$ level that concentration and time are linearly related.

2.12.2   Checking Regression Assumptions with Residual Plots

For statistical inferences to be valid in regression analysis, the following assumptions must be maintained:

1. The random error term, E_i, can be described by the normal probability model,
2. The E_i are independent, and
3. The E_i have the same variance, σ^2, for all i.
   In addition, for model (2.78)
4. The relationship between Y and X is reasonably defined by a straight line. (This is another way to state the E_i have a mean of zero.)

As was noted earlier, the E_i are unknown random errors. In order to monitor the assumptions, it is necessary to examine estimators of the E_i. These estimators are called residuals. The residual for the ith measurement where $X_i = X_P$ is defined as

$$e_i = Y_i - \hat{Y}_P \qquad (2.94)$$

where $\hat{Y}_P$ is defined in Eq. (2.79). Equation (2.94) is called the "raw residual." Most statistical software packages have alternative residuals that better serve the purpose of monitoring statistical assumptions. For example, Minitab offers standardized and deleted residuals in addition to raw residuals. The standardized and deleted residuals are recommended over raw residuals because they have properties that are more consistent with the properties of the E_i. For example, it is known that the raw residuals do not have equal variance when the E_i have equal variance. The variance of the raw residuals increases as the predictor value moves away from $\bar{X}$. However, both standardized and deleted residuals have a common variance when the E_i have a common variance. Vining (2011) provides calculations and mathematical definitions of these various residuals. It is important to select residuals that best mimic the behavior of the E_i. Typically, you should select the deleted residuals if they are available.

We recommend the following three steps for determining if the regression assumptions are reasonable.

1. Construct a scatterplot. A scatterplot of Y versus X is one of the most helpful tools for checking regression assumptions. Figure 2.16 presented the scatterplot for the example data provided in Table 2.25. By examining the scatterplot you should be able to decide if the regression relationship is linear as stated in assumption 4. When the regression function appears not to be linear, you may consider more complex nonlinear regression functions. Sections 2.12.4 and 2.12.6 provide some simple techniques that can be used if it appears the model is better described by a nonlinear model.
2. Plot the residuals. If the regression assumptions are satisfied, the deleted residuals will (roughly) behave like a simple random sample from a normal

population with mean 0 and variance σ^2. To help decide if the assumptions are satisfied, the following residual plots are useful:

(a) A normal quantile plot as described in Sect. 2.6.10. This plot will monitor assumption 1 concerning the appropriateness of the normal probability model.
(b) A scatterplot with the deleted residuals on the vertical axis and the predicted values (fitted values) on the horizontal axis. If the regression assumptions are satisfied, the points in this scatterplot should be randomly scattered about a horizontal line through zero. There should be no trend in the scatterplot, and the spread in the residuals should be constant across the plot.
(c) Plot the residuals in a time order if there is a possibility of a time effect. This might demonstrate a cyclical pattern that would demonstrate a lack of independence among the E_i (Note: even the deleted residuals are not independent, but the degree of dependence should be small enough not to matter if the E_i are independent.)

3. Examine all plots for anomalous points or outliers. Outliers are sample values that do not appear to be from the same population as the rest of the data. Outliers sometimes occur because mistakes were made during data collection, transcription, or entry. Each sample value that appears to be an outlier should be thoroughly investigated to determine the reason for its existence. If you cannot determine why the outlier exists, it may be wise to carry out two analyses—one with the outlier included and one with it excluded. If the two analyses lead to identical decisions, then the outlier does not impact the result. If the two analyses lead to different decisions, you must look further into the matter.

Figure 2.18 presents the residual plots described in this section for the data in Table 2.25.

There does not appear to be any evidence that the regression assumptions are unreasonable for this application.

A complete discussion of remedial measures when assumptions are violated is beyond the scope of this book. There are numerous textbooks on regression analysis that provide discussion of these remedial measures.

2.12.3 Multiple Regression Analysis

2.12.3.1 Model Definitions

Multiple regression considers models with more than one predictor variable. Predictor variables can be either quantitative or categorical. Consider the data presented in Table 2.29.

These data were collected during a process characterization experiment for a purification column used in a biopharmaceutical manufacturing process. The

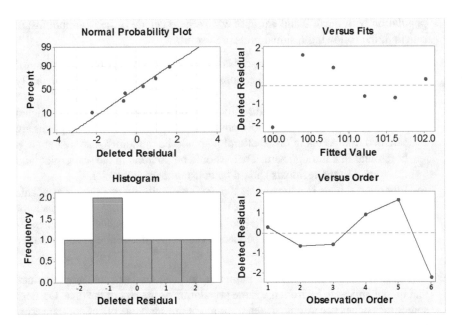

Fig. 2.18 Plots of deleted residuals

Table 2.29 Results of characterization study

Experiment	Load rate (g/L)	pH	Y (%)
1	2.5	4.7	94.01
2	2.5	5	94.91
3	2.5	5.3	99.65
4	7.5	4.7	75.7
5	7.5	5	81.54
6	7.5	5	84.92
7	7.5	5.3	91.37
8	12.5	4.7	67.11
9	12.5	5	80.62
10	12.5	5.3	92.36

response variable, Y, is a measure of product purity expressed as a percentage. There are two quantitative predictor variables, X_1 = Load rate measured in grams per liter and X_2 = pH.

The multiple regression model used to represent the data is

$$Y_i = \beta_0 + \beta_1 X_{1i} + \beta_2 X_{2i} + E_i \quad i = 1, \ldots, n; \qquad (2.95)$$

where Y_i is the measured response for experiment i, X_{1i} is the value of Load Rate for experiment i, X_{2i} is the value of pH for experiment i, β_0 is the y-intercept of the linear relationship, β_1 and β_2 are regression (slope) parameters, and E_i is the error term described for regression model (2.78). Essentially, model (2.95) is the same model as (2.78), but now the conditional mean of Y is a function of both X_1 and X_2.

One important difference between a regression model with one predictor and one with more predictors is the interpretation of the slope parameters β_1 and β_2. Similar to β_1 in the single predictor regression, β_1 and β_2 both describe rates of change. In particular, β_1 represents the change in the conditional mean of Y as X_1 increases by one unit *while holding X_2 constant*. Similarly, β_2 represents the change in the conditional mean of Y as X_2 increases by one unit and X_1 *is held constant*. Due to the conditional interpretation of these parameters, they are referred to as partial regression coefficients. When a partial regression coefficient is equal to zero, then the conditional mean of Y is not a function of the associated predictor variable. As in simple regression analysis, tests for evidence of a non-zero partial regression coefficients are used to determine if predictor variables are useful for predicting Y. Confidence intervals for this purpose are presented after a brief discussion concerning calculation of statistics in a regression model.

2.12.3.2 Regression Calculations

As a practical matter, it will be necessary to compute statistics for a multiple regression model using a statistical software package. However, it is instructive to write the formulas using the same format provided in Sect. 2.12.1, understanding that some of the quantities will be taken from computer output. Table 2.30 provides the essential information for this strategy.

Using the notation in Table 2.30, the following statistical intervals can be used with any number of predictor variables. Modification of Eq. (2.81) provides the 100 $(1 - \alpha)\%$ two-sided confidence interval on the conditional average of Y when $X_1 = X_{1P}, X_2 = X_{2P}, \ldots, X_m = X_{mP}$. This interval is

Table 2.30 Notation for multiple regression models

Verbal description	Notation used in Sect. 2.12.1	Notation with more than one predictor
Number of predictor variables	1	m
Estimated Y-intercept	b_0	b_0
Estimated regression coefficient	b_1	$b_1, b_2, \ldots, b_m$
Sample variance	S^2	S^2
Sample standard deviation	S	S
Error degrees of freedom	$n - 2$	$n - m - 1$
Estimated standard error of regression coefficient	k as defined in(2.92)	$k_1, k_2, \ldots, k_m$
Estimated standard error of fitted value $\hat{Y}_P$	$S \times d$	$S \times d$

$$L = \hat{Y}_P - t_{1-\alpha/2:n-m-1} \times S \times d$$
$$U = \hat{Y}_P + t_{1-\alpha/2:n-m-1} \times S \times d \qquad (2.96)$$
$$\hat{Y}_P = b_0 + b_1 X_{1P} + b_2 X_{2P} + \ldots + b_m X_{mP}$$

where d and S must be obtained using statistical software.

The $100(1 - \alpha)\%$ two-sided prediction interval is modified from interval (2.82) as

$$L = \hat{Y}_P - t_{1-\alpha/2:n-m-1} \times S \times \sqrt{1 + d^2}$$
$$U = \hat{Y}_P + t_{1-\alpha/2:n-m-1} \times S \times \sqrt{1 + d^2} \qquad (2.97)$$

The $100(1 - \alpha)\%$ two-sided tolerance interval that contains P% of all future values based on the approximate Wallis interval in (2.86) is

$$L = \hat{Y}_P - \sqrt{\frac{(n - m - 1) \times \chi^2_{P:1}\left(d^2\right)}{\chi^2_{\alpha:n-m-1}}} \times S$$
$$U = \hat{Y}_P + \sqrt{\frac{(n - m - 1) \times \chi^2_{P:1}\left(d^2\right)}{\chi^2_{\alpha:n-m-1}}} \times S. \qquad (2.98)$$

A $100(1 - \alpha)\%$ two-sided confidence interval on β_j for $j = 1, \ldots, m$ similar to (2.92) is

$$L = b_j - t_{1-\alpha/2:n-m-1} \times k_j$$
$$U = b_j + t_{1-\alpha/2:n-m-1} \times k_j \qquad (2.99)$$

where k_j is determined from a software program.

Consider the data from the experiment shown in Table 2.29. Table 2.31 reports the required statistics that were obtained from a statistical package assuming X_{1P} = 7.5 g/L and $X_{2P} = 5$.

Table 2.32 reports the statistical intervals using the formulas (2.96), (2.97), (2.98), and (2.99) for the data in Table 2.29. All intervals are two-sided intervals with 95% confidence. The tolerance interval contains 99% of the future values.

All of the intervals in Table 2.32 can be obtained directly from most software packages with the exception of the tolerance interval. An example printout from Minitab is shown in Table 2.33.

The 95% tolerance interval containing 99% of future values when $X_{1P} = 7.5$ and $X_{2P} = 5$ is computed with the information in Table 2.33 using Eq. (2.98). First, determine the non-centrality parameter $d^2 = (1.50513/4.75962)^2 = 0.100$, $\chi^2_{0.99:1}\left(d^2\right) = 7.260$, and $\chi^2_{0.05:7} = 2.167$. Then

Table 2.31 Example problem

Verbal description	Value with data from Table 2.29
Number of predictor variables	$m = 2$
Estimated Y-intercept	$b_0 = -31.0$
Estimated regression coefficients	$b_1 = -1.62$ $b_2 = 25.87$
Sample variance	$S^2 = 22.654$
Sample standard deviation	$S = 4.76$
Error degrees of freedom	$n - m - 1 = 7$
Estimated standard error of regression coefficients	$k_1 = 0.389$ $k_2 = 6.48$
Estimated standard error of fitted value $X_{1P} = 7.5, X_{2P} = 5$	$S \times d = 1.51$

Table 2.32 Computed statistical intervals

Verbal description	Interval with data from Table 2.29
Fitted value with $X_{1P} = 7.5, X_{2P} = 5$	86.22%
Confidence interval on β_1	-2.54 to -0.70
Confidence interval on β_2	10.55 to 41.18
Confidence interval on mean when $X_{1P} = 7.5$ and $X_{2P} = 5$	82.66 to 89.78%
Prediction interval when $X_{1P} = 7.5$ and $X_{2P} = 5$	74.41 to 98.02%
Tolerance interval when $X_{1P} = 7.5$ and $X_{2P} = 5$	63.17 to 109.27%

Table 2.33 Minitab output for multiple regression example

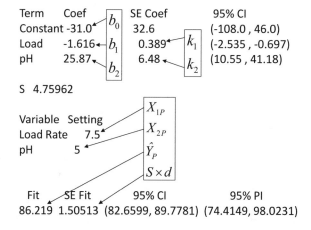

$$L = \hat{Y}_P - \sqrt{\frac{(n - m - 1) \times \chi^2_{P:1}(d^2)}{\chi^2_{\alpha:n-m-1}}} \times S$$

$$= 86.219 - \sqrt{\frac{7 \times 7.260}{2.167}} \times 4.76 = 63.17$$

$$U = \hat{Y}_P + \sqrt{\frac{(n - m - 1) \times \chi^2_{P:1}(d^2)}{\chi^2_{\alpha:n-m-1}}} \times S$$ (2.100)

$$= 86.219 + \sqrt{\frac{7 \times 7.260}{2.167}} \times 4.76 = 109.27$$

2.12.4 Incorporating Interaction and Quadratic Effects

Recall the definitions for the partial regression coefficients associated with the predictor variables X_1 and X_2. The partial regression coefficient β_1 provides the rate of change in the average of Y as X_1 increases one unit holding X_2 fixed. Notice that the definition does not stipulate a particular value at which X_2 is fixed. Regardless of what value of X_2 is assigned, the rate of change in the mean of Y for a one unit increase in X_1 is constant, and does not depend on the particular fixed value of X_2. However, situations arise where such an assumption is not reasonable. That is, the rate of change in the mean of Y for a one unit increase in X_1 varies as the value of X_2 changes. When such a situation occurs, it is stated that X_1 and X_2 interact.

It has been our experience that researchers often confuse the definitions of interaction and correlation. Correlation is a relationship between two variables (e.g., Y, X_1), whereas interaction involves at least three variables (e.g., Y, X_1, X_2). If two variables are being considered where one variable generally increases (decreases) as the other variable increases (decreases), then the two variables are positively correlated. If the two variables tend to move in opposite directions, then they are negatively correlated. Interaction is a condition that considers at least three variables. If two predictor variables X_1 and X_2 interact, this means the slope of the regression of Y on X_1 changes as X_2 changes. Similarly, the slope for the regression of Y on X_2 changes as X_1 changes. Figure 2.19 provides an illustration of an interaction effect between Load Rate (X_1) and pH (X_2) on Y. Note the slope of Y on X_1 is much steeper when pH $= 4.7$ than when pH $= 5.3$.

In order to include an interaction effect between X_1 and X_2 in a regression model, one creates a new predictor variable by multiplying the values of X_1 and X_2, say $X_3 = X_1 \times X_2$. Using the data in Table 2.29, a regression model that includes an interaction effect between X_1 and X_2 is

$$Y_i = \beta_0 + \beta_1 X_{1i} + \beta_2 X_{2i} + \beta_3 X_{3i} + E_i \quad i = 1, \ldots, n; \qquad (2.101)$$

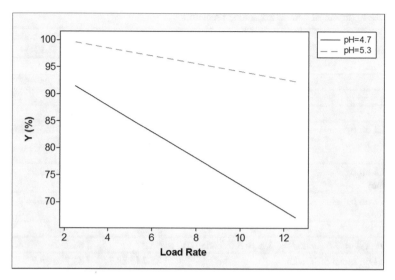

Fig. 2.19 Slopes change due to interaction between load rate and pH

where $X_{3i} = X_{1i} \times X_{2i}$. The prediction function associated with model (2.101) is written as

$$\hat{Y}_P = b_0 + b_1 X_{1P} + b_2 X_{2P} + b_3 X_{3P}$$
$$\hat{Y}_P = b_0 + b_1 X_{1P} + b_2 X_{2P} + b_3(X_{1P} \times X_{2P}) \qquad (2.102)$$
$$\hat{Y}_P = b_0 + (b_1 + b_3 X_{2P})X_{1P} + b_2 X_{2P}.$$

Notice that the slope coefficient on X_{1P} is now a function of X_{2P}, namely $b_1 + b_3 X_{2P}$. Thus, the rate of change in $\hat{Y}_P$ as X_{1P} increases one unit is now a function of X_{2P}. Similarly, the prediction function can be written as

$$\hat{Y}_P = b_0 + b_1 X_{1P} + b_2 X_{2P} + b_3 X_{3P}$$
$$\hat{Y}_P = b_0 + b_1 X_{1P} + b_2 X_{2P} + b_3(X_{1P} \times X_{2P}) \qquad (2.103)$$
$$\hat{Y}_P = b_0 + (b_2 + b_3 X_{1P})X_{2P} + b_1 X_{1P}$$

and one can see that the rate of change in $\hat{Y}_P$ as X_{2P} increases one unit is now a function of X_{1P}.

When fitting a regression model with an interaction effect, it is recommended that each predictor variable included in the interaction be coded prior to computing the interaction predictor variable. The appropriate coding formula is

Table 2.34 Impact of correlation among predictors

Statistic	Predictors not coded	Predictors coded using (2.104)
b_0	91.57	86.22
b_1	−17.96	−8.08
b_2	1.35	7.76
b_3	3.27	4.90
95% CI on β_1 (p-value)	−31.1 to −4.79 (0.016)	−11.3 to −4.86 (0.001)
95% CI on β_2 (p-value)	−21.1 to 23.8 (0.888)	4.54 to 10.98 (0.001)
95% CI on β_3 (p-value)	0.64 to 5.90 (0.023)	0.96 to 8.85 (0.023)
S	3.226	3.226
$\hat{Y}_P$ with $X_{1P} = 7.5$ (coded 0), $X_{2P} = 5.0$ (coded 0)	86.2	86.2
95% CI on mean of Y with $X_{1P} = 7.5$ (coded 0), $X_{2P} = 5.0$ (coded 0)	83.7 to 88.7	83.7 to 88.7
95% Prediction Interval with $X_{1P} = 7.5$ (coded 0), $X_{2P} = 5.0$ (coded 0)	77.9 to 94.5	77.9 to 94.5
95% Tolerance Interval with $X_{1P} = 7.5$ (coded 0), $X_{2P} = 5.0$ (coded 0) Equation (2.98)	69.6 to 102.9	69.6 to 102.9

$$X_{Coded} = \frac{2 \times X_{Original} - Max - Min}{Max - Min} \tag{2.104}$$

where *Max* is the maximum X value for the original (uncoded) data and *Min* is the minimum X value. By coding the data in this manner, the minimum value of X_{Coded} is −1 and the maximum value is +1. As an example, for pH in Table 2.29, the maximum value is 5.3 and the minimum is 4.7. The value of pH for the first experiment is 4.7, and so the coded value is

$$X_{Coded} = \frac{2 \times X_{Original} - Max - Min}{Max - Min}$$
$$X_{Coded} = \frac{2 \times 4.7 - 5.3 - 4.7}{5.3 - 4.7} = \frac{-0.6}{0.6} = -1. \tag{2.105}$$

The reason for coding using Eq. (2.104) is to mitigate a situation where predictor variables are correlated with each other. This condition is called collinearity. The primary problem caused by collinearity is that the estimated regression coefficients can have standard errors that are extremely large, creating point estimates that are "unstable." This instability leads to wide confidence intervals on the regression coefficients and associated p-values that are large even when there are strong statistical relationships between Y and the predictor variables. Since the predictor variable representing the interaction term (X_3) is a function of other predictor

Table 2.35 Correlation matrix of original data

	X_{1O}	X_{2O}	X_{3O}
X_{1O}	1	0	0.995
X_{2O}	–	1	0.090
X_{3O}	–	–	1

Table 2.36 Correlation matrix of coded data

	X_{1C}	X_{2C}	X_{3C}
X_{1C}	1	0	0
X_{2C}	–	1	0
X_{3C}	–	–	1

variables (X_1 and X_2), X_3 will likely be correlated with X_1 and X_2. By coding the predictors as described, the degree of correlation is reduced, if not eliminated entirely.

To demonstrate, Table 2.34 shows results from fitting interaction models to the data in Table 2.29 using both original and coded data.

Tables 2.35 and 2.36 report the correlation matrices between the original X values (X_{1O}, X_{2O}, X_{3O}) with the coded X values (X_{1C}, X_{2C}, X_{3C}).

There are several things to notice in Table 2.34. First, the confidence intervals on the regression coefficients for X_1 and X_2 are wider for the original data than for the coded data. Additionally, the associated p-values are greater for the original data. For example, using the original data the regression coefficient of X_2 does not appear to be "statistically significant." In contrast, when the model is fit with the coded data, the coefficient of X_2 is seen to be important ($p=0.001$). As seen in Table 2.35, there are non-zero correlations between X_{1O} and X_{3O}, and X_{2O} and X_{3O}. In contrast, Table 2.36 shows all pairwise correlations are zero.

One other important thing to notice is that the presence of collinearity *does not* impact S, $\hat{Y}_P$, or the three statistical intervals. This demonstrates that although collinearity can create problems in interpreting the individual impact of a predictor variable, it does not create a problem with predictions. However, since studies such as the one described in Table 2.29 are performed to discover relationships between Y and the predictor variables, one should attempt to control collinearity as much as possible.

Some care must be taken when interpreting the regression coefficients shown in Table 2.34. Consider the definition of the prediction function in Eq. (2.102) where $\hat{Y}_P = b_0 + (b_1 + b_3 X_{2P})X_{1P} + b_2 X_{2P}$. Using the coded data results in Table 2.34, the slope of the response Y on X_1 is $(b_1 + b_3 X_{2P}) = (-8.08 + 4.90 X_{2P})$. Suppose the original value of pH is $X_{2P} = 5.0$. Then the coded value is $X_{2P} = 0$, and the slope of Y on X_1 is -8.08. This means that as the coded value of X_1 increases 1 unit, say from -1 to 0, the average value of Y will decrease by 8.08%. Or in terms of the original data, when pH is equal to 5.0 and Load Rate increases from 2.5 to 7.5 g/L, the average value of Y will decrease by 8.08%. Similarly, when pH is equal to 5.0 and Load Rate increases from 7.5 to 12.5 g/L, the average value of Y will decrease

Table 2.37 Model fit with interaction and quadratic effects with predictors coded

Statistic	Value
b_0	83.17
b_1, b_2, b_3, b_4, b_5	-8.08, 7.76, 4.90, 4.66, 0.426
95% CI on β_1 (p-value)	-9.72 to -6.44 (0.000)
95% CI on β_2 (p-value)	6.12 to 9.40 (0.000)
95% CI on β_3 (p-value)	2.89 to 6.91 (0.003)
95% CI on β_4 (p-value)	2.03 to 7.29 (0.008)
95% CI on β_5 (p-value)	-2.20 to 3.06 (0.676)
S	1.4472
$\hat{Y}_P$ $X_{1P} = 7.5 (\text{coded } 0), X_{2P} = 5.0 (\text{coded } 0)$	83.17
95% CI on mean of Y with $X_{1P} = 7.5$ (coded 0), $X_{2P} = 5.0$ (coded 0)	80.77 to 85.57
95% Prediction Interval with $X_{1P} = 7.5$ (coded 0), $X_{2P} = 5.0$ (coded 0)	78.49 to 87.85
95% Tolerance Interval with $X_{1P} = 7.5$ (coded 0), $X_{2P} = 5.0$ (coded 0) Equation (2.98)	73.10 to 93.24

by 8.08%. Thus, if pH is equal to 5.0, as Load Rate increases across the entire range from 2.5 to 12.5 g/L, the average of Y will decrease by $2 \times 8.08 = 16.16\%$. Now if pH is set equal to $X_{2P} = 5.3$ and Load Rate increases from 2.5 to 7.5 g/L, then the average value of Y will change by only $(-8.08+4.90(1)) = -3.18\%$. That is, the slope of Y on X_1 flattens out as pH increases. This was displayed in Fig. 2.19.

One final thing to note in Table 2.34 is that the width of the three statistical intervals associated with $\hat{Y}_P$ is shorter than those computed for the model without the interaction shown in Table 2.32. This demonstrates the value of adding the interaction effect to the model.

One final addition to the model are quadratic effects for both X_1 and X_2. In many applications, the functional relationship between Y and a predictor variable may not be linear. In order to fit a quadratic term for the predictor variable X_1, one creates a new predictor variable, say X_4, by multiplying X_1 with itself. That is, $X_4 = X_1 \times X_1$. Similarly, one can define $X_5 = X_2 \times X_2$ to represent the quadratic effect for X_2. As with the predictor variables associated with interaction, the quadratic terms should be formed with coded values of X_1 and X_2. Table 2.37 reports an analysis of the data in Table 2.29 fitting the interaction term and both quadratic effects.

The p-values in Table 2.37 provide evidence of both an interaction effect between X_1 and X_2, and a quadratic effect for X_1 (Load Rate).

2.12.5 Incorporating Qualitative Predictor Variables

Regression models can also include qualitative factors. This is done by creating "indicator" or "dummy" variables to represent the factor levels. Consider the example used to introduce regression analysis in Table 2.25. In this example, concentration values were recorded for a single drug product lot as it degraded over time. The data set shown in Table 2.38 provides the same data but now with measurements from two additional lots.

The variable "Lot" in Table 2.38 is a qualitative factor with three categories: A, B, and C. In general, a qualitative variable has c categories. To represent this factor in the regression model, it requires the creation of $c - 1$ indicator variables. These indicator variables can be defined in several ways, but we apply coding to produce a range of $(-1,+1)$ as in the previous section. In particular, to represent a qualitative factor with c categories, define $c - 1$ indicator variables of the form

$$
I_i = \left\{ \begin{array}{ll} +1 & \text{if category } i \\ -1 & \text{if category } c \\ 0 & \text{otherwise} \end{array} \right\} \tag{2.106}
$$

for $i = 1, 2, \ldots, c - 1$.

You might wonder why only $c - 1$ indicator variables are required to represent c categories. The reason is that knowledge of $c - 1$ variables will uniquely define the last category. That is, use of all c indicator variables is unnecessary because the

Lot	Vial	Concentration (%) (Y)	Time in months (X)
A	1	102.1	0
A	2	101.4	6
A	3	101.0	12
A	4	101.1	18
A	5	100.8	24
A	6	99.6	30
B	7	100.0	0
B	8	100.0	6
B	9	100.2	12
B	10	98.8	18
B	11	99.8	24
B	12	99.0	30
C	13	97.6	0
C	14	98.3	6
C	15	98.1	12
C	16	97.1	18
C	17	96.5	24
C	18	96.0	30

Table 2.38 Concentration (%) over time for three lots

last one provides redundant information and will lead to a singularity in the solution of the regression equation. In our present example, $c = 3$ categories. Thus we define $c - 1 = 2$ indicator variables

$$I_1 = \begin{cases} +1 & \text{if Lot A} \\ -1 & \text{if Lot C} \\ 0 & \text{if Lot B} \end{cases}$$
$$I_2 = \begin{cases} +1 & \text{if Lot B} \\ -1 & \text{if Lot C} \\ 0 & \text{if Lot A.} \end{cases} \tag{2.107}$$

The assumed regression model is now

$$Y_i = \beta_0 + \beta_1 X_{1i} + \beta_2 I_{1i} + \beta_3 I_{2i} + E_i \quad i = 1, \ldots, n; \tag{2.108}$$

where X_1 is time in months. Note that we have now defined $c = 3$ regression models. For Lot A, $I_1 = +1$ and $I_2 = 0$ so model (2.108) becomes

$$Y_i = \beta_0 + \beta_1 X_{1i} + \beta_2(1) + \beta_3(0) + E_i$$
$$Y_i = \beta_0 + \beta_1 X_{1i} + \beta_2 + E_i \tag{2.109}$$
$$Y_i = (\beta_0 + \beta_2) + \beta_1 X_{1i} + E_i \ .$$

The model in Eq. (2.109) has a slope of β_1 and a y-intercept of $\beta_0 + \beta_2$. For Lot B, $I_1 = 0$ and $I_2 = 1$ so model (2.108) becomes

$$Y_i = \beta_0 + \beta_1 X_{1i} + \beta_2(0) + \beta_3(1) + E_i$$
$$Y_i = \beta_0 + \beta_1 X_{1i} + \beta_3 + E_i \tag{2.110}$$
$$Y_i = (\beta_0 + \beta_3) + \beta_1 X_{1i} + E_i \ .$$

The model in Eq. (2.110) has a slope of β_1 and a y-intercept of $\beta_0 + \beta_3$. Finally, for Lot C, $I_1 = -1$ and $I_2 = -1$ so model (2.108) becomes

$$Y_i = \beta_0 + \beta_1 X_{1i} + \beta_2(-1) + \beta_3(-1) + E_i$$
$$Y_i = \beta_0 + \beta_1 X_{1i} - \beta_2 - \beta_3 + E_i \tag{2.111}$$
$$Y_i = (\beta_0 - \beta_2 - \beta_3) + \beta_1 X_{1i} + E_i \ .$$

Model (2.111) is a regression with slope β_1 and y-intercept $\beta_0 - \beta_2 - \beta_3$. Thus, the regression model in (2.108) represents the data as three parallel lines, all with

Table 2.39   Concentration (%) over time with indicator variables

Lot	Vial	Concentration (%) (Y)	Time in months (X)	I_1	I_2
A	1	102.1	0	1	0
A	2	101.4	6	1	0
A	3	101.0	12	1	0
A	4	101.1	18	1	0
A	5	100.8	24	1	0
A	6	99.6	30	1	0
B	7	100.0	0	0	1
B	8	100.0	6	0	1
B	9	100.2	12	0	1
B	10	98.8	18	0	1
B	11	99.8	24	0	1
B	12	99.0	30	0	1
C	13	97.6	0	-1	-1
C	14	98.3	6	-1	-1
C	15	98.1	12	-1	-1
C	16	97.1	18	-1	-1
C	17	96.5	24	-1	-1
C	18	96.0	30	-1	-1

slope β_1. The intercepts for the three lots are $\beta_0 + \beta_2$, $\beta_0 + \beta_3$, and $\beta_0 - \beta_2 - \beta_3$, respectively.

Table 2.39 adds the two indicator variables to the data in Table 2.38.

Table 2.40 reports the results of this regression using the data in Table 2.39, and Fig. 2.20 shows a plot of the fitted regression line for each lot.

A model that allows different slopes for each lot can be obtained by using interaction effects between time and the indicator variables. In particular, the following regression model will allow a different slope within each lot.

$$Y_i = \beta_0 + \beta_1 X_{1i} + \beta_2 I_{1i} + \beta_3 I_{2i} + \beta_4 (X_{1i} \times I_{1i})$$
$$+ \beta_5 (X_{1i} \times I_{2i}) + E_i \quad i = 1, \ldots, n. \tag{2.112}$$

Fitting the interaction terms from the same data used to produce Table 2.40, Table 2.42 reports the results of the analysis. Since an interaction term has been added to the model, time has been coded using the formula in (2.104) as shown in Table 2.41. Figure 2.21 presents the plot of the model fit allowing different slopes for each lot.

Notice that since the intervals in Table 2.42 are wider than those in Table 2.40, and neither of the interaction effects have p-values less than 0.05, it appears an assumption of equal slope across lots is reasonable.

Table 2.40 Model with equal slopes across lots

Statistic	Value
b_0	100.15
b_1	-0.0565 (p < 0.001)
b_2	1.700 (p < 0.001)
b_3	0.333 (p $= 0.054$)
S	0.4744
Results for Lot A $(I_1 = 1, I_2 = 0)$ at $X_1 = 30$ months	
$\hat{Y}_P$	100.152
d	0.5345
95% confidence interval on mean of Y	99.61–100.70
95% prediction interval	99.00–101.31
95% tolerance interval that contains 99% using (2.98)	98.16–102.14

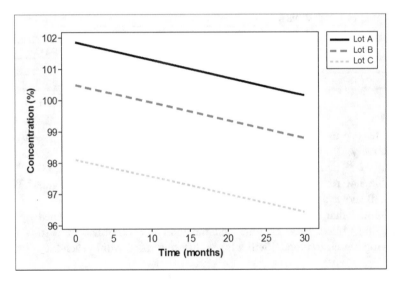

Fig. 2.20 Fitted model with qualitative predictor (Equal Slopes)

2.12.6 Nonlinear Models Using Variable Transformation

In some situations, the assumption of linearity between the response variable and a predictor is not reasonable. The incorporation of quadratic effects as described in Sect. 2.12.4 is one approach for modeling such data. Another approach is to perform a transformation on either the response or the predictor variable. Consider the data set shown in Table 2.43 which displays the response variable, Proportion Cell Viability (Y), and the predictor variable Cell Age at Time of Inoculation (X) in an experiment to examine the growth of cells in a bioreactor.

Table 2.41   Concentration (%) over time with time coded

Lot	Vial	Concentration (%) (Y)	Time in months (X)	I_1	I_2	Time (coded)
A	1	102.1	0	1	0	−1
A	2	101.4	6	1	0	−0.6
A	3	101.0	12	1	0	−0.2
A	4	101.1	18	1	0	0.2
A	5	100.8	24	1	0	0.6
A	6	99.6	30	1	0	1
B	7	100.0	0	0	1	−1
B	8	100.0	6	0	1	−0.6
B	9	100.2	12	0	1	−0.2
B	10	98.8	18	0	1	0.2
B	11	99.8	24	0	1	0.6
B	12	99.0	30	0	1	1
C	13	97.6	0	−1	−1	−1
C	14	98.3	6	−1	−1	−0.6
C	15	98.1	12	−1	−1	−0.2
C	16	97.1	18	−1	−1	0.2
C	17	96.5	24	−1	−1	0.6
C	18	96.0	30	−1	−1	1

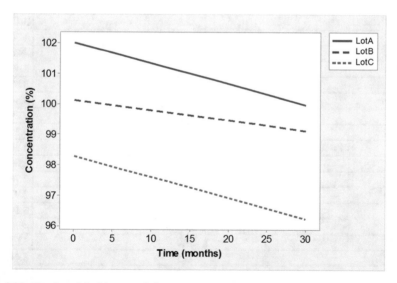

Fig. 2.21   Fitted model with unequal slopes

Table 2.42 Model with unequal slopes across lots

Statistic	Value
b_0	99.3
b_1 (Time)	-0.848 ($p < 0.001$)
b_2	1.700 ($p < 0.001$)
b_3	0.333 ($p = 0.055$)
b_4	-0.167 ($p = 0.481$)
b_5	0.348 ($p = 0.155$)
S	0.4694
Results for Lot A ($I_1 = 1, I_2 = 0$) at $X_1 = 30$ months	
$\hat{Y}_P$	99.986
d	0.7237
95% confidence interval on mean of Y	99.25–100.73
95% prediction interval	98.72–101.25
95% tolerance interval that contains 99% using (2.98)	97.81–102.16

Table 2.43 Data from cell growth study

Proportion cell viability (Y)	Cell age in days (X)	Logit (T)
0.931	13.0	2.602
0.94	18.6	2.751
0.946	22.1	2.863
0.957	32.3	3.102
0.991	96.7	4.701
0.946	24.0	2.863
0.968	43.5	3.409
0.967	46.0	3.377
0.979	63.5	3.842
0.995	118.5	5.293
0.995	126.0	5.293
0.997	138.5	5.806
0.937	16.0	2.699
0.942	21.6	2.787
0.949	25.1	2.923
0.957	35.3	3.102
0.991	99.7	4.701
0.954	27.0	3.032
0.969	46.5	3.442
0.968	49.0	3.409
0.982	66.5	3.999
0.995	121.5	5.293
0.995	129.0	5.293
0.997	141.5	5.806

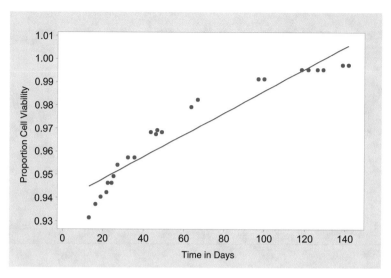

Fig. 2.22 Plot of cell viability against cell age

Notice from the plot in Fig. 2.22 that the points at both the left and right fall below a straight line, whereas those in the middle fall above the line. This suggests a nonlinear relationship between Y and X. Generally, the improvement in viability is slow for very young cells, followed by an increasing linear growth, and then a plateau at higher ages. Additionally, the response variable is such that a value less than 0 or greater than 1 is not possible. However, the straight line in Fig. 2.22 has predicted values that exceed 1 within the timeframe of interest.

A useful transformation for this situation is the logit transformation. The response variable Y is replaced with the transformed variable T where

$$T = \ln\left(\frac{Y}{1-Y}\right) \quad 0 \leq Y \leq 1. \tag{2.113}$$

The reverse transformation that allows one to re-express predicted values of T into predicted values of Y is

$$Y = \frac{e^T}{(1+e^T)}. \tag{2.114}$$

Table 2.43 shows values of T in the last column. For example, in the first row $T = \ln(0.931/(1 - 0.931)) = 2.602$.   Figure 2.23 displays the data for the transformed value T against cell age.

Note that the linear model of T on X is now a much better representation of the data. Suppose it is desired to predict Y when the predictor variable Cell Age is $X = 140$ days. Table 2.44 reports results of a simple linear regression after transforming Y to the values of T as shown in Table 2.43.

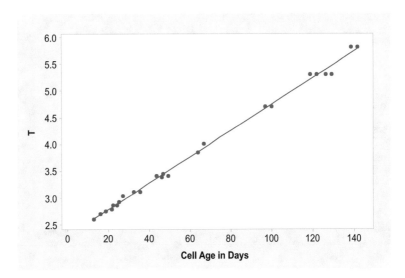

Fig. 2.23 Plot of T against cell age

Table 2.44 Regression fit of t on cell age (days)

Statistic	Value
b_0	2.293
b_1	0.0246
S	0.0646
$\hat{T}_P$ when $X = 140$	5.731
95% prediction interval	(5.586, 5.876)

Transforming these values back to the original measurements, the predicted value of Y at $X = 140$ days using (2.114) is $Y = \exp(5.731)/[1 + \exp(5.731)]$ $= 0.997$ with a lower prediction bound of $Y = \exp(5.586)/[1 + \exp(5.586)] = 0.996$ and an upper prediction bound of $Y = \exp(5.876)/[1 + \exp(5.876)] = 0.997$. Note that all values are less than 1 as desired.

There are many other transformations that might be considered in a particular application. The logit transformation is one of several transformations that are commonly used with proportions. Other useful transformations for proportions or percentages are the square root and arcsine transformations. An exponential model is also useful for some growth patterns. Logarithms are useful for transforming non-normal data with a long tail to more normal appearing data.

2.12.7 Mixed Models

All of the regression models considered to this point contain only a single error term. However, as was encountered with the dependent data in Sect. 2.7, it is often

Table 2.45 Concentration
(%) over time for three lots

Lot	Vial	Concentration (%) (Y)	Time in months (X)
A	1	102.1	0
A	2	101.4	6
A	3	101.0	12
A	4	101.1	18
A	5	100.8	24
A	6	99.6	30
B	7	100.0	0
B	8	100.0	6
B	9	100.2	12
B	10	98.8	18
B	11	99.8	24
B	12	99.0	30
C	13	97.6	0
C	14	98.3	6
C	15	98.1	12
C	16	97.1	18
C	17	96.5	24
C	18	96.0	30

necessary to model more than a single random error. To demonstrate, consider the previously described stability data shown in Table 2.45.

In Sect. 2.12.5 these data were modeled using indicator variables to represent "Lot" as a fixed effect. This means that the inferences collected from the sample pertain only to the three specific lots shown in Table 2.45. More realistically, it is desired to make an inference on the *process* that produces the lots, rather than this single fixed set of three lots. In fact, regulatory expectations are that companies provide assurance that their manufacturing processes are continuously producing lots that meet quality standards. (see Chap. 8 for more discussion on stability models).

In order to allow inferences concerning the ongoing process, it is necessary to consider the three lots tested in the sample as a representative sample from a conceptual population of future lots that will be produced by the process. In statistical terms, this means the factor lot is treated as a random effect. The statistical model is

$$Y_{ij} = \mu + L_i + \beta \times t_{ij} + E_{ij} \quad i = 1, \ldots, n; \, j = 1, \ldots, T_i; \qquad (2.115)$$

where Y_{ij} is a product quality attribute for lot i at time point j, μ is the average y-intercept across all lots, β is the average slope across all lots, L_i is a random variable that allows the y-intercept to vary from μ for a given lot, L_i has a normal distribution with mean 0 and variance σ_L^2, t_{ij} is the time point for measurement j of lot i, E_{ij} is a random normal error term created by measurement error and model misspecification with mean 0 and variance σ_E^2, n is the number of sampled lots, T_i is the number of time points obtained in lot i, and L_i and E_{ij} are jointly independent.

Table 2.46 Mixed model with random intercept

Parameters	Description	Computed estimate from Table 2.45
μ	Overall average across all lots	100.15
β	Average slope across all lots	-0.0565
σ_L^2	Variance from lot-to-lot	3.5303
σ_E^2	Variance of analytical method	0.2251

The model in (2.115) shows a process where individual lots have varying y-intercepts and is sometimes referred to as a random intercepts model. This model is appropriate when it can be assumed degradation rates (slopes) are equal to β across all lots. Such a model is reasonable when the process is well controlled, the reaction kinetics governing the stability properties are consistent from lot-to-lot, and variations in slope estimates from lot-to-lot are attributable primarily to analytical method variability. However, in some situations, it may be necessary to allow slopes to vary across lots. A model that allows such variation is

$$Y_{ij} = \mu + L_i + (\beta + B_i) \times t_{ij} + E_{ij} \quad i = 1, \ldots, n; \; j = 1, \ldots, T_i; \qquad (2.116)$$

where β is the average slope across all lots and B_i is a normal random variable that allows the slope to vary from lot-to-lot with mean 0 and variance σ_B^2.

Note that model (2.116) has three random error terms, L_i, B_i, and E_{ij}. As before, upper case Latin letters represent random effects (e.g., E_{ij}) and Greek symbols represent fixed effects (e.g., β). A model that contains both fixed and random effects is called a mixed model (hierarchical model). In a mixed model, one generally desires inferences on both the fixed effects and the covariances and variances associated with the random effects.

Table 2.46 reports the parameters of interest associated with model (2.115) using uncoded time data. Also shown in the table are the estimates using the data in Table 2.45. Minitab can be used for computations for some mixed models, but more generally JMP, SAS, or R code is recommended.

Determination of appropriate statistical intervals becomes a bit more complex in a mixed model because there are more intervals to consider. For example, suppose one asks the question, "What is the predicted value for a future lot at 30 months?" The predicted value using model (2.115) is $\hat{\mu} + \hat{\beta} \times t_{ij}$ where the "hats" on the parameters denote the estimates shown in Table 2.46. So in this case, the answer to the question is $100.15 + (-0.0565) \times 30 = 98.452\%$. However, suppose the question is "What is the predicted value for Lot A at 30 months?" In this case, the predicted value now includes an estimate of a random effect and is obtained from the equation $\hat{\mu} + \hat{L}_1 + \hat{\beta} \times t_{ij} = 100.135\%$. A predicted value that contains estimates of both fixed and random effects is called a best linear unbiased predictor or BLUP. In the fixed effects model of Sect. 2.12.5, the predicted mean for Lot A at time 30 was 100.152% as shown in Table 2.40. This example demonstrates that model assumptions concerning whether factors are fixed or random make a difference in the estimates. When using mixed models, the complexity necessarily

Table 2.47 ANOVA for mixed model

Source of variation	Degrees of freedom	Mean square	Expected mean square
Concentration	$n_1 = c - 1$	S_1^2	$\theta_1 = \sigma_E^2 + r\sigma_A^2 + ar\dfrac{\sum_{i=1}^{c}(\mu_i - \mu)^2}{c-1}$
Between runs	$n_2 = c(a - 1)$	S_2^2	$\theta_2 = \sigma_E^2 + r\sigma_A^2$
Within runs	$n_3 = ac(r - 1)$	S_3^2	$\theta_3 = \sigma_E^2$

increases because there are more quantities (and questions) of interest. Statistical intervals can also be computed with mixed models, but the formulas are more complex than those presented to this point, and statistical software is needed to compute them.

For some mixed models, simple results can be obtained by considering the analysis of variance (ANOVA) table associated with the model. As an example, consider the mixed model often used in method validation

$$Y_{ijk} = \mu_i + A_{j(i)} + E_{ijk} \; i = 1, \ldots, c; \; j = 1, \ldots, a; \; k = 1, \ldots, r; \qquad (2.117)$$

where Y_{ij} is the measured value for the kth replicate of the jth analytical run for concentration level i. The random error $A_{j(i)}$ represents between run variability. It is assumed to have a mean of zero and a variance σ_A^2. The random error E_{ijk} is the within run variability which has an assumed mean of zero and variance σ_E^2. The parameter μ_i represents the mean of Y_{ijk} for concentration level i.

The ANOVA table for model (2.117) is shown in Table 2.47. It can be computed in any of the software programs previously cited.

The factor concentration is a fixed effect because the inference pertains to the particular concentration range examined in the study. The factor "Between runs" is a random effect since runs are not unique and change from application to application. The "Within runs" source of variation represents the method's repeatability. Inferences are desired for the parameters shown in the expected mean square (EMS) column of the ANOVA table. For example, the total variation of the analytical method is the sum of the two variance components, $\sigma_A^2 + \sigma_E^2$. This sum is called the intermediate precision.

A useful approximation for constructing statistical intervals on sums of variance components is the Satterthwaite approximation. This approximation is used to form an approximate chi-squared random variable that can be used for constructing statistical intervals. To demonstrate, suppose it is desired to compute a 95% confidence interval on the intermediate precision variance $\sigma_A^2 + \sigma_E^2$. From the ANOVA in Table 2.47, consider the two expected mean squares $\theta_2 = \sigma_E^2 + r\sigma_A^2$ and $\theta_3 = \sigma_E^2$. The intermediate precision variance, $\gamma = \sigma_A^2 + \sigma_E^2$, can be written as the function of these expected mean squares

$$\gamma = k_2\theta_2 + k_3\theta_3,$$

$$k_2 = \frac{1}{r}, \text{ and} \tag{2.118}$$

$$k_3 = 1 - k_2.$$

The estimator for γ is obtained by replacing the EMS values with their respective mean squares, S_2^2 and S_3^2. This provides the point estimator

$$\hat{\gamma} = k_2 S_2^2 + k_3 S_3^2. \tag{2.119}$$

The Satterthwaite approximation determines a value m such that $(m\hat{\gamma})/\gamma$ has an approximate chi-squared distribution where

$$m = \frac{\hat{\gamma}^2}{\frac{\left(k_2 S_2^2\right)^2}{c(a-1)} + \frac{\left(k_3 S_3^2\right)^2}{ac(r-1)}}. \tag{2.120}$$

In practice, m will often be non-integer. This causes no problem if using a software package that allows non-integer degrees of freedom for the chi-squared distribution. However, if performing calculations in Excel, one should round the value of m to the nearest integer.

Once m has been determined, statistical intervals can be computed using $\hat{\gamma}$ and m. For example, Eq. (2.10) can be used to construct a confidence interval on γ by replacing S^2 with $\hat{\gamma}$ and n-1 with m. Consider the present example and suppose $S_2^2 = 30$, $S_3^2 = 15$, $c = 3$, $a = 4$, and $r = 3$. The estimator for the intermediate precision is

$$\hat{\gamma} = k_2 S_2^2 + k_3 S_3^2 = \frac{1 \times 30}{3} + \frac{2 \times 15}{3} = 20. \tag{2.121}$$

The value of m is obtained as

$$m = \frac{\hat{\gamma}^2}{\frac{\left(k_2 S_2^2\right)^2}{c(a-1)} + \frac{\left(k_3 S_3^2\right)^2}{ac(r-1)}}$$

$$m = \frac{(20)^2}{\frac{\left(\frac{30}{3}\right)^2}{3(4-1)} + \frac{\left(\frac{2 \times 15}{3}\right)^2}{4 \times 3 \times (3-1)}} \tag{2.122}$$

$$m = 26.18$$

Rounding m to 26, the 95% confidence interval on γ is

$$L = \frac{m\hat{\gamma}}{\chi^2_{1-\alpha/2:m}} = \frac{26 \times 20}{41.92} = 12.4$$
$$U = \frac{m\hat{\gamma}}{\chi^2_{\alpha/2:m}} = \frac{26 \times 20}{13.84} = 37.6. \tag{2.123}$$

The Satterthwaite approximation can be applied for any sum of expected mean squares. If $\gamma = \sum_{q=1}^{Q} k_q\theta_q$ is a sum of Q expected means squares, the Satterthwaite approximation is

$$\hat{\gamma} = \sum_{q=1}^{Q} k_q S_q^2$$
$$m = \frac{\hat{\gamma}^2}{\sum_{q=1}^{Q} \frac{\left(k_q S_q^2\right)^2}{df_q}} \tag{2.124}$$

where df_q is the degrees of freedom associated with the qth mean square. If differences or ratios of expected mean squares are needed, then other methods should be applied (see Chap. 8 of Burdick et al. (2005) for a discussion of these methods).

Some useful references describing mixed model analysis include Brown and Prescott (2006), Gelman and Hill (2007), and Littell et al. (2006). A few applications of mixed models are provided in this book, but with emphasis on the interpretation of the statistical computations.

2.13 Bayesian Models

Everything described in the chapter to this point is based on the classical frequentist model. Under the frequentist model, parameters that describe a population or process are assumed to be fixed and unknown. A sample is selected from the population, and the unknown parameters are estimated using numerical functions of the sample called statistics. Unlike the fixed parameters of interest, sample statistics are random variables because their values change each time a new sample is drawn from the population. In the classical frequentist model, all probability statements refer to the theoretical sampling distribution of random statistics that would be obtained in a hypothetical infinite series of resampling and re-estimation steps. Frequentist probabilities are thus based on long-run relative frequencies. Computation of these probabilities requires a sampling distribution for the sample statistics and an assumed value for the fixed parameter. Very briefly, the classical frequentist approach is concerned with the

$$\text{Probability}(\text{Random Sample Statistic} | \text{Fixed Parameter Value}). \qquad (2.125)$$

Statement (2.125) reads as "The probability that a given sample statistic will take on some value or values, conditional on a set of fixed unknown model parameters."

An alternative inference approach is the Bayesian model. In contrast to the classical approach, parameters in a Bayesian model are assumed to be random, rather than fixed. Thus, in addition to a probability distribution that describes random sample statistics, there is also a probability distribution that describes uncertainty in the unknown random parameter. The probability of interest in the Bayesian approach is

$$\text{Probability}(\text{Random Parameter Value} | \text{Observed Sample Statistic}). \qquad (2.126)$$

Statement (2.126) reads as "The probability that a given unknown parameter will take on some value or values, conditional on the observed sample statistic". The mathematics of (2.126) is discussed in the next section.

To contrast the two approaches, consider the client meeting described in Sect. 2.2. The company is building a new manufacturing site in Ireland, and wants to know the average purity of the product produced at the Ireland plant.

The frequentist approach assumes the true average purity is fixed and unknown. To estimate the average, lots are sampled from the Ireland plant, sample means and sample variances are computed, and the resulting values are used to construct a 100 $(1 - \alpha)\%$ confidence interval on the unknown average purity. The confidence coefficient $100(1 - \alpha)\%$ expresses the confidence that this procedure will provide an interval that truly contains the average purity for the Ireland site. Based on the resulting interval, a decision is made concerning the quality of the manufacturing process. Notice that the confidence coefficient is a fixed probability that refers not to the unknown average purity, but to the statistical methodology used to construct the interval. From the classical frequentist perspective, a computed confidence interval will be correct $100(1 - \alpha)\%$ of the time. Whether the confidence interval is correct for any particular application is unknown because the true mean potency is unknown.

To conceptualize this problem in a Bayesian framework, assume that the company has performed many manufacturing transfers over the years. They know that average purities vary across manufacturing sites in accordance with a defined probability distribution. It is desired to combine this information with the sample collected in Ireland to provide information concerning the average purity at the Ireland site. This combined information is used to form the expression (2.126). The Bayesian analogue of a frequentist confidence interval is called a credible interval. Although similar to a confidence interval, the interpretation of a credible interval is much different. Bayesian inference is not based on the long-run frequentist behavior of statistics. Rather, a credible interval is conditional on the particular data set that was observed and is therefore fixed, and not random. In this example, the probability that the true unknown average purity at the Ireland site lies within a fixed credible interval based on (2.126) is $100(1 - \alpha)\%$. This interval is assessed to make a decision concerning the quality of the Ireland manufacturing site. Notice that the probability in (2.126) refers directly to the parameter of interest, the average purity of the product from the Ireland site. More generally, Bayesian procedures can be used to estimate the probability that the true average purity of the product made in Ireland lies in any interval that might be of interest.

Importantly, the coverage properties of credible intervals are also of interest because one wants to understand the long-run reliability of any statistical procedure. Unlike many (but not all) frequentist interval procedures, the coverage properties of Bayesian credible intervals are usually not discoverable from theory and must be determined by computer simulation.

The selection of either a Bayesian approach or a frequentist approach depends on many factors. Some considerations when making this decision are offered in Sect. 2.13.5. Excellent introductions to Bayesian statistics are provided in the books by Bolstad (2007) and Kruschke (2015).

2.13.1   Expressing Prior Information on Model Parameters

Both frequentist and Bayesian procedures make use of probability density functions (pdfs). A pdf is a function that gives the probability density (or relative likelihood) as a function of values assigned by a random variable. The greater the probability density, the more likely a variable is to have the corresponding value. Univariate pdfs describe the probability density of scalar variables and multivariate pdfs describe the probability density of vector variables. The normal probability distribution is the primary pdf used for many statistical methods.

Some notation is needed to describe the Bayesian procedure. Let the bold-face letter $\mathbf{y}$ represent a vector that will contain a random sample of data values to be collected from a population or process. For example, this could be a set of concentration values for a sample of manufactured lots in Ireland. Before the data $\mathbf{y}$ are analyzed, Bayesian procedures require that a pdf be selected to define knowledge concerning $\mathbf{y}$ before the analysis. This pdf is called the "prior" pdf. This prior pdf is denoted as the bold-face function $\mathbf{p}(\boldsymbol{\theta})$ where the bold Greek letter theta represents parameter values that define the pdf. For example, if the prior pdf is a normal distribution, then θ would be a vector that contains the mean and the variance parameters. The prior pdf encapsulates any empirical or theoretical knowledge one may have about the likely values of the parameters prior to data analysis. In the Irish manufacturing example, it would include empirical information from other process transfers. The idea is illustrated in Fig. 2.24.

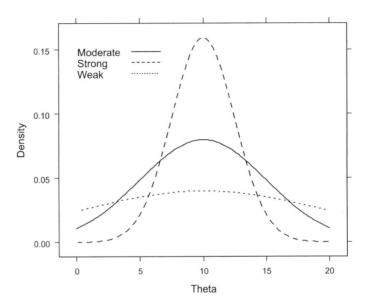

Fig. 2.24 Quantifying prior knowledge using a prior pdf

If one has a great amount of prior knowledge about a parameter, one might define a very narrow pdf that limits the possible values of the parameter. This is referred to as a strong prior. On the other hand, if one has very little information concerning the parameter values, a wider prior pdf is assigned. Such pdfs are described as either moderate or weak priors. If one has no prior knowledge, it is customary to assign a very wide and flat pdf that is often referred to as a non-informative prior. Often, the use of a non-informative prior pdf yields Bayesian credible intervals that correspond closely with statistical intervals developed using the frequentist approach. However, remember that the interpretation of any probability statements is different between these two intervals.

2.13.2 Bayes Rule and the Posterior Distribution

From a Bayesian perspective, all knowledge of the prior pdf and the sampled data set is encapsulated in their combined "posterior" pdf. This posterior pdf is obtained by combining the prior pdf with new information from the collected sample.

More formally, the probability of obtaining the observed data given a function of the parameter values is called the likelihood. The likelihood is a pdf denoted as $\mathbf{p}(\mathbf{y}|\boldsymbol{\theta})$. Note that the likelihood forms the basis of the frequentist relationship shown in (2.125). The prior pdf $\mathbf{p}(\boldsymbol{\theta})$ and the likelihood $\mathbf{p}(\mathbf{y}|\boldsymbol{\theta})$ are combined to produce the posterior pdf $\mathbf{p}(\boldsymbol{\theta}|\mathbf{y})$ using Bayes rule. According to Bayes rule, the posterior pdf at a given parameter value is proportional to the product of the prior pdf and the likelihood. In statistical shorthand,

$$\mathbf{p}(\boldsymbol{\theta}|\mathbf{y}) \propto \mathbf{p}(\boldsymbol{\theta}) \times \mathbf{p}(\mathbf{y}|\boldsymbol{\theta}). \tag{2.127}$$

Equation (2.127) uses two conventions commonly applied in Bayesian expositions. First, the symbol $\mathbf{p}(\cdot)$ is used instead of the traditional $f(\cdot)$ to represent a pdf. Second, the symbol $\propto$ (proportional to) is used in place of an equality sign. This indicates that in general, specified constants of the density functions are not important to the derivation of the posterior distribution.

To put this in a context, consider $\mathbf{y}$ to represent a single data value selected from a normal population with an unknown mean μ and a known variance σ^2. The likelihood is represented as $\mathbf{p}(\mathbf{y}|\boldsymbol{\theta})$ where θ is the scalar μ. The actual formula for the likelihood is

$$\mathbf{p}(y|\mu) = \frac{1}{\sqrt{2\pi\sigma^2}} e^{-\frac{(y-\mu)^2}{2\sigma^2}} \propto e^{-\frac{(y-\mu)^2}{2\sigma^2}}. \tag{2.128}$$

Now assume the prior pdf on μ is a normal distribution with mean μ_0 and variance σ_0^2. That is, the prior density is written as

$$p(\mu) \propto e^{-\frac{(\mu-\mu_0)^2}{2\sigma_0^2}} \tag{2.129}$$

Using Eq. (2.127), the posterior density is

$$\mathbf{p}(\mu|y) \propto \mathbf{p}(\mu) \times \mathbf{p}(y|\mu)$$
$$\propto e^{-\frac{(\mu-\mu_0)^2}{2\sigma_0^2}} \times e^{-\frac{(y-\mu)^2}{2\sigma^2}} \tag{2.130}$$

After performing some algebra, $\mathbf{p}(\mu|y)$ is shown to be a normal distribution with mean μ_P and variance σ_P^2 where

$$\mu_P = \frac{\frac{1}{\sigma_0^2} \times \mu_0 + \frac{1}{\sigma^2} \times y}{\frac{1}{\sigma_0^2} + \frac{1}{\sigma^2}} \tag{2.131}$$

$$\sigma_P^2 = \left(\frac{1}{\sigma_0^2} + \frac{1}{\sigma^2} \right)^{-1}.$$

Once the posterior distribution, $\mathbf{p}(\boldsymbol{\theta}|\mathbf{y})$, is estimated using (2.127), it can be used to obtain credible intervals, make predictions of future data, or test hypotheses of interest. The relationship between the three pdfs is illustrated in Fig. 2.25.

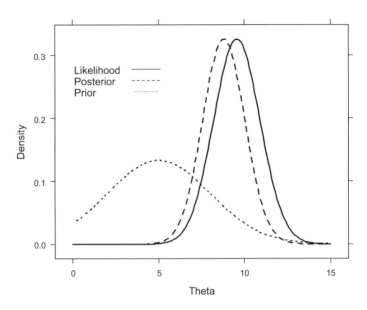

Fig. 2.25 The relationship between prior, likelihood, and posterior pdf

Prior to observing any data, knowledge about the parameter θ with possible values plotted on the horizontal axis is represented by the prior pdf $\mathbf{p(\theta)}$. The prior pdf in Fig. 2.25 encapsulates prior evidence that the likely value of θ lies somewhere between 0 and 10 with the most likely value being 5. Based on the collected data, the likelihood $\mathbf{p(y|\theta)}$ suggests that the value for θ may be between 7 and 13 with the most likely value being about 10. The likelihood is narrower than the prior indicating that the evidence collected in the data set is stronger than what was believed prior to the data collection. The posterior pdf $\mathbf{p(\theta|y)}$ represents a compromise between the likelihood and prior. Because the weight of evidence is stronger with the data (likelihood), the posterior is very close to the likelihood and somewhat removed from the prior.

2.13.3   An Example

To demonstrate how Bayesian inference can be used in a CMC setting, consider the data collected in the Ireland manufacturing example and shown in Table 2.1. The sample mean is $\bar{Y} = 94.305\%$. Noel tells Tom he can assume virtually all the variation in the measured values is due to the analytical method. The method is well characterized and it is known the standard deviation is 0.8. Thus, for a sample mean based on a sample of $n = 8$ observations, the standard deviation is $\sigma = 0.8/\sqrt{8} = 0.28\%$. In discussion to determine prior information, Noel shares that based on past transfers, he believes there is a 99% chance that the true value of

Table 2.48 Probabilities based on posterior distribution

Possible value in Ireland	Probability that Ireland mean Exceeds value in column 1
95%	0.007
94.5%	0.265
94%	0.882
93.5%	0.999

μ is between 91 and 99%. This subjective assessment is described by a normal distribution with mean $\mu_0 = 95\%$ and $\sigma_0 = 1.55\%$. (Based on a standard normal curve, 99% has a Z-score 2.58. Thus $\sigma_0 = (99 - 95)/2.58 = 1.55\%$. Similarly, the same result can be obtained using the lower value of the range, 91%.) Using (2.127), the posterior distribution for the mean is shown to be a normal distribution with mean μ_P and variance σ_P^2 where

$$\mu_P = \frac{\frac{1}{\sigma_0^2} \times \mu_0 + \frac{1}{\sigma^2} \times \bar{Y}}{\frac{1}{\sigma_0^2} + \frac{1}{\sigma^2}} = \frac{\frac{1}{(1.55)^2} \times 95 + \frac{1}{(0.28)^2} \times 94.305}{\frac{1}{(1.55)^2} + \frac{1}{(0.28)^2}} = 94.32$$

$$\sigma_P^2 = \left(\frac{1}{\sigma_0^2} + \frac{1}{\sigma^2}\right)^{-1} = \left(\frac{1}{(1.55)^2} + \frac{1}{(0.28)^2}\right)^{-1} = 0.0759.$$

(2.132)

Thus the posterior distribution has a mean of 94.32% and a standard deviation of $\sqrt{0.0759} = 0.276\%$. Table 2.48 reports the probability using this distribution that the mean in Ireland exceeds various values.

One question that Noel wanted answered is whether the mean at the site in Ireland exceeds 93%. Table 2.48 indicates that this is indeed the case.

2.13.4 Software for Estimation of the Posterior Distribution

Except in some very simple situations, application of (2.127) to perform a Bayesian analysis requires computer software. Modern Bayesian software, such as Bayesian inference Using Gibbs Sampling (BUGS), the Windows version WinBUGS, Just Another Gibbs Sampler (JAGS), or Stan, are freely available and have been widely used for over a decade. WinBUGS comes with a wide range of examples and many books are available to describe how to use it to implement Bayesian procedures.

In using such software, it is typically necessary to specify $\mathbf{y}$, $\mathbf{p}(\boldsymbol{\theta})$, and $\mathbf{p}(\mathbf{y}|\boldsymbol{\theta})$ using whatever functions are available in the particular software package. The software will handle the computation of $\mathbf{p}(\boldsymbol{\theta}|\mathbf{y})$ behind the scenes. Usually $\mathbf{p}(\boldsymbol{\theta}|\mathbf{y})$ is provided not as an analytical pdf expression, but as a sample "drawn" from this posterior pdf. Generally the sample size will consist of thousands or millions of draws. From this sample of draws, the posterior distribution of any fixed or random function of θ can be easily obtained simply by calculating the function for each draw.

2.13.5 Bayesian Analysis in Practice Today

There are many applications that may be formulated using either a frequentist or a Bayesian approach. Generally speaking, the Bayesian approach is preferred when there is information that can be used to construct an informative prior. The Bayesian approach is necessary if one wishes to make a probability statement concerning a parametric value. However, it seems most of the applications in the CMC world presently involve frequentist formulations. One may ask why this is the situation, given the potential benefits that can be derived from a Bayesian approach. Some possible roadblocks to greater implementation of Bayesian methods include the following:

1. The subjectivity of prior knowledge gives the perception to some that Bayesian approaches allow one to demonstrate a preconceived position regardless of the data. This may seem to be a major deterrent in a regulated industry.
2. As with all scientific analyses, care and understanding are needed to ensure that results are meaningful and appropriately derived. Thus, statistical knowledge is needed to apply Bayesian procedures correctly. Although this knowledge can be acquired by non-statisticians, many of these scientists are not inclined to learn Bayesian methods without first understanding the benefits that might result.
3. User-friendly statistical packages are more developed for the frequentist approach.

With regard to the first objection, it is true that the accuracy of Bayesian probabilities depend on both mechanistic understanding of random processes and one's prior knowledge of how the mechanism behaves. Because prior knowledge and expression of that knowledge vary among individuals, Bayesian probabilities are subjective in nature.

However, Bayesian probabilities can be just as objective as frequency-based probabilities. The same fundamental laws of probability apply to both. Given that prior knowledge is justified and properly expressed, a Bayesian procedure is no less rigorous or appropriate than a corresponding frequentist procedure. Moreover, the use of prior knowledge can render a Bayesian analysis more efficient (i.e., provide tighter intervals) than a corresponding frequentist analysis, or provide a solution where none exists in the frequency world. As an example, Bayesian approaches developed by Joseph et al. (1995) and Suess et al. (2000) are used to analyze screening tests when a gold standard is absent. Frequentist approaches are not useful here due to the limited information that cause traditional models to be "under-specified." Because Bayesian procedures provide probability statements about unknown parameters, they are arguably more often in alignment with questions posed by clients. That is, clients often want to know the probability of some event occurring in the future, and a Bayesian analysis more directly answers such a question.

Although some will criticize Bayesian methods for the subjectivity inherent in defining prior information, it is well known that subjectivity is an inherent part of most regulatory and business decisions. This is because we generally make better decisions when we consider the totality of evidence that bears on a problem, including evidence beyond the particular limited data set in hand. Additionally, subjectivity is hardly absent in frequentist statistical methods where one is required to make subjective decisions concerning model assumptions, the choice of confidence levels, selected decision limits, pooling of data, and weighting rules that may be included in an analysis.

Bayesian approaches are commonly found in the pharmaceutical industry. Examples of such applications in clinical activities are provided in the book by Berry et al. (2010). We also provide some examples in this book in Chaps. 6 (analytical methods) and 8 (stability testing). Thus, the argument that Bayesian methods are not appropriate in a regulated industry does not hold in the pharmaceutical industry.

Roadblocks 2 and 3 are slowly dissolving. Bayesian methods are becoming part of every statistician's tool box, and scientists outside the field of statistics have more opportunities to experience their benefits.

As greater demand arises for Bayesian methods, statistical software will appear to meet the demand. Prior to the existence of modern computer software, Bayesian analysis was difficult to handle since the computations are very involved. The advent of powerful techniques such as Gibbs sampling (see, e.g., Casella and George 1992) has facilitated the use of the Bayesian approach since now practitioners can concentrate on the analysis rather than on the difficult calculation of posterior densities.

References

ASTM E29 (2013) Standard practice for using significant digits in test data to determine conformance with specifications. ASTM International, West Conshohocken

Berger RL, Hsu JC (1996) Bioequivalence trials, intersection-union tests and equivalence confidence sets. Stat Sci 11(4):283–302

Berry SM, Carlin BP, Lee JJ, Muller P (2010) Bayesian adaptive methods for clinical trials. CRC Press, Taylor & Francis Group, Boca Raton

Bolstad WM (2007) Introduction to Bayesian statistics, 2nd edn. Wiley, New York

Borman PJ, Chatfield MJ (2015) Avoid the perils of using rounded data. J Pharm Biomed Anal 115:502–508

Boylan GL, Cho RR (2012) The normal probability plot as a tool for understanding data: a shape analysis from the perspective of skewness, kurtosis, and variability. Qual Reliab Eng Int 28:249–264

Brown H, Prescott R (2006) Applied mixed models in medicine, 2nd edn. Wiley, New York

Burdick RK, Graybill FA (1992) Confidence intervals on variance components. Marcel Dekker, New York

Burdick RK, Borror CM, Montgomery DC (2005) Design and analysis of gauge R&R experiments: making decisions with confidence intervals in random and mixed ANOVA models. ASA-SIAM Series on Statistics and Applied Probability. SIAM, Philadelphia

Burgess C (2013) Rounding results for comparison with specification. Pharm Technol 37(4):122–124

Casella G, George EI (1992) Explaining the Gibbs sampler. Am Stat 46:167–174

Cleveland WS (1985) The elements of graphing data. Wadsworth Advanced Books and Software, Monterey

Dong X, Tsong Y, Shen M, Zhong J (2015) Using tolerance intervals for assessment of pharmaceutical quality. J Biopharm Stat 25(2):317–327

Eberhardt KR, Mee RW, Reeve CP (1989) Computing factors for exact two-sided tolerance limits for a normal distribution. Commun Stat Simul Comput 18(1):397–413

Gelman A, Hill J (2007) Data analysis using regression and multilevel/hierarchical models. Cambridge University Press, New York

Gitlow H, Awad H (2013) Intro students need both confidence and tolerance intervals. Am Stat 67 (4):229–234

Graybill FA, Wang C-M (1980) Confidence intervals on nonnegative linear combinations of variances. J Am Stat Assoc 75(372):869–873

Hahn GJ, Meeker WQ (1991) Statistical intervals: a guide for practitioners. Wiley, New York

Hannig J, Iyer H, Patterson P (2006) Fiducial generalized confidence intervals. J Am Stat Assoc 101(473):254–269

Hoffman D, Kringle R (2005) Two-sided tolerance intervals for balanced and unbalanced random effects models. J Biopharm Stat 15:283–293

Howe WG (1969) Two-sided tolerance limits for normal populations--some improvements. J Am Stat Assoc 64:610–620

International Conference on Harmonization (2004) Q5E Comparability of biotechnological/biological products subject to changes in their manufacturing process

Joseph L, Gyorkos TW, Coupal L (1995) Bayesian estimation of disease prevalence and the parameters of diagnostic tests in the absence of a gold standard. Am J Epidemiol 141 (3):263–272

Kelley K (2007) Confidence intervals for standardized effect sizes: theory, application, and implementation. J Stat Softw 20(8):1–24. http://www.jstatsoft.org/

Krishnamoorthy K, Mathew T (2009) Statistical tolerance regions: theory, applications, and computation. Wiley, New York

Krishnamoorthy K, Xia Y, Xie F (2011) A simple approximate procedure for constructing binomial and Poisson tolerance intervals. Commun Stat Theory Methods 40(12):2243–2258

Kruschke JK (2015) Doing Bayesian data analysis, 2nd edn. Academic Press, Burlington, MA

Limentani GB, Ringo MC, Ye F, Bergquist ML, McSorley EO (2005) Beyond the t-test: statistical equivalence testing. Anal Chem 77(11):221–226

Littell RC, Milliken GA, Stroup WW, Wolfinger RD, Schabenberger O (2006) SAS for mixed models, 2nd edn. SAS Institute Inc, Cary

Mee RW (1984) β-expectation and β-content tolerance limits for balanced one-way ANOVA random model. Technometrics 26(3):251–254

Mee RW (1988) Estimation of the percentage of a normal distribution lying outside a specified interval. Commun Stat Theory Methods 17(5):1465–1479

Myers RH, Montgomery DC, Vining GG (2002) Generalized linear models with applications in engineering and the sciences. Wiley, New York

Nijhuis MB, Van den Heuvel ER (2007) Closed-form confidence intervals on measures of precision for an inter-laboratory study. J Biopharm Stat 17:123–142

NIST/SEMATECH e-handbook of statistical methods. http://www.itl.nist.gov/div898/handbook/. Accessed 6 Jan 2016

Office of Regulatory Affairs (ORA) Laboratory Manual. http://www.fda.gov/ScienceResearch/FieldScience/LaboratoryManual/ucm174283.htm. Accessed 6 Jan 2016

Persson T, Rootzen H (1977) Simple and highly efficient estimators for a type I censored normal sample. Biometrika 64:123–128

Quiroz J, Strong J, Zhang Z (2016) Risk management for moisture related effects in dry manufacturing processes: a statistical approach. Pharm Dev Technol 21:147–151

Suess EA, Fraser C, Trumbo BE (2000) Elementary uses of the Gibbs sampler: applications to medical screening tests. STATS Winter 27:3–10

Thomas JD, Hultquist RA (1978) Interval estimation for the unbalanced case of the one-way random effects model. Ann Stat 6:582–587

Torbeck LD (2010) %RSD: friend or foe? Pharm Technol 34(1):37–38

Tufte ER (1983) The visual display of quantitative information. Graphics Press, Cheshire

USP 39-NF 34 (2016a) General Chapter <1010> Analytical data—interpretation and treatment. US Pharmacopeial Convention, Rockville

USP 39-NF 34 (2016b) General Notices 7.20: Rounding Rules. US Pharmacopeial Convention, Rockville

Vangel MG (1996) Confidence intervals for a normal coefficient of variation. Am Stat 15(1):21–26

Vining G (2010) Technical advice: quantile plots to check assumptions. Qual Eng 22(4):364–367

Vining G (2011) Technical advice: residual plots to check assumptions. Qual Eng 23(1):105–110

Vukovinsky KE, Pfahler LB (2014) The role of the normal data distribution in pharmaceutical development and manufacturing. Pharm Technol 38(10)

Yu Q-L, Burdick RK (1995) Confidence intervals on variance components in regression models with balanced (Q-1)-fold nested error structure. Commun Stat Theory Methods 24:1151–1167

Chapter 3
Process Design: Stage 1 of the FDA Process Validation Guidance

Keywords Critical process parameter (CPP) • Critical quality attribute (CQA) • Experimental design • Factorial design • Knowledge space • Normal operating range • Process capability • Process design • Process robustness • Quality by design • Quality target product profile (QTPP) • Region of goodness • Risk assessment

3.1 Introduction

This is the first of three chapters that describe statistical approaches related to the three stages of process validation described in the FDA Process Validation Guidance for Industry (2011). The three stages are

1. Process Design (Chap. 3),
2. Process Qualification (Chap. 4), and
3. Continued Process Verification (Chap. 5).

The three-stage process validation guidance aligns process validation activities to the product life cycle concept. Along with existing FDA guidance, it links the quality of the product with ensuring quality of the process, from product and process design through mature manufacturing. The FDA process validation guidance supports process improvement and innovation through sound science and includes concepts from other FDA supported guidance, including the International Conference on Harmonization (ICH) chapters Q8(R2) Pharmaceutical Development (2009), Q9 Quality Risk Management (2005), and Q10 Pharmaceutical Quality System (2008).

A goal of quality assurance is to produce a product that is fit for its intended use. Very broadly, within the guidance, process validation is defined as the collection and evaluation of knowledge, from the process design stage through commercial production, which establishes scientific evidence that a process consistently delivers quality product. This knowledge and understanding is the basis for establishing an approach to control a manufacturing process that results in products with the desired quality attributes. Across the three stages, the statistics contribution is to iteratively work on understanding sources of variability and the impact of variability on the process and product attributes, build quality into the product and process, and detect and control the variation in a manner commensurate with the product risk and the patient needs.

This three-step approach assumes the following conditions:

1. Quality, safety, and efficacy are designed or built into the product.
2. Quality cannot be adequately assured merely by in-process and finished product inspection or testing.
3. Each step of a manufacturing process is controlled to ensure that the finished product meets all quality attribute specifications.

In slightly more detail, the three stages are

Stage 1—Process Design: The commercial manufacturing process is defined during this stage based on knowledge gained through development and scale-up activities. The knowledge can be of several forms: fundamental science, mechanistic or physics-based models, data-driven models based on previous compounds, and experimental understanding of the product being developed. In the process design stage various tools are employed to understand inputs to the process (parameters and material attributes) and their effect on the outputs (quality attributes). Throughout this development stage, decisions are made on how to establish and control the process to ensure quality in the output. This design stage can be in accordance with ICH Q8(R2) and ICH Q11 (2012) and as such may be a key change in the focus of activity for many companies.

Stage 2—Process Qualification: Following a process design stage where sufficient understanding has been gained to provide a high degree of assurance in the manufacturing process, the process design is evaluated to determine if the process is capable of reproducible commercial manufacturing. This stage has two elements: (1) design of the facility and qualification of the equipment and utilities and (2) process performance qualification (PPQ). This later element was historically called process validation, and most often conducted by executing three lots within predetermined limits.

Stage 3—Continued Process Verification: After establishing and confirming the process, manufacturers should maintain the process in a state of control over the life of the process, even as materials, equipment, production environment, personnel, and manufacturing procedures change. The goal of the third validation stage is continued assurance that the process remains in a state of control (the validated state) during commercial manufacturing. Systems for detecting departures from the product quality are helpful to accomplish this goal. Data are collected and analyzed to demonstrate that the production process remains in a state of control and to identify any opportunities for improvement.

Stages 2 and 3 are discussed in Chaps. 4 and 5, respectively.

3.2 More on PV-1

Inspection is too late, the quality or lack thereof, is already in the product. Inspection does not improve the quality, nor guarantee quality. As Harold F. Dodge said, "You cannot inspect quality into a product."

"Quality cannot be inspected into a product or service; it must be built into it." W.E. Deming in *Out of the Crisis* (2000).

Joseph M. Juran, renowned quality guru, characterized the development process as a hatchery for new quality issues and coined the term "quality by design" to describe the comprehensive discipline required to pro-actively establish quality (Juran 1992).

The pharmaceutical industry has traditionally been highly dependent on end-product testing and inspection. However, this has changed and continues to develop. The appropriate balance of a holistic quality approach versus end-product testing is now common across the industry. Concepts from ICH Q8–Q11 and the FDA guidance for process validation facilitate a move from an inspection-based to a design-based system. The focus of PV-1 is to design a product and associated processing by identifying and controlling process inputs so that the resulting output is of acceptable quality (defined as "what the patient needs") and well-controlled. The result of PV-1 is to create a manufacturing process with an appropriate risk-based control strategy.

The PV-1 process progresses in the following manner:

1. Develop a Quality Target Product Profile (QTPP).
2. Iteratively design the active pharmaceutical ingredient (API) process, formulation, analytical methods, and final drug product process to achieve the QTPP.
3. Define the "region of goodness" for each process and process input.
4. Determine critical parameters and propose a control strategy.
5. Transition to Manufacturing, PV-2.

The next sections of this chapter overview and connect statistically related tools used in process design. These tools are used to identify good and flexible operating regions, help to determine critical parameters, and propose a control strategy.

Figure 3.1 presents terminology and provides a high level summary of the QbD development process. When starting to develop any product or process, the unknown is called the "unexplored space." The subset labeled "knowledge space" consists of prior learnings on similar products, first principles understanding of the present process, and empirical information from experiments and other data analyses. After assessing what is known and unknown, the task is to identify and prioritize the knowledge necessary to produce a high quality, safe, and efficacious product. Risk assessment helps in the prioritization and both statistical and non-statistical tools are used to obtain the knowledge. Following an iterative development cycle, the knowledge and scientific experience might lead to several defined regions:

1. A process set point, where if needed, represents where the process is nominally operated.

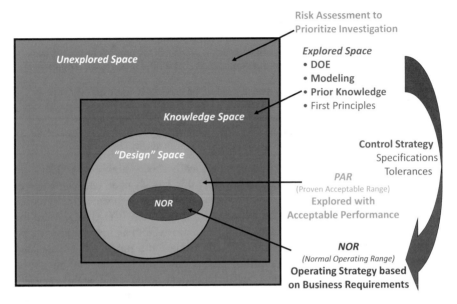

Fig. 3.1 Structure of PV-1 spaces and terminology

2. A normal operating range (NOR) which accounts for variability in the set point.
3. A proven acceptable range (PAR) is a region of goodness which allows for variability in incoming raw materials or otherwise permits flexibility in assuring quality.

Based on knowledge gained through development, parameters and process elements which must be controlled and monitored are identified via the "control strategy." This control strategy allows the manufacturing process to stay within a region of goodness.

The following definitions are useful in navigating the PV-1 landscape.

1. Attribute: A characteristic or inherent property of a feature. This term is used in two contexts. The first is as a reference to raw material or excipient features, called material attributes. The other is in reference to the features of the drug substance or drug product. These attributes are significant in defining product safety and efficacy, and are termed critical quality attributes.
2. Control Strategy: A planned set of controls, derived from current product and process understanding that ensures a consistent level of process performance and product quality. The controls can include parameters and attributes related to drug substance and drug product materials, facility and equipment operating conditions, in-process controls, finished product specifications, and the associated methods and frequency of monitoring and control (ICH Q10 2008).
3. Critical Process Parameter (CPP): A process parameter whose variability has an impact on a critical quality attribute and must be monitored or controlled to ensure the process produces the desired quality (ICH Q8(R2)).

4. Critical Quality Attribute (CQA): A physical, chemical, biological, or micro-biological property or characteristic that should be within an appropriate limit, range, or distribution to ensure the desired product quality (ICH Q8(R2)).

5. Design Space: The multidimensional combinations and interaction of input variables (e.g., material attributes) and process parameters that have been demonstrated to provide assurance of quality. Movement of a process within the design space is not considered to be a change. Movement out of the design space is considered to be a change and would normally initiate a regulatory postapproval change process. Design space is proposed by the applicant and is subject to regulatory assessment and approval (ICH Q8(R2)).

6. Knowledge Management: Systematic approach to acquiring, analyzing, storing, and disseminating information related to products, manufacturing processes, and components. Sources of knowledge include prior knowledge (public domain or internally documented), pharmaceutical development studies, technology transfer activities, process validation studies over the product life cycle, manufacturing experience, innovation, continual improvement, and change management activities (ICH Q10).

7. Life cycle: All phases in the life of a product from the initial development through marketing until the product's discontinuation (ICH Q8(R2)).

8. Normal Operating Range (NOR): A defined range within (or equal to) the Proven Acceptable Range. It defines the standard target and range under which a process operates.

9. Parameter: A measurable or quantifiable characteristic of a system or process (ASTM E2363).

10. Process Design (PV-1): Defining the commercial manufacturing process based on knowledge gained through development and scale-up activities.

11. Process Qualification (PV-2): Confirming that the manufacturing process as designed is capable of reproducible commercial manufacturing.

12. Process Validation: The collection and evaluation of data, from PV-1 through PV-3, which establishes scientific evidence that a process is capable of consistently delivering quality products.

13. Process Capability: Ability of a process to manufacture and fulfill product requirements. In statistical terms, process capability is measured by comparing the variability and targeting of each attribute to its required specification. The capability is summarized by a numerical index C_{pk} (see Chap. 5 for information on this topic). A process must demonstrate a state of statistical control for process capability to be meaningful.

14. Process Parameter: A process variable (e.g., temperature, compression force) or input to a process that has the potential to be changed and may impact the process output. To ensure the output meets the specification, ranges of process parameter values are controlled using operating limits.

15. Process Robustness: The ability of a manufacturing process to tolerate the variability of raw materials, process equipment, operating conditions, environmental conditions, and human factors. Robustness is an attribute of both

process and product design (Glodek et al. 2006). Robustness increases with the ability of a process to tolerate variability without negative impact on quality.

16. Proven Acceptable Range: A characterized range of a process parameter for which operation within this range, while keeping other parameters constant, will result in producing a material meeting relevant quality criteria (ICH Q8 (R2)).

17. Quality: The suitability of either a drug substance or drug product for its intended use. This term includes such attributes as the identity, strength, and purity (ICH Q6A 1999), ICH Q8(R2)).

18. Quality by Design: A systematic approach to process development that begins with predefined objectives and emphasizes product and process understanding based on sound science and quality risk management (ICH Q8(R2)).

19. Quality Risk Management: A systematic process for the assessment, control, communication, and review of risks to the quality of the drug (medicinal) product across the product life cycle (ICH Q9).

20. Quality Target Product Profile (QTPP, pronounced Q-tip): A prospective summary of the quality characteristics of a drug product that ideally will be achieved to ensure the desired quality, taking into account safety and efficacy of the drug product (ICH Q8(R2)).

21. Risk: The combination of the probability of occurrence of harm and the severity of that harm (ICH Q9).

22. Risk Assessment: A systematic process of organizing information to support a risk decision to be made within a risk management process. It consists of the identification of hazards and the analysis and evaluation of risks associated with exposure to those hazards (ICH Q9).

23. State of Control: A condition in which the set of controls consistently provides assurance of continued process performance and product quality (ICH Q10).

3.3 Iterative Process Design

Once the QTPP has been developed, product and process design can begin. Process design is the activity of defining the commercial manufacturing process that will be reflected in master production and control records. The goal of this stage is to design a process suitable for routine commercial manufacturing that can consistently deliver a product that meets the acceptance criteria of its quality attributes.

Process design is iterative and can include all processes associated with the product: API process, formulation, analytical methods, and final product processes. Not all processes are developed in the same manner. For example, the API synthesis process is not developed in the same fashion as a formulation process or an analytical method.

Step 1: Form a team.

A systematic team-based approach to development is the most efficient manner to develop a robust process. This team should include expertise from a variety of disciplines including process engineering, industrial pharmacy, analytical chemistry, microbiology, statistics, manufacturing, and quality assurance.

Step 2: Define the process.

Typically, a manufacturing process is defined by a series of unit operations or a series of synthesis steps. Prior to initiation of any studies, the team needs to agree which unit operations, reactions, or steps are included in the process. To aid in this definition, the team creates a map or flowchart of the process.

A process is a combination of people, machines, methods, measurement systems, environment, and raw materials that produces the intended output. Figure 3.2 displays a process flow diagram for a dry granulation process. Once the process has been defined, meaningful groupings of the unit operations are developed to form the basis for experimentation. Figure 3.3 provides a schematic of these grouping or "focus areas." The parameters (inputs) and attributes (outputs) for each focus area are discussed and studied in detail. The team discusses a focus area and identifies the attributes and parameters that could potentially affect each attribute. Figure 3.4 provides an Ishikawa diagram (also known as a cause and effect or fishbone diagram) which is helpful in mapping potential sources of parameter variability by categories (e.g., machine, method, manpower, and environment) that could influence attributes.

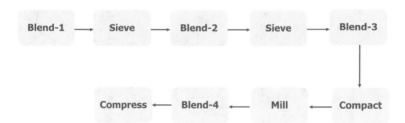

Fig. 3.2 Process flow diagram for a dry granulated product

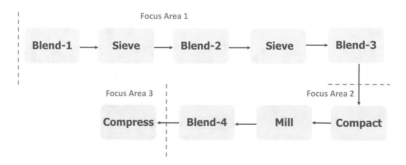

Fig. 3.3 Process map with experimental focus areas

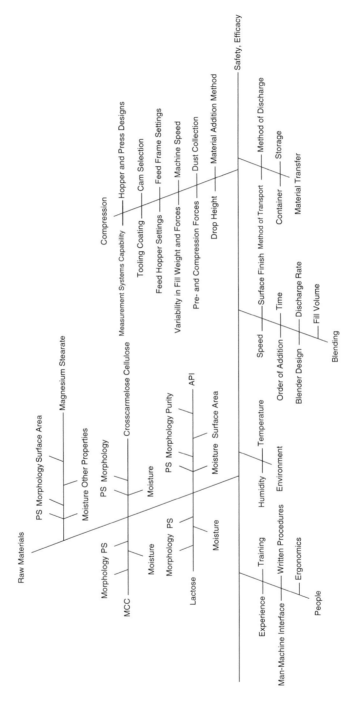

Fig. 3.4 Ishikawa cause and effect (fishbone) diagram (Glodek et al. 2006)

Table 3.1 Risk assessment matrix

Attribute Rank	10	7	10	10	7		
Attribute Parameter	Genotoxic Impurity	Tablet Potency	Drug Release Rate	Shelf Life	Content Uniformity	Score	Strategy
Sampling Method	9	5	5	5	9	288	MSA
Excipient Attribute	10	5	5	9	1	282	DoE
Drug Particle Size	1	9	9	1	9	236	Model
Roll Force	1	9	5	1	9	196	DoE
Screen Size	1	5	1	1	5	100	DoE
Blend Speed	1	5	1	1	5	100	Model
Blend Size	1	5	1	1	5	100	Model
Roll Gap Width	1	5	1	1	5	100	DoE
Compression Force	1	1	5	1	1	84	DoE

Step 3: Prioritize Team Actions

All attributes and parameters should be evaluated in terms of their roles in the process, and on their impact on the final product or in-process material. They should also be reevaluated as new information becomes available.

Following the relationship building of the Ishikawa diagram, the team will select attributes that best define the process. It is typical to perform a risk assessment to prioritize actions taken by the team in developing the process. As per the Pareto principle, it is important to identify the significant parameters for further study. Not all parameters will have an impact and prior knowledge of this improves the impact of the planned studies. Prioritization establishes a risk-based approach to development. Table 3.1 provides the results of a risk assessment.

- The attributes from the Ishikawa diagram are listed across the top and the team of knowledgeable experts rates their importance from 1 to 10, with 10 being the most important in impacting the final product quality.
- The parameters are listed down the left side and the hypothesized or known strength of the relationship between the attribute and parameter is supplied in each box.
- The score is determined for each parameter by multiplying the attribute score and the parameter strength and summing across the attributes. For example, the score for the excipient attribute is $10 \times 10 + 5 \times 7 + 5 \times 10 + 9 \times 10 + 1 \times 7 = 282$.
- The score is sorted from high to low and the strategy to study each parameter is determined.
- The team's actions and the work performed in developing the product are prioritized based on importance as indicated by the total score.

Step 4: Take Action to Understand and Solidify Functional Relationship

Throughout the product life cycle, various studies can be initiated to discover, observe, correlate, or confirm information about the product and process. Knowledge exists in many forms including fundamental knowledge, data-driven models from experimentation, data-driven models on related compounds or equipment, and experimental studies meant to establish or confirm relationships. Studies to gain knowledge are planned for areas where information does not exist. These studies should be planned and conducted according to sound scientific principles and appropriately documented.

Ultimately, the result of the functional understanding is coined as a knowledge space. The region defined as the design space is a subset of the knowledge space. Operation within the design space will ensure product quality. Note that this is clearly not the traditional statistical definition, as a design space in statistics refers to the study range. The design space is meant to be defined in a multifactor fashion and is optional from a regulatory perspective. Another region traditionally defined to represent a region of goodness is the proven acceptable range (PAR). This range has traditionally, although not exclusively, been set in a univariate manner. Rather than compare and contrast a design space with a PAR, suffice it to say each can be called a "region of goodness." This term is used in future discussion in this chapter to cover both regions. For more on the topic of design space see the papers by Peterson (2004, 2010), Vukovinsky et al. (2010a, b, c), and Stockdale and Cheng (2009).

An initial set point is established within the region of goodness to define the nominal operating condition. Around that point, a normal operating range (NOR) is defined that considers expected operational variability. In the ICH literature, the NOR is permitted to vary within the region of goodness. For example, incoming raw material variability might necessitate a change in set point and NOR or additional process understanding at scale could be used to establish a new set point within the region of goodness.

Step 5: Confirm

Once an NOR is determined, the selected operating conditions or ranges are confirmed. In many cases the NOR has been determined based on experimental design and predictive modeling, but it hasn't been run at either development or full scale. The paper champion needs to be realized and the knowledge confirmed. The initial confirmation might be at development scale (Garcia et al. 2012). The ultimate confirmation, process qualification, is usually conducted at the manufacturing facility where the product will be produced.

Step 6: Document Control Plan

Justification of the controls should be sufficiently documented and internally reviewed to verify and preserve their value for use or adaptation later in the process life cycle. Process knowledge and understanding is the basis for establishing an approach to process control for each unit operation. Strategies for process control can be designed to reduce input variation, adjust for input variation during manufacturing, or combine both approaches. Manufacturing controls mitigate variability to assure quality of the product. Controls can consist of material analysis and equipment monitoring at significant processing points (21 CFR 211§ 211.110 (c)). Decisions regarding the type and extent of process controls can be aided by earlier risk assessments, then enhanced and improved as process experience is gained. The degree of control over attributes and parameters should be commensurate with their risk to the process. In other words, a higher degree of control is appropriate for attributes and parameters that pose a higher risk. The planned commercial production and control records, which contain the operational limits and overall strategy for process control, should be carried forward to the next stage for confirmation.

Step 7: Iterate as Needed

Typically, all development decisions are not made in one shot. This is an iterative process that continues as new information becomes available.

3.4 PV-1 Statistical Tools

Knowledge is defined as facts, information, and skills acquired through experience or education. It is the theoretical or practical understanding of a subject. Knowledge does not need to be recreated ab initio for every product being developed, but should be created where necessary. That is, in the design process, teams leverage relevant existing data along with fundamental knowledge to make initial decisions, perform risk assessments that identify gaps, and take actions to gain more knowledge. Figure 3.5 summarizes statistical tools that are important in PV-1. These tools include data-based decision making, data collection and experimental design,

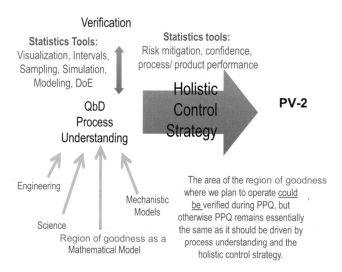

Fig. 3.5 Application of statistical tools in PV-1

descriptive data analysis, and complex modeling. These tools are used to gather, summarize, or quantify knowledge in PV-1.

Some of the PV-1 statistical tools are

1. Visualization: It is said that a picture is worth a thousand words. The benefit of effective and simple display of information cannot be overstated and the ability to take a set of data, summarize the information, and visually display this information is both an art and a science. A primary goal of data visualization is to communicate information clearly and efficiently in order to induce the viewer to think about the substance being displayed without distorting or misrepresenting the information. There are many graphical tools available in spreadsheet and statistical software programs. It is necessary to learn these tools in order to present a meaningful data analysis. Section 2.4 provides more discussion on this topic.

2. Simple Descriptive Statistics: Descriptive statistics is the discipline of quantitatively describing a set of data. This usually includes a description of the central tendency of the data (mean, geometric mean, median, or mode) and a measure of the dispersion or variability in the data (range, standard deviation, or variance). The data summary can be displayed visually as a boxplot by itself or with other groups of similar data as a comparison. Section 2.4 provides more discussion on a boxplot.

3. Statistical Intervals (Confidence, Prediction, and Tolerance): Statistical intervals are the most useful tools for quantifying uncertainty. Section 2.5 discusses these tools in detail.

4. Sampling Plans: In pharmaceutical development and manufacturing, sampling is used in many applications. Included are sampling processes used for making batch release decisions, demonstrating homogeneity of drug substance and drug product, accepting batches of raw material, and selection of units for environmental monitoring. Examples of sample plans are discussed throughout this book.

5. Monte Carlo Simulation: Simulation is most useful for studying future events that can be predicted from historical data and theorized or established models. The impact of considered changes can be simulated to obtain an understanding of future outcomes under various possible scenarios. Simulation applications are provided throughout this book.

6. Measurement System Analysis (MSA): These analyses are referred to as repeatability and reproducibility (R&R) in some industries. They involve the design and analysis of experimental data to understand, quantify, and reduce the variability in the measurement system (analytical method). Variability in the measurement system is normally reduced to categories of bias, linearity, stability, repeatability, and reproducibility. These types of data analysis are critical for the development of useful analytical methods, and are discussed in Chap. 6.

7. Hypothesis Testing: Hypothesis testing is a formal statistical process of comparison and inference. Such tests are often required by regulatory agencies in many evaluations. This topic is discussed in Sect. 2.10.

8. Models and Modeling: Prior to running experiments, information based on either first-principles or data-driven models should be exercised to help inform relationships.
9. Data-driven Modeling: Data-driven models are developed through fitting models to data. In PV-1, there is often data related to the process or compound being developed. Sometimes, as is the case with material property data, chemical structure data, and processing data, small to large data sets exist and data-driven models are developed to best express relationships. In the case of a material property data base, relationships between material properties and product attributes would be examined and data-driven models developed to predict product properties based on the material attributes. These data-driven models permit a decrease in experimentation, or at least provide a starting point for further experimentation. Common modeling techniques include simple linear regression, partial least squares, regression trees, and machine learning algorithms.
10. First-principle or Fundamental Models: First-principle, engineering, physics, or fundamental models explain relationships between parameters, material attributes, or manufacturing factors and product attributes. These models seek to predict product attributes directly from established laws of science.
11. Design of Experiments (DoE): DoE is a highly used tool in investigating unknown relationships within the framework of PV-1. DoE provides a systematic approach to study prioritized factors and establish a relationship with quality or in-process attributes. More information on DoE is provided in Sect. 3.5.

3.5 Design of Experiments

> To call in the statistician after the experiment is done may be no more than asking him to perform a post-mortem examination: he may be able to say what the experiment died of.
>
> R. A. Fisher (1890–1962)

DoE has become a bedrock of the framework of PV-1. Why has this become such an integral part of the process? The strength of DoE is in the application of a systematic approach to data-based decision making along with the selection of a study design. Because of the complexity of most processes, several factors are usually studied in a series of experiments. Historically, students learn to vary one-factor-at-a-time (OFAT) and this practice is applied on the job in research, development, and manufacturing. A reason provided in support of this approach is that if more than one factor is changed, the experimenter will not be able to determine which factor was responsible for the change in the response. In reality, the proper selection of experimental runs combined with the proper analysis removes this source of concern. In addition, there are two major deficiencies with an OFAT study. The first is that there are often interactions between the parameters under study. An interaction means that the effect a process parameter has on the response may depend on the levels of another process parameter. Statistical experimental designs that permit the estimation of interactions will allow for their study, whereas OFAT studies do not. The other deficiency of OFAT is that data previously collected to study other factors is set aside and new data are collected. The structure of the statistical experimental design allows all the data from the entire study to be used to draw conclusions on each factor. This results in savings of both time and money over the OFAT process. In fact, the statistical design approach provides a proper design structure that when combined with the analysis method maximizes the amount of information for the minimum number of runs (i.e., the knowledge development process is highly efficient)

Many textbooks and papers have been written on this subject, and the reader is encouraged to have some of these books in a personal library. Three books are Box et al. (2005), Montgomery (2012), and Morris (2011). Since so much is available on the topic, there is no intention to provide comprehensive technical details in this book. Rather, the focus in this chapter is on the high level application of DoE within a PV-1/QbD environment.

Underlying all processes are mathematical and statistical models, the behavior of which is interrogated via experimentation. Designing an efficient process with an effective process control strategy is dependent on the process knowledge and understanding obtained. DoE studies can help develop process knowledge by revealing relationships between process parameters and the resulting quality attributes measured on process outputs. Efficiently determining an approximate equation representing the underlying physical equation is best accomplished by DoE and an effective experimental strategy. The experimental design process is only one element of QbD, and it closely follows the QbD process as shown in Fig. 3.6.

Fig. 3.6 Experimental design flowchart

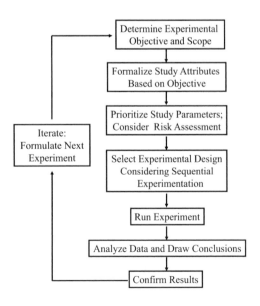

As shown in Fig. 3.6, the experimental process consists of the following seven steps:

1. Determine the experimental objective and scope: Before initiating any series of experiments, define the purpose of the study. Is the goal to improve yield, increase selectivity or the reaction, achieve a particular dissolution profile while minimizing content variability, or optimize a potency method? It is very important to be clear about the purpose of the experiment and the decisions to be based on the study. Under the QbD paradigm, agreement on the goal and alignment on the experimental objective is especially important. Everyone on the team needs to be progressing toward the same goal.

2. Formalize study attributes based on the objective: Agree on the attributes (responses or outputs) to study in the experiment. Attributes should be aligned with the experimental objective. In addition, an important consideration at this stage is inclusion of attributes that are not primary to the objective. Selected attributes should include those directly related to the experimental objective, and those not of primary interest, but with an ability to impact the study later in the process. In addition, selection of a measurement system, how responses will be measured, and the required level of precision should all be considered before running the experiment.

3. Prioritize study parameters which may affect the responses: Define the factors (e.g., process parameters, material attributes, and starting materials) that are hypothesized to have an impact on the quality attribute responses. Scientifically analyze the issue at hand and the process under study in order to determine all the possible factors influencing the situation. This could be achieved by examining literature reports or other prior knowledge, employing fundamental or mechanistic understanding, or through preliminary laboratory experiments. Judicious selection of factors is important in keeping the number of experiments manageable. Consider the appropriateness of a single large design, or a series of reduced sequential designs. A risk assessment can be used to prioritize variables for DoE studies. There are two ways to usefully categorize process performance parameters. One is traditionally statistical, and the other is of special consideration within the pharmaceutical industry. In each case, there are special issues with each parameter type that should be addressed prior to designing and running the experiment.

   a. *Control and noise factors*: Selecting factors to control is not always an easy decision. In fact, even if a factor is not one of the process parameters to be controlled during manufacturing (so-called noise factors), it might be beneficial to control the noise factor during experimentation. For example, humidity at the site might be an uncontrollable factor during manufacturing. However, if an effect on response attributes can be demonstrated during experimentation which might drive improvement of the manufacturing process to mitigate the effect of humidity. Thus, it is important to consider all factors that can impact values of the quality attributes in designing your experiment.

b. *Scale dependent vs. scale independent or scalable*: Parameters that are scale independent or scalable can be studied at a smaller scale than full scale commercial manufacturing and results are applicable for full scale. With scale dependent parameters, there is a dependency of the results based on the scale of the equipment. A strategy is needed to assess the DoE results at scale. Examples of scale independent factors include pressure, temperature, and Gerties roller compactors. Examples of scale dependent factors include mixing rpm and high sheer granulation.

In addition to defining process parameters, it is necessary to define the experimental domain by assigning the upper and lower limit ranges to all continuous variables. For discrete variables, one must define categories. Probably the most difficult component of designing an experiment is selection of the levels or ranges for each parameter. Consider an experiment with the process performance parameter "revolutions per minute (rpm)". How does one decide to set the low and high study levels to (75, 125), as opposed to (50, 100) or (50, 150), or (75, 150)? Selection of such ranges depends on the experimental objective and the overall development strategy. In general, the range limit span should be as wide as is practical, but neither too large nor too narrow. If limits are too narrow, there is a risk of not seeing the parameter effect. If the range is too wide, the parameter will be characterized on a macro level, but may not provide information on the micro level and hide effects of other parameters. It is often helpful to examine existing experimental data, fundamental knowledge, and similar compounds or processes of interest.

4. Select experimental design and consider sequential experimentation: Experimental design is based on the following principles:

   a. *Randomization*: Statistical methods require that observations be independently distributed random variables, and randomization helps make this assumption valid. Randomizing helps "average out" uncontrolled noise variables (lurking or extraneous variables). There are situations where the experiment is not run in a completely randomized fashion due to practical situations. However, this should be by design and the data analyzed in a manner consistent with the design.

   b. *Blocking*: Blocking removes unwanted variability and allows focus on the factors of interest. Pairing is a special type of blocking. As an example, consider a comparison of the bias for two analytical methods. It is expected that there will be variation among the test samples measured in the experiment. For that reason, a paired design requires that each method be used to measure each test sample. In this manner, variation among test samples will not manifest in differences between the measured values from the two analytical methods.

   c. *Replication*: Replication allows the estimate of experimental error to be obtained. This estimate is the basic unit of measurement for determining whether observed differences in the data are *statistically* different.

Replication represents the between run variability. True replication means that the process is completely restarted for each replicate run. Often times, the person running the experiment will merely take repeated measures from a single experimental setup. This is not a true replication, but rather repetition. The manner in which one employs these three principles is impacted by the realities of the scale of the design, and time and other resource constraints. If is easy to get confused by the many statistical designs that are described in the literature.

Design selection is largely dependent on the objective, goals, process knowledge, and the stage of the experimentation. Sections 3.5.1 and 3.5.2 provide high level descriptions of some common experimental designs.

5. Run the experiment: The details in this step are often ignored. It is extremely important that the one who carries out the experiment understands the underlying details of the experiment. In general, run the experiments in a random order to distribute unknown sources of variability and minimize the effect of systematic errors on the observations. It may be tempting to re-order for convenience, but running an experiment in such a non-random pattern can create problems with the data analysis. Some designs can account for planned restrictions on randomization (e.g., split-plot or hard-to-change factors). However, any such restrictions should be built into the experimental design.

   It is important for the person running the experiment to understand the process. The person should understand the difference between experimental factors and factors which should stay fixed during the entire DoE. It may be tempting for those who truly understand the process to make slight tweaks to try and "save" a run. Such adjustments are not allowed, and risk destroying the study conclusions. Review the experimental protocol and determine if there is something that should be changed before running the experiment. Record actual levels of the process parameters and note if they deviate from the planned levels. Record other unexpected events.

6. Analyze the data and draw conclusions: If care was taken in setting up the experimental plan and the experiments were executed as expected, then the data analysis will be relatively straightforward. It might be necessary to meet with a subject matter expert should any of the data appear as outliers, or if the analysis results have influential observations or the models don't appear to be feasible.

7. Confirm results and document: Verify as necessary the decisions made based on the experimental data and the model. It is very important to perform confirmation runs as necessary to verify the best predicted condition. Often times, the best model predicted condition or region has not been run in the experiment, so physical confirmation is highly recommended. Plan the next sequential design, if appropriate.

Results of the experimental effort must be documented. Recommended documentation should include a brief description of the background information that led to the experimentation, the objective, study process parameters (including names, levels, and units), response attributes (including measurement method), details of replication and randomization, the design matrix, a summary of the data, the statistical methods and software used in the analysis, and summary results.

3.5.1 Full and Fractional Factorial Experiments

The factorial family provides powerful and flexible designs for collecting information on main effects and interactions in a minimum number of runs. They are highly flexible and can be used in the screening and interaction phases and as a base for optimization. The ability to add onto these designs facilitates sequential experimentation and enhanced refinement of knowledge.

To execute, factors for experimentation are selected and a fixed number of "levels" (usually high-low) are defined for each parameter. A full factorial design considers all possible combinations of the levels of each input factor. This design permits estimation of the main effects and interactions. In general, assume there are ℓ_1 levels for the first factor, ℓ_2 levels for the second factor, and ℓ_k levels for the kth and last factor. The complete arrangement of $\ell_1 \times \ell_2 \times \ldots \times \ell_k$ experimental combinations is called a full factorial design (e.g., a $2 \times 2 \times 3$ full factorial design yields 12 experimental runs). A full factorial design including five factors varying each factor across two levels is written as 2^5, and has 32 experimental runs.

The 2^5 full factorial design permits estimation of the five main effects, 10 - two-way interactions, 10 three-way interactions, five four-way interactions, and one five-way interaction. The remaining experimental run is used to estimate an overall mean.

It is usually not required to estimate all multifactor interactions, and so a specific fraction of the full factorial is selected. This necessarily results in a reduction of the number of experimental runs. This so-called fractional factorial design is a mathematically correct subset of the full factorial that permits estimation of main effects and some subset of interaction effects. Some loss in experimental information (i.e., resolution) generally results by fractionating, but knowledge of the desired information can be used a priori to select an appropriate fraction. For example, a half fraction of the 2^5 full factorial design includes $2^{5-1} = 16$ runs. This design permits estimation of the main effects and all of the 10 two-factor interactions. Figure 3.7 displays a 2-level full factorial with 3 factors (A, B, C) on the left, and a half fraction of that design on the right.

Understanding the structure of a factorial experiment is important as a base to understanding all designed experiments. An experiment to study the effect of factors A and B on attributes of interest would consist of the four unique runs: (A low, B low), (A low, B high), (A high, B low), and (A high, B high). This is denoted as a 2^2 full factorial experiment, and is shown in Table 3.2. A full factorial experiment to study three factors at two levels, 2^3, has eight unique runs as shown in Table 3.3.

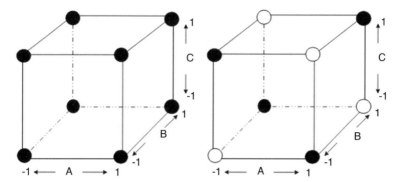

Fig. 3.7 Two level three full factorial (*left*) and half fraction factorial (*right*)

Table 3.2 2^2 full factorial experiment

Standard order	A	B
1	Low	Low
2	Low	High
3	High	Low
4	High	High

Table 3.3 2^3 full factorial experiment

Standard order	A	B	C
1	Low	Low	Low
2	Low	Low	High
3	Low	High	Low
4	Low	High	High
5	High	Low	Low
6	High	Low	High
7	High	High	Low
8	High	High	High

As noted earlier, it is generally accepted that even complicated relationships between parameters and attributes can have a large proportion of the relationship explained by linear effects. Less can be explained by interactions, and even less from nonlinear or quadratic effects. Hence, 2-level factorial experiments are all that is required in many cases.

An example is now provided to demonstrate the power of the two-level factorial structure. In the development of a wet granulation process, it is desired to study the impact of impeller speed, binder level, and binder addition rate on the average particle size (D50). Since it is desired to look at all possible combinations of the three process parameters, the selected design is the 2^3 full factorial experiment shown in Table 3.4. The experiment was run in a random order and three replicated center point conditions were run in addition to the eight factorial

Table 3.4 Wet granulation design with particle size data

Standard order	Impeller speed (A)	Binder level (B)	Binder addition rate (C)	D50
1	Low	Low	Low	156.5
2	Low	Low	High	146.3
3	Low	High	Low	198.8
4	Low	High	High	209.5
5	High	Low	Low	158.4
6	High	Low	High	161.6
7	High	High	Low	136.4
8	High	High	High	142.7

Table 3.5 2^3 design illustrating all estimable effects

Standard order	Impeller speed (A)	Binder level (B)	Binder addition rate (C)	A × B	A × C	B × C	A × B × C	D50
1	−1	−1	−1	1	1	1	−1	156.5
2	−1	−1	1	1	−1	−1	1	146.3
3	−1	1	−1	−1	1	−1	1	198.8
4	−1	1	1	−1	−1	1	−1	209.5
5	1	−1	−1	−1	−1	1	1	158.4
6	1	−1	1	−1	1	−1	−1	161.6
7	1	1	−1	1	−1	−1	−1	136.4
8	1	1	1	1	1	1	1	142.7

runs (not shown in the table). The results are in Table 3.4 in standard order without the center points to simplify the analysis and more effectively demonstrate the power of the analysis.

In the analysis of this particular experiment, it is possible to obtain the linear effects of the parameters (A, B, C), the two-way interactions (A × B, A × C, B × C), and the three-way interaction (A × B × C). Now, replace the word "low" in Table 3.4 with a "−1" and the word "high" with a "1", as shown in Table 3.5. Notice that the sum of multiplied values in the same row of any two columns is equal to zero. In matrix algebra nomenclature, such columns are said to be linearly independent. An experimental design in which all columns are linearly independent is said to be an orthogonal design. An orthogonal design permits estimation of all of the effects individually without interference from any other effects. Notice that each column in Table 3.5 is unique. All eight values of D50 will be used to estimate all seven effects.

Table 3.6 displays this same information in an alternate form. White space in Table 3.6 indicates the correct −1 or 1 positioning of the data for each effect.

Data analysis will normally be conducted using a computer program. For this example, a simple analysis representation which will match a computer analysis is shown in Table 3.7. Each D50 value is placed in the column of its row

Table 3.6 Alternative representation of Table 3.5

Random Order Trial Number	Standard Order Trial Number	Response (Observed Value)	Impeller (A)		Binder Level (B)		Binder AddRate (C)		AB		AC		BC		ABC	
			-1	1	-1	1	-1	1	-1	1	-1	1	-1	1	-1	1
	1	156.5														
	2	146.3														
	3	198.8														
	4	209.5														
	5	158.4														
	6	161.6														
	7	136.4														
	8	142.7														
Total																
Number of Values		8	4	4	4	4	4	4	4	4	4	4	4	4	4	4
Average																
Effect																

corresponding to the level of the performance parameter for which it was collected. For example, in the first row, A is at level -1, and so D50 in the first row is placed in the -1 column of A. From this display, patterns may become apparent, and certainly, data from standard order trial numbers 3 and 4 appear greater than the rest of the data.

To perform the analysis, add each column of values and place the sum in the row labeled "Total." For example, the sum for Impeller (A) at -1 is $156.5 + 146.3 + 198.8 + 209.5 = 711.1$. Next average each column by dividing the column total by the total number of observations included in the total (4 in this example). For Impeller (A) at -1, the average is $711.1/4 = 177.78$. Comparing the average between the low and high level of each factor, it is observed that some of the differences are large (e.g., 177.78 versus 149.78 for Impeller (A)) and some of the differences are small (e.g., 162.53 versus 165.03 for Binder Addition Rate (C)). The difference between the $+1$ average and the -1 average is summarized into a factor effect shown in the last row of the table. By this method, the effect of A is found by $149.78 - 177.78 = -28$. From this row, it can be seen that A $\times$ B, A, and B are the largest effects.

Table 3.8 presents results of a regression model as described in Sect. 2.12 that is fit to include the three large effects (A,B, A $\times$ B) using the data in Table 3.5.

Note the intercept term is the overall average of all eight values of D50. The regression estimate (slope) for each parameter is equal to the effect value in Table 3.7 divided by two. Recall that the effect is the difference from low (-1) to high $(+1)$, whereas the slope is the difference for one unit change (e.g., from -1 to 0 or from 0 to $+1$). Since the overall average of the attribute D50 represents a

Table 3.7 Data analysis, for example

Random Order Trial Number	Standard Order Trial Number	Response (Observed Value)	Impeller (A)		Binder Level (B)		Binder AddRate (C)		AB		AC		BC		ABC	
			-1	1	-1	1	-1	1	-1	1	-1	1	-1	1	-1	1
	1	156.5	156.5		156.5		156.5			156.5		156.5		156.5	156.5	
	2	146.3	146.3		146.3			146.3		146.3	146.3		146.3			146.3
	3	198.8		198.8		198.8	198.8		198.8			198.8	198.8			198.8
	4	209.5		209.5		209.5		209.5	209.5		209.5			209.5	209.5	
	5	158.4		158.4	158.4		158.4		158.4		158.4			158.4		158.4
	6	161.6		161.6	161.6			161.6	161.6			161.6	161.6		161.6	
	7	136.4		136.4		136.4	136.4			136.4	136.4		136.4		136.4	
	8	142.7		142.7		142.7		142.7		142.7		142.7		142.7	142.7	
Total		1310.2	711.1	599.1	622.8	687.4	650.1	660.1	728.3	581.9	650.6	659.6	643.1	667.1	664	646.2
Number of Values		8	4	4	4	4	4	4	4	4	4	4	4	4	4	4
Average		163.775	177.78	149.78	155.7	171.85	162.53	165.03	182.08	145.48	162.65	164.9	160.78	166.78	166	161.55
Effect				-28		16.15		2.5		-36.6		2.25		6		-4.45

Table 3.8 Estimates of regression slopes

Term	Estimate	Prob > |t|
Intercept	163.775	<0.0001
Impeller speed (A)	−14	0.0024
Binder level (B)	8.075	0.0169
Binder level (B)*Impeller speed (A)	−18.3	0.0009

Fig. 3.8 Examples of interaction plots

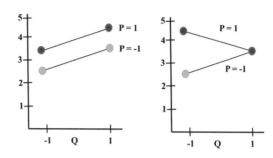

model baseline, the estimates describe the amount of change from baseline as a given factor moves from −1 (low) to +1 (high).

As discussed in Sect. 2.12, existence of an interaction A × B means that the effect of A on the response attribute depends on the selected level for B. This means information from two levels of each parameter is required to decompose the shape of the interaction. In the case of the example, the existence of the A × B interaction points to its strength but not to the functional nature. The functional nature can be described using an interaction plot.

Two interaction plots are provided in Fig. 3.8 for factors P and Q. The vertical axis represents the response attribute, and the horizontal axis shows the two levels of Q. There is one line on the plot for each level of P. The circles represent the average of the response attribute at the given combination of P and Q.

No interaction exists between P and Q in the plot on the left. This is because the lines are parallel, and the amount of change in the response attribute as Q changes from −1 to +1 is constant for both values of P. The y-intercept is different for the two lines, but the rate of change (slope) is identical. Thus, the change in the response attribute as a function of Q is not dependent on the setting for P. Such effects are said to be additive rather than interactive. Similarly, the change in the response as a function of P is not dependent on the setting for Q. (This can be demonstrated by placing P on the horizontal axis, and drawing a line for each level of Q.)

The interaction plot on the right of Fig. 3.8 indicates a strong interaction between P and Q. Notice that as Q moves from −1 to +1 when P = 1, there is a decrease in the average response. However, if P = −1, movement of Q from −1 to +1 results in an increase in the average of the response. This is an important discovery when interactions of this magnitude exist, and provides important information to be considered in process development.

Table 3.9 Calculations of interaction example

Run	Response, No Interaction	P (-1)	P (1)	Q (-1)	Q (1)	P*Q Low, No Interaction	P*Q High, No Interaction	Response, Yes Interaction	P (-1)	P (1)	Q (-1)	Q (1)	P*Q Low, Yes Interaction	P*Q High, Yes Interaction
1	2.5	2.5		2.5			2.5	2.5	2.5		2.5			2.5
2	3.5	3.5			3.5	3.5		3.5	3.5			3.5	3.5	
3	3.5		3.5	3.5		3.5		4.5		4.5	4.5		4.5	
4	4.5		4.5		4.5		4.5	3.5		3.5		3.5		3.5
Total	14	6	8	6	8	7	7	14	6	8	7	7	8	6
Number of Values	4	2	2	2	2	2	2	4	2	2	2	2	2	2
Average	3.5	3	4	3	4	3.5	3.5	3.5	3	4	3.5	3.5	4	3
Effect		1		1		0			1		0		1	

The analysis of the data from these two graphs is provided in Table 3.9. The no interaction graph on the left of Fig. 3.8 shows visually that as Q changes from low to high, there is a 1 unit change in the response. As P changes from low to high, there is also a 1 unit change in the response and there is no dependency between P and Q. Note the sum that defines the effect of the interaction is 0. This must be true when there is no interaction, and the lines when graphed will be parallel. On the other hand, the interaction graph on the right shows there is not consistent behavior in the effect of Q changing between P low and P high. Table 3.9 calculates the effects given this situation. In this case, the effect of P changing from low to high is 1, the effect of Q changing from low to high is 0, and the effect of the interaction is 1. The weight of importance in correctly understanding the situation has shifted from an individual parameter effect to the interaction effect.

3.5.2   Other Experimental Designs

Other experimental designs are now briefly discussed.

1. *Plackett–Burman Designs (PBD)*: The PBD design is used in screening where one has a large set of candidate factors and it is necessary to select a small set of the most important factors. Unlike the factorial design structure, the PBD design is constructed in multiples of four rather than powers of two. For example, a PBD design with 12 runs may be used for an experiment containing up to 11 factors. These very economical screening designs are most normally used when only main effects are of interest and are most useful if you can safely assume that interactions are not significant. Another useful application is in ruggedness testing or confirmation within a region of goodness where there should not be an effect on the attributes of interest. Alias structures can be very messy in some situations and it is advised that someone with experience in experimental design be consulted in selecting an appropriate design. Because these designs are used for screening, follow-on designs are usually conducted with the process parameters identified as significant. PBDs are difficult to augment except under specific circumstances when combined with a computer optimal design.

2. *Central Composite Designs (CCD)*: The CCD is used in optimization or to map a region of interest in more detail. These are response surface designs to which a full quadratic model can be fit. The CCD is part of the factorial family of designs and contains a factorial or fractional factorial design that is augmented with both center and axial points. As the name implies, axial points appear on the axis outside of the cube defined by the full factorial corner points. If the distance from the center of the design to a factorial point is defined to be ±1 unit for each factor, then the distance from the center of the design to an axial point is ±α with |α| is greater than or equal to 1. The precise value of the distance depends on the properties desired for the design and on the number of factors included in the design. The axial points require that each design factor can be changed across either three or five levels. Similarly, the number of center points depends on preferred design properties.

3. *Box–Behnken Designs (BBD)*: The BBD is a three-level design used for fitting response surfaces. BBDs are experimental designs used to fit a model which includes main effects, two-factor interactions, and quadratic effects. They are formed by combining 2^k factorials with incomplete block designs. In the experiment, each factor is placed at one of three equally spaced values, usually coded −1, 0, +1. The design itself is structured as a series of two level (full or fractional) factorial designs (−1, +1) in usually 2–3 factors while the other factors are kept at the center (0) values. In this design, several center points are run. The structure of the BBD provides a convenience of not running at extremes, should the extreme be a concern. However, the predictive ability is not generally as good as the CCD. Like a CCD, these designs can be augmented. The augmentation for BBD permits estimation of cubic and quartic effects. In the case of 3–4 factors, the BBD will require a fewer number of experiments than the CCD.

4. *Split-Plot Fractional Factorial Designs*: Split-plot designs are required when there are constraints on randomization of the experimental runs. For example, the temperature of an incubator cannot be randomly changed across units placed in the same incubator. Factors such as temperature in this example are referred to as hard to change factors. Hard to change factors appear often in CMC applications, and proper analysis of the data requires proper recognition of these factors in the experimental design.

5. *Mixture Designs*: Mixture designs include factors that are compounds or ingredients in a mixture. The objective of these experiments is to determine the optimal proportion of each ingredient in order to accomplish some objectives.

6. *Computer Optimal Designs*: Computer optimal designs allow alternatives that are not considered in the more classical designs. In particular, they allow definition of an experimental range that is not defined by a cube or sphere. They also allow selection of specific models that might include a pre-selected subset of interactions. They also allow the opportunity to select designs with small sample sizes relative to the number of parameters to be estimated. These designs are popular as they can reduce the required number of experiments and are helpful in augmenting experimental runs to a previously designed study.

They can also be helpful in tricky situations, such as when there are an uneven number of levels of the experimental factors, when certain combinations of the factors cannot be run, or when multiple level discrete factors combine with continuous and mixture factors. Care should be given in employing this design as the design is only optimal if the pre-specified model is current. This requires understanding of the underlying mathematics of statistical experimental design and practical knowledge of the process under study. Two popular criteria include both D-optimal and I-optimal designs. The D-optimality criterion minimizes the joint confidence region of the regression coefficients, and I-optimality minimizes the average prediction variance over the design space.

There are many other designs that are useful in special applications, and new designs to be developed. Some of these other designs include saturated designs (designs where the number of parameters is equal to the number of data points), definitive screening designs, and hybrid designs. Information on these designs can be found in the statistical literature.

3.5.3 Experimental Strategy

Determination of an experimental strategy is both an art and a science. If research studies are sequential in nature or cover multiple unit operations, it may be advantageous to break up a study into parts. Strategy depends on prior knowledge, available time, and material and equipment availability. Regardless of the particular intricacies of a situation, it is best to make decisions as expeditiously and efficiently as possible. To do so, a hierarchical effect principle is employed. Many processes involve complicated relationships between process parameters and attributes. In general, the large portion of the relationship can be explained by the linear effect, less by interactions between parameters, and less again by a nonlinear or quadratic effect. It takes two experiments (low, high) to estimate a linear trend, four experiments to estimate an interaction between two parameters, and three to five experiments (depending on the nonlinearity) to estimate curvature. This generality is consistent with a strategy to first understand linear relationships and interactions, and then examine curvature as needed.

Consider two such examples:

- HPLC process parameters are known to be linear in their effect on certain attributes. The signature for a new piece of HPLC equipment may be unknown, but the underlying trend in the parameter to attribute relationships could be well known based on prior fundamental and experimental knowledge. A couple of familiarization runs plus a screening design combined with prior knowledge might be all that is needed to establish the functional relationship between the process parameter and attribute for the compound being developed.
- In developing understanding around an active ingredient's synthetic route with minimal prior knowledge, one might require all stages of experimentation. As a

first familiarization step, a small number of experiments at the extremes could be run to gain knowledge on the compound and equipment. A screening experiment could then be run to identify the significant few from the trivial many parameters. Once the important 3–5 parameters are determined, a factorial or central composite design is run to estimate interactions and quantify nonlinearities.

In each of these examples, a scientist works to understand the particular strategy and integrate all prior knowledge and tools in order to set up the most efficient experimental strategy. Table 3.10 presents the four stages of strategy:

1. Familiarization,
2. Screening,
3. Interaction, and
4. Optimization.

Each stage is now described in more detail.

1. *Familiarization*: As the name implies, the basic purpose of this phase is to better understand the problem at hand. The experimenter should keep in mind that engaging in a full DoE without a basic understanding of the system practically assures a study of limited value. If the system is well known, this step can be skipped. There are no set guidelines or specific requirements for executing familiarization runs as part of an experimental design (with perhaps the

Table 3.10 Overview of experimental strategy by level of understanding

	Little System Knowledge		Detailed Understanding	
	Familiarization Phase	Screening Phase	Interaction Phase	Optimization Phase
# Parameters (Factors)	2 - 15	5 - 15	2 - 8	2 - 5
# Experiments	2 – few	11 – 19	7 – 35	11 - 31
Questions to be Answered	If there is little knowledge about the study environment, a couple of runs to establish ranges and investigate the system is valuable. Need to be aware of when to stop and move on to formal efforts.	What factors should be further studied? Is there a workable solution? Should the ranges be adjusted in the next study? Is there an area or direction of goodness? Are any responses nonlinear? What is the control strategy?	Are the center points repeatable and reproducible? Which factors interact? Is there a potential solution? Is there an area or direction of goodness? Are any responses nonlinear? What is the control strategy?	Are the center points repeatable and reproducible? Are nonlinear responses adequately modelled? Is there an area or direction of goodness? What is the control strategy?

exception of the initial runs in a sequential simplex). However, a familiarization phase is essential. The following outcomes would generally describe a successful completion of this stage:

- Any new equipment has been tested and enjoys a degree of reliability.
- Potential performance parameters have been identified with some degree of certainty.
- A range for the performance parameters has been defined that appears practical from a process point of view (i.e., not difficult to control and are scalable) and provide results that are not extraordinarily atypical.
- At least several replicate runs have been completed to estimate the system variability.

2. *Screening*: The main purpose of a screening design is to select a small number of performance parameters from a large set of potential parameters in a minimal number of experiments. Many times, one can identify several potential performance parameters after only a few experiments. At this very early stage, the relative impact of these parameters on the quality attributes may be based more on prior knowledge than on empirical experimentation. Since there is a severe penalty in terms of the number of experiments required to complete a full factorial design, the wise experimenter will embark on a full DoE with only those process parameters that are truly important in this stage.

Screening designs are obtained by using fractional factorial designs, Plackett–Burman designs, or computer optimal designs. One drawback to screening designs is they have a complicated confounding of interactions. However, any process parameters that affect the attributes to an extent greater than the experimental error will be identified. Although some modeling can be done with the data, the basic idea is that once a screening design is completed, the experimenter will eliminate the superfluous variables and embark on a more detailed study of the important process parameters using higher resolution factorial designs.

Screening designs can also be used for purposes that don't require additional experimentation. One such case is the confirmation of an area of robustness. In demonstrating robustness, key process parameters are studied across their recommended manufacturing ranges and the responses or quality attributes are measured. The effects of raw materials and environmental or human factors may be considered in the experiment as noise factors. Noise factors are controlled at the time of the DoE and then left uncontrolled in process operation. Robustness demonstration can be conducted in conjunction with the optimization phase. Variability reduction activities take place in manufacturing, but are best performed early in the development life cycle.

3. *Interaction*: The interaction phase permits study of interactions between the input process parameters. During this phase fewer input factors are studied as compared to the screening phase, because knowledge of interaction effects requires more testing than knowledge of main effects. There may also be a greater level of process understanding. The several process parameters being

studied are believed more likely significant than the many studied in the screening phase. If there are no more than 5–6 process parameters, and only 2-factor interactions are of interest, the screening and interaction phase may be conducted at the same time.

Possible designs considered in this phase include fractional to full factorial designs, depending on the level of interaction required in the study. Data from the screening design can be used in a "fold-over" study (Box et al.) to reduce the total number of runs. A computer optimal design may be used if the region is of unusual shape, if a known model exists, or if design modifications are required unexpectedly in the process of running the experiment (e.g., design repair).

4. *Optimization*: Optimization refers to examination of nonlinear effects, usually quadratic effects, about a smaller region of interest (e.g., the NOR). This typically occurs following the interaction and the screening phases. Designs used in this phase include central composite designs, optimal designs, and Box–Behnken designs. The most popular designs in this phase are central composite designs as in many cases information from experiments included in the screening and interaction phase can be reused and included in the study design and analysis. Note that the screening, interaction, and optimization phases do not need to be sequential and can be conducted simultaneously.

Strategic questions to answer that are crucial to proper execution of an experiment include the following:

1. How will the responses be measured? What measurement system will be used? What is the expected variability?
2. What performance parameter factors are hypothesized to have the largest effect on the quality attributes of interest?
3. Are there any known interactions between factors? Increased prior knowledge can help in decreasing the required experiments.
4. How will the rest of the parameters and material attributes be controlled or blocked during the experiment?
5. Are there noise factors which cannot be controlled? How can their effect be minimized? Should blocking be used to minimize the effect of the hard to control sources of variability?
6. Can the entire experiment be randomized or is this not practical? Should there be a partial randomization scheme?
7. How many replicates are needed for each attribute in an experimental run? It is often both acceptable and necessary to perform unreplicated experiments, but it is important to understand the considerations of these experiments. For example, consider tablet potency as a quality attribute for a study. How will potency be measured? Certainly not by a single replicate assay injection from a single tablet. More likely, it could be measured as the average of two replicate HPLC injections from a composite of five tablets. In general, knowledge of past estimates of variability for similar compounds or similar processes will help inform the replicate strategy. Ultimately, this subject is important to ensure

sufficient statistical power to detect differences which are meaningful to the experimental objective.

8. Will center points be used to estimate variability? For example, in a 16 run factorial design, it is usually of benefit to run at least three center points at the beginning, middle, and end of the experiment. These center points are used to assess variability across the experiment and also to judge nonlinearity or curvature in the experimental space.

9. Is it expected that center points will be in the center of the experimental design, or will those points be at a manufacturing set point that may be off-center in the experimental space? What is the effect to the properties of the experimental design if the points are off-center?

3.6 Nominating a Parameter as Critical

The assessment of critical quality attributes (CQAs) and the control of critical process parameters (CPPs) that affect these attributes is an important component of the overall control strategy for drug substance and drug product manufacturing. There are many different approaches for assessing process parameter criticality, and although the determination of criticality is not primarily a statistics function, statistics can play a part in helping to identify CPPs.

One particular challenge involves assessing when a relationship between a process parameter and a CQA represents a significant impact on that CQA. For example, Fig. 3.9 provides two statistically significant relationships between a CQA and a process parameter across the explored space. Both equations are statistically significant, however, it is clear that the blue equation has a practically more meaningful relationship than the green equation. The blue equation has a chance of producing product outside specification if operated within the range, whereas, the green equation does not. Assessing impact based solely on statistical significance (p-value) is not appropriate, because statistical significance does not take into

Fig. 3.9 Statistically significant equation comparison

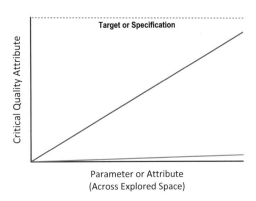

account the strength of the relationship relative to the relevant quality requirements. Ignoring this fact can lead to the inclusion of relatively unimportant process parameters as critical elements of the control strategy. Including these unimportant process parameters as CPPs is undesirable as it effectively dilutes the focus on process parameters that are truly important for ensuring product quality. It can also place an unnecessary burden on manufacturing operations resulting in an increased cost.

An alternative two-step procedure is provided by Wang et al. (2016).

Step 1: Perform a process risk evaluation for each relevant CQA.

For each CQA, evaluate the data set responses across the investigated range without focusing on any single or particular parameter. A Z-score assessment is employed to determine how close the results are to the specification or targeted response for the attribute. The Z-score is calculated as

$$Z^* = min\left(\frac{U - \bar{x}}{S}, \frac{\bar{x} - L}{S}\right)$$

$\bar{x}$ = average of data across the explored space,

S = standard deviation of data across the explored space,   (3.1)

U = upper target or specification, and

L = lower target or specification.

It is not necessary to have both an upper and lower limit to calculate Z^*. In the case of a one-sided specification, Z^* is simply the single value corresponding to the specification of interest.

Figure 3.10 provides an illustration for a one-sided Z^*. For this case, if "specification #1" is the upper specification limit, then the Z^* for this data is expected to be small, indicating that the data is at risk of being greater than the upper specification limit at some operating conditions in the explored space. Alternatively, for "specification #2," the data is far from the specification limit indicating that there is no risk of being beyond the specification for a well-controlled process operating within the explored space.

In Fig. 3.10 cutoff values for Z^* of 2 and 6 were selected as decision points in the analysis. Say that potency is a CQA and the analysis of the DoE data found a significant relationship between potency and milling speed, roll force, and compression force. The prediction equation is

$$\text{Potency} = 98 + 2.5 \times \text{Millspeed} - 0.5 \times \text{Roll force} + 1.5 \\ \times \text{Compression force} \tag{3.2}$$

- If Z^* is less than 2, then all of the parameters in the significant model are CPPs. For the potency example, mill speed, roll force, and compression force are all significant.

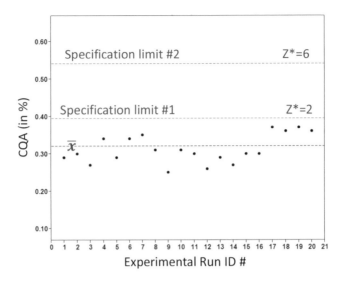

Fig. 3.10   Potential Z-score cutoff values for determining significance

- If Z^* is greater than 6, then none of the parameters in (3.2) are CPPs. For the potency example, the response is performing similar to the green line in Fig. 3.9. Hence, there is no risk across the explored region and no CPPs.
- If Z^* is between 2 and 6, then go to Step 2.

Step 2: Assess the criticality of individual parameters as necessary.

This step is performed if Z^* is between 2 and 6. Here, the fitted statistical model is utilized to quantify individual parameter effects against the proposed specification. This is termed the 20% rule for this application. For the potency CQA, if the specification is 95–105%, then the specification width is 10%. This specification width is multiplied by 20% to yield 2% for this example. As the mill speed coefficient in (3.2) is greater than 2%, the mill speed is a practically meaningful CPP. The other two parameters, roll force and compression force, are not CPPs.

3.7   Determining a Region of Goodness

A significant outcome of the DoE is determination of a region of goodness to operate the process. For example, two responses, impurity 1 and impurity 2, were studied in a two-factor (B, C) full factorial design with replicated center points. From the experimental data, models were developed and summarized in a contour plot (see Fig. 3.11).

It is desired to minimize impurity. The arrows in Fig. 3.11 show the direction of this minimization, or the so-called direction of goodness. For impurity 1, the

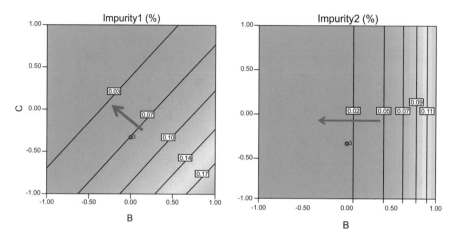

Fig. 3.11   Contour plots of impurity 1 and 2

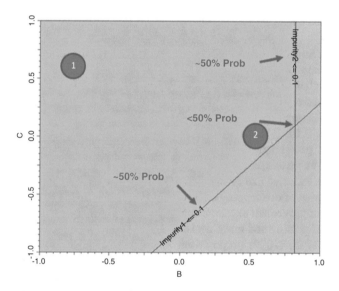

Fig. 3.12   Region of goodness defined by pass/fail mindset

combination of factor B at the low level with factor C at the high level is the best combination to minimize impurity 1. For impurity 2, factor B at the low level is the best, and factor C has no impact.

Assume the specification for each impurity is 0.10%. Examination of Fig. 3.11 shows the region and boundary where each impurity is less than 0.1%. It is common to summarize this information by providing pass (orange) and fail (gray) regions as shown in Fig. 3.12. The orange region represents an area where both 0.10%

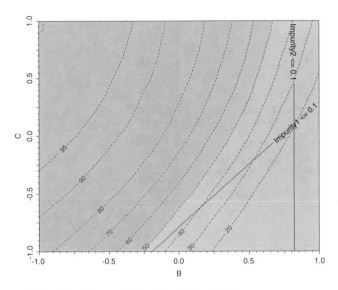

Fig. 3.13 Probability of passing considering the relationship between responses

specification limits are simultaneously met. However, across this region, there are a range of success probabilities. That is, based on Fig. 3.11, it is expected that there will be a higher probability to pass the specification of 0.1% in the upper left quadrant of the region than in the rest of the space. The predicted value at point #1 in Fig. 3.12 for impurity 1 and 2 is 0.01. The predicted value at point #2 for impurity 2 is close to 0.10. It makes sense that although the orange region will produce product that passes specifications, the probability of passing a specification of 0.10% must be greater at point #1 than point #2. In general, at the impurity limit of 0.10% it is expected that the probability of passing the specification is about 50% and in the region of overlapping requirements there is less than a 50% probability of passing. An improvement to examining the pass/fail plot in Fig. 3.12 is to assess the probability of passing and make decisions based on this probability.

Peterson et al. (2009) proposed an approach to calculate the probability of simultaneously passing all relevant specifications using seemingly unrelated regression (SUR) and has also, although unpublished, outlined a parametric bootstrap simulation approach to calculating this probability across the space of interest.

Using a bootstrap method, the probability to simultaneously pass multiple specifications is provided in Fig. 3.13. The levels on these contours now show the probability of passing while taking into account the predictive distribution, not simply the average prediction.

This more descriptive probability can be used to make better informed decisions about the process. That is not to say that the method is perfect or cannot be improved. A Bayesian approach to specify a range of parameter and variability estimates might help stabilize predictions in some cases.

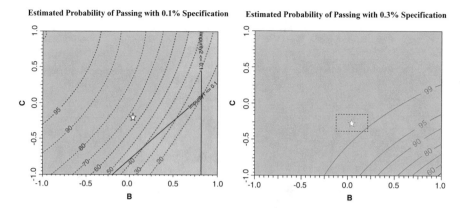

Fig. 3.14 Example using the probability of passing specification

Continuing the earlier example of the two impurities, assume the process was initially set to operate at the point indicated by a star in Fig. 3.14. Following the probability calculation, it is determined that there is a 70% chance of passing both specifications simultaneously. There could be several solutions that may improve or remediate this probability, and knowledge of the estimated probability is a step in proposing the process.

- It may be that the process is improved by downstream processing. So although there may be a cost associated with a 70% probability of passing at this stage, the probability will be improved in the future.
- It may be that the experiment was performed sub-scale. There is a known improvement to the probability when performed at scale.
- The set points of parameters B and C may need to be adjusted to improve the probability of passing.
- The initial specifications on the impurities of 0.1% may need to be increased. The effect of increasing the specification to 0.3% is provided in the right-hand side of Fig. 3.15.
- Finally, true process variability may be greater or less than the magnitude realized in the experimental data. The simulation can be performed again with the more appropriate error structure.

3.8   Process Capability and Process Robustness

The process capability index, abbreviated broadly as C_{pk}, is a widely used summary statistic describing the ability of a process to produce output within specification limits. The index plays a prominent role in PV-3, and is discussed in Chap. 5 of this book. An assessment of capability is also useful in PV-1. Obtaining a meaningful estimate of process capability early in a product's life cycle is difficult because

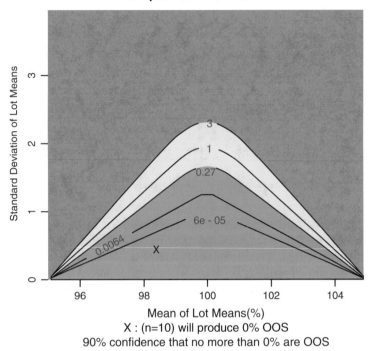

Fig. 3.15 Robustness contour

many lots are needed to provide a meaningful capability index. For these indices to have predictive meaning, the process must have demonstrated adequate statistical control prior to their calculations. This effort requires at least 25 lots.

Within the last decade, the concept and industrial practices of QbD have led to greater process understanding in R&D leading to increased knowledge of process capability that is not specifically captured by the small number of lots manufactured early in a product's life cycle. Although there may only be a couple produced lots, the scientific understanding, fundamental knowledge, and development experience is substantial and provides an opportunity to assess process capability. A proposal for a robustness calculation, meant to distinguish an early estimate of control and capability developed within a QbD framework from the rigorous assessment of control and capability implied by a capability statistic was proposed by Vukovinsky et al. (2017). This contour-based tool calculates the percent out-of-specification (% OOS), based on the mean, standard deviation, and specification of an attribute. The contours provide a clear visualization of the ability of the process to meet the specification, making it a useful tool for products in development as well as new and marketed products. Figure 3.15 provides an example contour plot for potency, based on a sample size of only 10 lots resulting in a mean potency of about 98.4%, a standard deviation between lots of 0.5%, and a specification of 95–105%.

These %OOS contour regions use the following coloring scheme:

- Green: less than a 0.27% OOS rate (good performance).
- Yellow: greater than or equal to 0.27% OOS rate and less than 3% (further discussion required).
- Red: greater than 3% OOS rate (requires improvement).

The OOS% contour levels of 0.27, 0.006, and 6e-5 displayed on the plot are approximately related to C_{pk} values of 1, 1.33, and 1.67, respectively. Associating the green contour with 0.27% OOS implies a minimum C_{pk} of one in transition to manufacturing.

Once a process robustness contour plot is constructed, the relative location of the present process within the colored contour is examined to assess the product performance. In Fig. 3.15 the "X" represents the location of the attribute of interest. The ultimate goal for the product should be emphasized more than the color zone containing the "X". The relative location provides information concerning the sensitivity of the attribute to change in the sample mean and sample standard deviation and can guide the search for potential improvements in product performance or the need to modify data-driven specifications. As with any summary statistic, there is variability in the %OOS estimates. This variability is described in Fig. 3.15 footnote as an upper confidence estimate on the %OOS. This estimated upper bound is based on the data, and can be quite wide for a small sample size. The fundamental, scientific, and experimental understanding of the process gained through the design process along with the calculated bound should be considered in process decisions.

Once constructed, the contour plots should support an active discussion about the product performance amongst a cross-functional team. In general, data external to the summarized lot data, estimates of variability components from methods and processes, or knowledge from modeling efforts on similar products can be used to assess potential future process behavior and expectations. All of these discussions can use the robustness contour as a foundation.

3.9 Control Strategy Implementation

ICH Q8(R2) documents a "Minimal Approach" to Control Strategy which is contrasted with the "Enhanced, Quality by Design Approach." Here, criticality of parameters is determined following scientific investigation through the QbD process.

The concept of criticality can be used to describe any material attribute, characteristic of a drug substance, component, raw material, drug product or device, process attribute, parameter, condition, or factor in the manufacture of a drug product. The assignment of attributes or parameters as critical or non-critical is an important outcome of the development process and provides the foundation for the control strategy. Critical Process Parameters (CPPs), the relationship between Critical Quality Attributes (CQAs) and Critical Process Parameters (CPPs), and the ranges for CPPs (PAR and NOR) are documented as a control plan. The control strategy provides a plan to prevent operating in regions of limited process knowledge or those that are known to cause product failure.

Underlying the criticality assignment process is the concept that the primary assessment and designation of criticality should be made relative to the impact that quality attributes or process parameters have on the safety, efficacy, and quality of the product. The material in Sect. 3.6 provides one option to determine criticality. Once criticality is determined, a control strategy that focuses on the most appropriate control points and methods is developed.

ICH Q10 defines a control strategy as

a planned set of controls derived from current product and process understanding that assures process performance and product quality. The controls can include parameters and attributes related to drug substance and drug product materials and components, facility and equipment operating conditions, in process controls, finished product specifications and the associated methods and frequency of monitoring and control.

QbD also introduced the concept of a traditional versus a dynamic control strategy. In a traditional control strategy, any variability in process inputs (such as quality of the feed material or raw materials) results in variability in the quality of the product because the manufacturing controls are fixed. In a dynamic control strategy, the manufacturing controls can be altered (within the region of goodness) to remove or reduce the variability caused by process inputs.

A holistic control strategy mitigates any risk from a single unit operation. The control strategy includes the process definition, control limits of process parameters, and release limits, amongst other considerations. It is important in determining the manufacturing process that specifications be set appropriately (see Chap. 7).

A statistically related example illustrates the translation from an equation derived from a DoE to a control strategy. Here, dissolution (*Diss*) is found to be a function of API particle size (*API*), magnesium stearate surface area (*MgSt*), lubrication time (*LubT*), and compression force (*Crush F*).

$$
\begin{aligned}
Diss = {} & 108.9 - 11.96 \times API - 7.556 \times 10^{-5} \times MgSt - 0.1849 \times LubT \\
& - 3.783 \times 10^{-2} \times CrushF - 2.557 \times 10^{-5} \times MgSt \times LubT.
\end{aligned}
\tag{3.3}
$$

Assume these parameters are both statistically significant and their effect on dissolution is practically meaningful. Equation (3.3) describes the current understanding and can be used to define meaningful limits on the parameter specifications and process controls. Using this information, quality is built into the process by managing the process inputs. Although there may not be direct control of *Diss*, it might be controlled upstream by one of the variables on the right of Eq. (3.3).

- API: To control dissolution, it is important to maintain the D90 API particle size within a certain range. Here, the predicted equation is used to determine the range of 5–30 µm and the high shear wet milling equipment is set to achieve a value within this range.
- MgSt: The surface area of the magnesium stearate (lubricant) particles is controlled to ensure dissolution. This assurance is performed upon receipt of the MgSt from the supplier.
- LubT: Lubrication time is controlled between 1 and 8 min via automated equipment.
- CrushF: Tablet hardness is controlled by the crushing force at the time of compression to a targeted amount and within an acceptable range.

All decisions concerning the CQA are documented within a control plan.

3.10 Preparation for Stage PV-2

After the control strategy has been defined and the product and process ranges are established, product and process qualification (PV-2) is performed to demonstrate that the process will deliver a product of acceptable quality if operated within the region of goodness. This will also confirm whether the small and/or pilot-scale systems used to establish the region of goodness can accurately model the performance of the manufacturing scale process. PV-2 is really a confirmation of the understanding and control strategy. Following PV-2, the regulatory filing is compiled, which includes the acceptable ranges for all critical operating parameters that define the manufacturing process.

References

ASTM E2363 (2014) Standard Terminology Relating to Process Analytical Technology in the Pharmaceutical Industry. ASTM International, West Conshohocken, PA

Box GEP, Hunter JS, Hunter WG (2005) Statistics for experimenters: design, innovation, and discovery, 2nd edn. Wiley, NY

Code of Federal Regulations (CFR), Title 21, Food and Drugs Administration (FDA), Part 210. http://www.accessdata.fda.gov/scripts/cdrh/cfdocs/cfcfr/CFRSearch.cfm?CFRPart=210

Deming WE (2000) Out of the crisis, Reprint edition. The MIT Press

FDA. CDER (2011) Process validation: general principles and practices, guidance for industry

Garcia T, McCurdy V, Watson T, Ende M, Butterell P, Vukovinsky K, Cheuh A, Coffman J, Cooper S, Schuemmelfeder B (2012) Verification of design space developed at subscale. J Pharm Innov 7:13–18

Glodek M, Liebowitz S, McCarthy R, McNally G, Oksanen C, Schultz T, Sundararajan M, Vorkapich R, Vukovinsky K, Watts C (2006) Process robustness – a PQRI white paper. Pharm Eng 26:6

ICH (1999) Q6A Specifications: test procedures and acceptance criteria for new drug substances and new drug products: chemical substances

ICH (2005) Q9 Quality risk management

ICH (2008) Q10 Pharmaceutical quality system

ICH (2009) Q8 (R2) Pharmaceutical development

ICH (2012) Q11 Development and manufacture of drug substances (chemical entities and bio-technological/biological entities)

Juran JM (1992) Juran on quality by design: the new steps for planning quality into goods and services. Free Press

Montgomery DC (2012) Design and analysis of experiments, 8th edn. Wiley, NY

Morris M (2011) Design of experiments: an introduction based on linear models. Chapman & Hall/CRC

Peterson JJ (2004) A posterior predictive approach to multiple response surface optimization. J Qual Technol 36:139–153

Peterson JJ (2010) What your ICH Q8 design space needs: a multivariate predictive distribution. Pharm Manuf 8:23–28

Peterson JJ, Miró-Quesada G, del Castillo E (2009) A Bayesian reliability approach to multiple response optimization with seemingly unrelated regression models. Qual Technol Quant Manage 6(4):353–369

Stockdale GW, Cheng A (2009) Finding design space and a reliable operating region using a multivariate Bayesian approach with experimental design. Qual Technol Quant Manage 6:391–408

Vukovinsky KE, Altan S, Bergum J, Pfahler L, Senderak E, Sethuraman S (2010a) Statistical considerations in design space development, I of III. Pharm Technol 34(7)

Vukovinsky KE, Altan S, Bergum J, Pfahler L, Senderak E, Sethuraman S (2010b) Statistical considerations in design space development, II of III. Pharm Technol 34(8)

Vukovinsky KE, Altan S, Bergum J, Pfahler L, Senderak E, Sethuraman S (2010c) Statistical considerations in design space development, III of III. Pharm Technol 34(9)

Vukovinsky KE, Li F, Hertz D (2017) Estimating process capability in development and for low volume manufacturing: process robustness assessment using %OOS contour plots. Pharm Eng Jan/Feb 2017

Wang K, Ide N, Dirat O, Subashi A, Thomson N, Vukovinsky K, Watson T (2016) Statistical tools to aid in the assessment of critical process parameters. Pharm Technol 40(3):36–44

Chapter 4
Process Qualification: Stage 2 of the FDA Process Validation Guidance

Keywords Data analysis • Effects of scale • Equivalence testing • Inter-batch variation • Intra-batch variation • Lot homogeneity • Lot-to-lot variation • Process characterization • Process performance qualification (PPQ) • Profiler • PPQ batch size • Risk management solution

4.1 Introduction

In Chap. 3, the first stage of the FDA's 2011 process validation guidance was described demonstrating various options to drive process understanding using experimental design. The resulting process design experiments yield information that can be used to define future operating ranges for the new process. The second stage of process validation is process qualification.

There are two elements associated with this stage of process validation. As described in the guidance, process qualification is the point where commercial manufacturing is demonstrated to be reproducible. This stage consists of two elements. The first element is the design of a facility and qualification of utilities and equipment. This precedes the second element, Process Performance Qualification (PPQ). Without validating the facility, utilities, and equipment, it is not possible to determine if the process can operate within the proposed operating ranges defined in Stage 1. Although an important step for Stage 2, it is not discussed in this chapter. Rather, this chapter focuses on the second element, PPQ. By definition, "PPQ will confirm the process design and demonstrate that the commercial manufacturing process performs as expected." By confirming that the process performs as expected, there is a high level of assurance that the manufacturing process consistently produces the active pharmaceutical ingredient (API) and drug product that meet the specifications related to identity, strength, purity, and potency. This requirement is also consistent with ICH Q7 (2000) and the EMA's guidance on process validation (EMA 2014). To demonstrate that a process consistently delivers product that is safe and efficacious, it is imperative to understand variability of the product over time versus the specifications. Gaining such understanding requires knowledge of multiple lots.

© Springer International Publishing AG 2017
R.K. Burdick et al., *Statistical Applications for Chemistry, Manufacturing and Controls (CMC) in the Pharmaceutical Industry*, Statistics for Biology and Health, DOI 10.1007/978-3-319-50186-4_4

Statistical analysis is at the core of any demonstration of process consistency. There are three statistical issues that are particularly germane to this stage of process validation that will be described in this chapter:

1. Evaluation of the effects of scale.
2. Demonstration of both inter-batch and intra-batch consistency.
3. Determination of the number of lots to be run in a PPQ campaign.

Section 4.3 describes a process for evaluating effects of scale. Sections 4.4 and 4.5 provide approaches for demonstrating both inter- and intra-batch consistency. Finally, Sect. 4.6 describes two approaches for selecting the appropriate number of lots to run in a PPQ campaign. The first approach considers a risk-based decision making tool that uses qualitative knowledge gained during Stage 1 and from previous experience with similar products, scales, and processes. The second approach considers the use of statistics to select the number of lots based on an understanding of lot-to-lot variability and the established sampling plans for process validation.

4.2 Process Performance Qualification

In the regulations, process validation is a mandatory step performed by the manufacturer to confirm that the process consistently delivers safe and efficacious API and drug product for commercial use. Although the requirement to validate a process has not changed over time, the activities that define process validation have evolved. Both the EMA and FDA promote a risk-based approach to process validation providing the manufacturer the flexibility to incorporate knowledge from process development studies and relevant scientific knowledge from similar products, processes, and scales. The FDA has matured the definition of process validation further by moving process validation from a single event to a series of events that extend through the lifecyle of a product. The new term used by the FDA to capture the prior definition of process validation is Process Performance Qualification (PPQ). PPQ is called process validation by the EMA and ICH Q7.

For PPQ, the FDA places emphasis on incorporating process variability knowledge into the justification of the sampling plan, the acceptance criteria selected for each of the process steps, and the number of lots required to execute the PPQ campaign. The level of monitoring and testing should be designed to confirm that there is uniform product quality both within manufactured lots (intra-batch variability) and among manufactured lots (inter-batch variability). The following sections will describe the statistical techniques used to address these areas.

4.3 Evaluation of Effects of Scale

The ultimate objective of process qualification is to demonstrate an understanding of the manufacturing process that allows an assurance that a process will operate as intended with respect to attributes of product identity, strength, quality, purity, and potency. Because of the costs and time associated with a full scale manufacturing process, smaller scale processes are used to develop understanding concerning the full scale process. Such processes are referred to as small-scale studies, and these studies can themselves vary by size of scale. Combining data across different scales provides opportunities for discovering scale-up effects and also leverages learnings from small scale concerning relationships between quality attributes (QAs) and process parameters (PPs). In statistical jargon, the QAs are the responses (Y) and the PPs are the predictors (X). However, combining information across scales presents several challenges.

1. Process means can shift when moving from one scale to another scale.
2. Process variances may change across scales.
3. Functional relationships between QAs and PPs may change across scales.

If proper statistical adjustments are not made, combining such data may lead to an erroneous analysis. Scale must be modeled so that scale shifts can be properly identified in the analysis.

Changes in functional relationships (i.e., changes between QAs and PPs) are the most difficult to detect across changes in scale. This is because unless one actually varies a PP, it is not possible to estimate a functional relationship with the QA. Large-scale runs are very expensive and it is generally not possible to manufacture lots under different settings of the PPs. Thus, functional relationships can only be inferred from small-scale experiments. In most practical cases, the relationships discovered at smaller scale transfer to larger scales. However, it is important to remember that transitivity is never guaranteed.

The data set in Table 4.1 reports results of six small-scale experiments conducted in a laboratory. There are two PPs: pH and Temperature ($^\circ$C). The QA is Yield (%).

The six experiments consist of four corner points and two center points for a two-factor factorial experiment. (See Chap. 3 for more on factorial experiments). There is interest in estimating both main effects and the interaction. Figure 4.1 provides an interaction plot of the data in Table 4.1. The plot suggests there is an interaction between pH and Temperature.

The PPs are coded as described in Eq. (2.106), and shown in Table 4.2.

Table 4.3 reports the analysis of variance table that reports estimated effects and p-values. The small p-values confirm the statistical significance of both main effects and interaction.

The data in Table 4.1 are now augmented with four medium-scale process runs conducted in a pilot plant. The augmented data are shown in Table 4.4. Note that a third PP, Scale, has been added to the data set. Scale is a qualitative PP. Here, Scale $= -1$ denotes small-scale, and Scale $= +1$ denotes medium-scale. (See Sect. 2.12.5 for coding a qualitative PP).

Table 4.1 Small-scale experiment

pH	Temp (°C)	Yield (%)
4	30	88.5
5	20	53.9
5	30	55.8
4	20	56.8
4.5	25	67.8
4.5	25	64.4

Fig. 4.1 Interaction plot for small-scale data

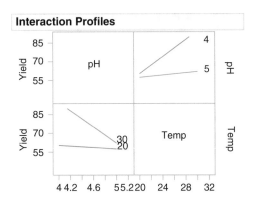

Table 4.2 Small-scale experiment with coded PPs

pH	Temp	Yield (%)
−1	1	88.5
1	−1	53.9
1	1	55.8
−1	−1	56.8
0	0	67.8
0	0	64.4

Table 4.3 Estimated effects for laboratory experiment

Term	Estimate	Prob > |t|
Intercept	64.533	0.0003
pH	−8.9	0.0201
Temp	8.4	0.0225
pH*Temp	−7.45	0.0283

Note that all pH levels for the medium runs are set to 0 (4.5 uncoded), but that Temperature was varied to include both −1 (20°C) and +1 (30°C). For this reason, no new information is provided for the pH and Yield relationship, although an interaction between Temperature and Scale can now be estimated. Estimation of this interaction allows one to determine if the functional relationship between Yield and Temperture changes across scales. Table 4.5 reports the results of a model that adds the Scale main effect and the Scale × Temperature interaction to the previous analysis.

Note that since pH was not manipulated in the medium-scale experiment, the effect estimates for pH and pH*Temp are the same in Table 4.3 and Table 4.5. The p-values

Table 4.4 Small-scale combined with medium-scale

pH	Temp	Scale	Yield (%)
−1	1	−1	88.5
1	−1	−1	53.9
1	1	−1	55.8
−1	−1	−1	56.8
0	0	−1	67.8
0	0	−1	64.4
0	1	1	81.7
0	−1	1	48.1
0	1	1	80.6
0	−1	1	69.5

Table 4.5 Estimates for combined small-scale and medium-scale data

Term	Estimate	Prob > |t|
Intercept	67.254	<0.0001
pH	−8.9	0.0843
Temp	9.788	0.0237
pH*Temp	−7.45	0.1283
Scale	2.721	0.3400
Temp*Scale	1.388	0.6409

Table 4.6 Combined data for all three scales

pH	Temp	Scale (medium)	Scale (large)	Yield (%)
−1	1	−1	−1	88.5
1	−1	−1	−1	53.9
1	1	−1	−1	55.8
−1	−1	−1	−1	56.8
0	0	−1	−1	67.8
0	0	−1	−1	64.4
0	1	1	0	81.7
0	−1	1	0	48.1
0	1	1	0	80.6
0	−1	1	0	69.5
0	0	0	1	88.2
0	0	0	1	68.1

for these effects have changed because the degrees of freedom and the point estimate for the model error have changed. The effect estimate for Temperature is different in the two tables because Temperature was manipulated in the medium-scale data. There is no statistical evidence ($p = 0.6409$) that temperature interacts with scale. It therefore seems appropriate to drop this interaction from the model. The functional relationship between Temperature and Yield is not a function of Scale.

Finally, Table 4.6 augments the data in Table 4.4 with two runs conducted at full manufacturing scale. Note that both the quantitative PPs are set to 0 for each run.

Table 4.7 Estimates for all data combined

Term	Estimate	Prob > |t|
Intercept	70.886	<0.0001
pH	−8.9	0.0883
Temp	9.788	0.0195
pH*Temp	−7.45	0.1398
Scale (large)	7.264	0.1606
Scale (medium)	−0.911	0.8170

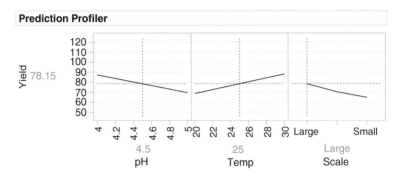

Fig. 4.2 Profiler of example data

The qualitative factor Scale (Small, Medium, and Large) is coded as described in Eq. (2.107) where small-scale is always coded −1. The resulting effect estimates are shown in Table 4.7.

The data have now been successfully combined. Figure 4.2 presents a visualization using the JMP[(c)] profiler of the final estimates when run at Large-scale at the pH and Temperature targets of 4.5 and 25°C, respectively.

The value 78.15% on the left most axis of Fig. 4.2 indicates the long-run average of lots manufactured at Large-scale when pH is held at 4.5 and Temperature is held at 25°C. There is also interest in the range of individual lots around the long-run average. An approach to provide that information is presented in Sect. 4.4.

It is important to remember there is an interaction between pH and Temperature. To demonstrate, compare Fig. 4.2 with Fig. 4.3 where Temperature has been increased to 28°C. The slope of pH is steeper in Fig. 4.3 than in Fig. 4.2. In Fig. 4.2 it runs from a yield of about 88% at pH = 4 to 70% at pH = 5 (88 % − 70 % = − 18%). In Fig. 4.3, the change in Yield runs from 98% at pH = 4 to 71% at pH = 5 (98 % − 71 % = − 27%).

The analysis to this point has examined the effect of scale on the process average and functional relationships (for those PPs varied under different scales). It is also important to examine the effect of scale on the process variance. One simple way to do this is to plot the residuals from the final model in Table 4.7 against the factor Scale. Figure 4.4 presents this plot.

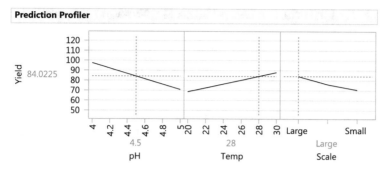

Fig. 4.3 Profiler with temperature = 28

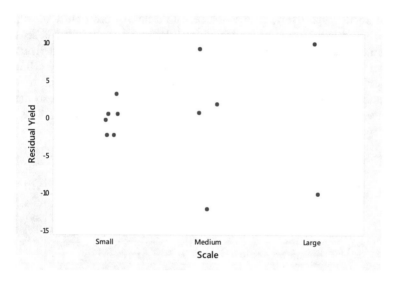

Fig. 4.4 Plot of residuals versus scale

Table 4.8 Standard deviation of residuals

Scale	Standard deviation of residuals
Small	2.04
Medium and large combined	9.37

Figure 4.4 suggests a greater variance in the model for the medium and large scales. Table 4.8 reports standard deviations of the residuals for the small scale and the medium and large scales combined.

This difference in the standard deviations is perhaps exaggerated due to the smallness of the data set. However, we will assume this difference is real and use it to establish criteria for the PPQ runs as described in the next section.

4.4   PPQ Criteria for Inter-Batch (Lot-to-Lot) Variability

One use of the combined data set in Sect. 4.3 is establishment of PPQ acceptance criteria that can be used to estimate inter-batch variability. Use of this data set allows one to form criteria based on the expected ranges of the QA when the PPs are run in an expected range.

Computer simulation based on results from the combined data set of manufacturing and laboratory experince is a useful approach for defining such criteria. Figure 4.5 shows the prediction profiler from Fig. 4.2 with some new information provided.

Figure 4.5 provides the results of a simulation produced in the following manner:

1. The PPs pH and Temperature are allowed to vary around the target values of 4.5 and 25°C, respectively, in accordance with a normal probability distribution. The mean is set equal to the target, and the standard deviation is set at the characterized range divided by five. For example, the characterized pH range is from 4 to 5, and so the value of the standard deviation is $(5 - 4)/5 = 0.2$. This calculation is based on the fact that 98.8% of the values within a normal distribution fall within 2.5 standard deviations of the mean, and so the total range from minimum to maximum is five standard deviations. This is an

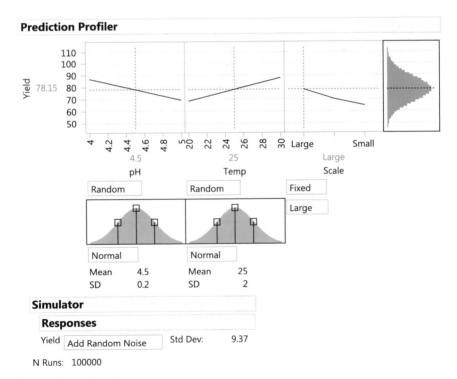

Fig. 4.5 JMP simulator

Distributions

Yield

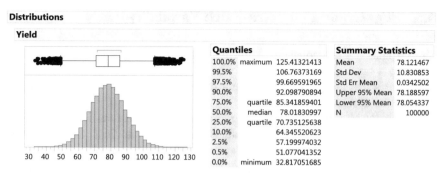

Quantiles			Summary Statistics	
100.0%	maximum	125.41321413	Mean	78.121467
99.5%		106.76373169	Std Dev	10.830853
97.5%		99.669591965	Std Err Mean	0.0342502
90.0%		92.098790894	Upper 95% Mean	78.188597
75.0%	quartile	85.341859401	Lower 95% Mean	78.054337
50.0%	median	78.01830997	N	100000
25.0%	quartile	70.735125638		
10.0%		64.345520623		
2.5%		57.199974032		
0.5%		51.077041352		
0.0%	minimum	32.817051685		

Fig. 4.6 Simulated values for yield

expected source of variation as it is likely PPs will vary from their target values. Because it is of interest to simulate the output from a large-scale (manufacturing run), Scale is held fixed at Large.

2. As presented in Table 4.8, there will be variability in values of the QA even when PPs are fixed. Based on Table 4.8, it is estimated that the standard deviation of this noise level for a large lot is 9.37. This value is shown at the bottom of Fig. 4.5.

3. The following operation is now repeated 100,000 times (N Runs in Fig. 4.5):

   a. One value for pH and one value for Temperature are simulated from the normal distributions defined in Fig. 4.5. The computed regression equation is used to predict the QA, Yield.

   b. A random normal error with mean zero and standard deviation 9.37 is then simulated and added to the predicted value from part a. This represents one simulated value of Yield for a manufacturing run.

   c. After performing steps a. and b. 100,000 times, the histogram of the simulated responses is plotted to the right of the profiler.

Figure 4.6 presents a numerical summary of the profiler histogram of simulated Yield values shown in the far right of the upper row of Fig. 4.5.

The interval of simulated Yield values that contain the middle 99% of all values is from 51.1 to 106.8%. This range can be considered as an expected range for the process and used as a criterion for the PPQ.

There are certainly other approaches for establishing criteria for the PPQ. The described approach is scientifically based because it makes use of knowledge acquired during stage 1 of process validation. It also takes into account expected variation in the process.

4.5 PPQ Criteria for Intra-Batch Variability

Drug product (DP) homogeneity refers to the sameness of QAs across the units that make up a batch (lot). Testing performed at release and during stability studies necessitates that homgeneity assessment of the DP batches be performed for the justification of release sample size.

Doymaz et al. (2015) provide the following statistical test for demonstration of batch homogeneity. Consider the operation of filling DP into final containers (vials or syringes). The QA of interest is protein concentration measured as mcg/mL. One can conceptually divide the filling process into three regions: beginning, middle, and end as shown in Fig. 4.7. Three independent measurements are made in each region. If the process is homogeneous, then the average protein concentrations of the three regions should be comparable. In fact, there should be no variation in the means except for that resulting from analytical method error. (Doymaz et al. describe a situation where variation in fill weight might also be expected).

To demonstrate homogenity, one can formulate a hypothesis set that compares region means:

H_0 : The absolute value of at least one pair-wise mean difference $\geq$ EAC

H_a : The absolute value of all pair-wise mean differences $<$ EAC   (4.1)

Let the three position means be represented as μ_B, μ_M, and μ_E, for beginning, middle, and end, respectively. Using this notation, the alternative hypothesis in (4.1) is

$$
\begin{aligned}
H_a : |\mu_B - \mu_M| &< \text{EAC} \\
|\mu_B - \mu_E| &< \text{EAC} \\
|\mu_M - \mu_E| &< \text{EAC.}
\end{aligned}
$$ (4.2)

The definition of the EAC should be based on the precision of the analytical method. For example, a reasonable defintion for EAC is $3 \times \sigma_M$ where σ_M is the standard deviation of the analytical method. The value of σ_M can typically be obtained from method qualification or from historical measurements of reference standards.

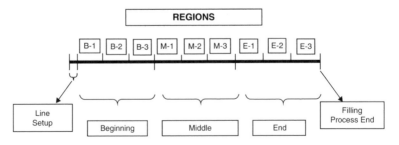

Fig. 4.7 Sampling plan for demonstrating lot homogeneity

Table 4.9 ANOVA for mixed model

Source of variation	Degrees of freedom	Mean square
Region	$n_1 = a - 1 = 2$	$S_1^2 = 1.96$
Error with regions	$n_2 = a(r - 1) = 18$	$S_2^2 = 0.907$

Since all three conditions in (4.2) must be satisfied in order to reject the null hypothesis, computation of three 90% two-sided confidence intervals (one for each pairwise difference) provides an equivalence test with an approximate test size of 0.05. The calculations are most easily performed using a one-factor analysis of variance table. Table 4.9 presents a summary of such an analysis where there are $a = 3$ regions and $r = 7$ samples taken at each region.

The sample means for the three regions are $\bar{Y}_B = 49.12 \text{ mcg/mL}$, $\bar{Y}_M = 50.15 \text{ mcg/mL}$, and $\bar{Y}_E = 49.84 \text{ mcg/mL}$. The mean squared error $S_2^2 = 0.907$ represents the pooled variance. Thus, using Eq. (2.56), the 90% two-sided confidence interval on the difference $\mu_B - \mu_M$ is

$$L = \bar{Y}_B - \bar{Y}_M - t_{0.95:a(r-1)} \sqrt{S_2^2 \left(\frac{1}{r} + \frac{1}{r} \right)}$$

$$= 49.12 - 50.15 - 1.73 \sqrt{0.907 \left(\frac{1}{7} + \frac{1}{7} \right)} = -1.91 \text{ mcg/mL} \qquad (4.3)$$

$$U = \bar{Y}_B - \bar{Y}_M + t_{0.95:a(r-1)} \sqrt{S_2^2 \left(\frac{1}{r} + \frac{1}{r} \right)} = -0.15 \text{ mcg/mL}$$

The computed 90% two-sided intervals for $\mu_B - \mu_E$ and $\mu_M - \mu_E$ are $(-1.60, 0.16)$ and $(-0.57, 1.19)$, respectively.

The analytical method has a reported %RSD of 2 at a target protein concentration of 50 mcg/mL (see Sect. 2.6.4 for %RSD definition). Thus, the standard deviation at the target concentration is $50 \times 0.02 = 1$ mcg/mL. An appropriate EAC for this demonstration is thus $3 \times 1 = 3$ mcg/mL. Since all three confidence intervals fall within the range from -3 mcg/mL to $+3$mcg/mL, homogeneity has been demonstrated.

Power is an important consideration in determining the sample size to select from each region. Computer simulation, as demonstrated by Doymaz et al., is a useful tool for this purpose.

4.6 Determining the Number of PPQ Lots

The final issue discussed in this chapter concerns determination of the number of batches needed for the PPQ campaign. Historically, traditional process validation was considered a one-time event, and the standard rule was to manufacture three batches. This rule required that three consecutive commercial scale batches meet the predefined acceptance criteria in order to declare the process validated. This "test into compliance" mindset has evolved in the FDA's 2011 process validation guidance. The shift has moved from the traditional three batch campaign to a campaign size that is dependent on process understanding and relevant historical knowledge.

Per the FDA's 2011 guidance document, "the number of samples should be adequate to provide sufficient statistical confidence of quality both within a batch and between batches. The confidence level selected can be based on risk analysis as it relates to the particular attribute under examination." From this excerpt, the FDA calls out the need to justify the number of samples for PPQ. In this section, the number of samples is defined as the number of lots (batches) in the PPQ campaign. Determining the number of lots required to execute PPQ is driven by multiple factors which together build the body of evidence driving the amount of data collected from lots manufactured during PPQ. Complexity of the process or product knowledge gained during Stage 1, and relevant knowledge of similar products and scales will inform the process engineer of potential gaps in process knowledge. These gaps will need to be filled by either collecting data from the PPQ campaign, or from statistical techniques that use mathematical models and simulation. Where process knowledge gaps are high, and the data are insufficient to fill them, the PPQ campaign will require a greater number of lots than if the converse is true.

Two approaches for determing the number of PPQ lots are discussed in this section. The first approach considers variations on risk-based decision making tools with ordinal scales that describe residual risk. The description of residual risk is designed to determine the number of lots based on qualitative process understanding. The second approach uses statistical methods to justify the number of PPQ lots. Regardless of the selected approach, there are jurisdictions that may still require a minimum of three lots for process validation, and this must be considered in the filing strategy.

4.6.1   Risk-Based Decision Approach to Determine Number of PPQ Lots

A risk-based approach to decision making in the GMP environment is not a new concept. ICH Q9 (2005) provides the framework to risk management and these principles can be extended to process validation. Early examples of incorporating risk-based decision making into process validation decisions include selection of equipment in cleaning validation (PDA 1998; Chao 2005), assessment of qualification, validation and change control (O'Donnell and Greene 2006a, b), and selection of process steps to include in process validation (Sidor and Lewus 2007). A risk-based decision making tool has been used to determine the number of PPQ lots in Levy (2012), Bryder et al. (2012) and Sidor (2013, 2014).

In Bryder et al., the risk-based approach is divided into three steps. Step 1 assesses product knowledge and process understanding, and Step 2 assesses the control strategy. Step 3 is the assessment of residual risk based on the outcome of Steps 1 and 2. The residual risk qualitatively describes gaps in knowledge and/or the control strategy. The residual risk is mapped into five levels from Severe to Minimal.

Depending on the residual risk, Bryder et al. present three options that drive the selection of the number of PPQ lots. The first option provides a qualitative description of rationale and experiences. Based on the qualitative description, the number of lots is determined. This is similar to the approach proposed by Sidor (2014) where quantitative scores are assigned to both process and scientific understanding. The primary difference between the two qualitative decision making tools is that Sidor's tool computes a cumulative score of process and scientific understanding. This score is translated to the number of lots using a decision making table where a range of scores define residual risk. Neither approach considers known information about process variability. This is the primary limitation for qualitative risk-based decision making tools. The other two options proposed in Bryder et al. involve statistical methods driven by the outcome of the qualitative tool. These methods are described in Sect. 4.6.2.

All approaches described above are ultimately based on a qualitative risk-based tool describing process and scientific understanding. There are several considerations in developing a risk-based tool. The tool is based on indicator characteristics and factors that describe the information collected during Stage 1 of the validation process or characterize additional relevant scientific knowledge. The information is scored based on its presence or absence using a predefined scoring system and aggregated to compute a final score that quantifies the residual risk. Cut-points based on the range of possible final scores will be organized into a decision making table. The number of lots required for PPQ will be determined using the final score and its position relative to the cut-points of the decision making table.

The first step in risk-based tool development is to describe the characteristics and factors to be used in the tool. If there are elements that describe knowledge or scientific understanding, these statements must be as objective as possible. Chao, ASTM E2475 (2010), and Bryder et al. offer examples of qualitative characteristics to consider for tool development. It is recommended to avoid adjectives in these statements since adjectives can lead to interpretation. If an adjective must be used, it is helpful to define the term. For example, an outcome of Stage 1 may result in *limited* understanding of critical raw materials. Here it would be useful to define the word *limited*. The word *limited* could be defined as "three or fewer raw material lots were used in Stage 1 experiments"or "the raw materials used during Stage 1 did not represent the range of the raw material specificaton." These modifed statements minimize interpretation error and drive consistent application among users of the tool. In addition, the choice of characteristics and factors describing the knowledge for each indicator characteristic depends on whether the tool is used for API/bulk drug substance or drug product. Characteristics unique to understanding API or drug product need to be accounted for because the knowledge gathered for bulk

Table 4.10 Limited example of a risk-based decision making tool for indicator characteristics (false statement = 0 and true statement = 1)

Indicator characteristic	Factor	Result	Scoring
Product knowledge	Analytical comparability has been demonstrated at commercial scale	True	1
	Drug product experience exists with a range of API age	False	0
Process knowledge	Process tracking is active in clinical manufacturing	False	0
Raw materials	All raw materials have been used for the same application at commercial scale	True	1
Total score			2

drug substance may differ from drug product. For example, process complexity with multiple seed trains may be of value for a bulk drug substance tool, whereas process understanding given different drug product run sizes may be appropriate for the drug product tool. An example with limited information for a tool's framework is shown in Table 4.10. Additional examples are provided by Levy and Bryder et al.

Once the indicator characteristics and the corresponding factors have been defined, a scoring mechanism needs to be developed. This can be a simple binary score as shown in Table 4.10, or a more complex scoring mechanism that provides greater values for more valuable or critical factors. The validity of the scoring needs to be tested before it can be tied to an action threshold table. Bryder et al. describe a low, medium, and high risk scenario for each factor. Based on the outcome of the risk for each factor, a qualitative description of residual risk is determined. One aspect that is not directly provided by Bryder et al. is a defined scoring system that determines residual risk. However, the example presented in Appendix 3 of their paper provides a risk priority number that implies a scoring of the different factors. Levy and Sidor also note the importance of weighting each factor based on its relative importance. In this manner, the residual risk will increase appropriately if there are gaps within the most critical information.

The final step in developing a risk-based decision making tool is construction of the decision making table. The decision making table describes the actions to be taken based on the tool's score. For the PPQ tool, the action is the selection of the number of PPQ lots. The decision making table consists of multiple cut-points based on a range of possible tool scores. An example of a decision making table is shown in Table 4.11. In this example, the overall tool score is translated to a percentage of false factors as described in Table 4.10. The score for the example presented in Table 4.10 is computed as the total number of false factors divided by the total number of factors.

Table 4.11 Example of a decision making table to determine the number of lots for PPQ

Total score (% false statements)	Number of lots for PPQ	Residual risk
≥80%	>6	Very high
>60 to <80%	5–6	High
>40 to ≤60%	3–4	Moderate
≤40%	1–2	Low

$$\left(\frac{2 \text{ false factors}}{4 \text{ factors}}\right) \times 100\% = 50\% \tag{4.4}$$

Using the decision making table in Table 4.11, a score of 50% results in moderate residual risk and the need for 3–4 lots in the PPQ campaign.

Although a risk-based decision making tool has some disadvantages, one major benefit is going through the process to develop the tool. Process engineers and other team members are required to articulate and discuss characteristics that lead to process understanding, and this improves and codifies product knowledge. A second advantage is that a risk-based tool can be developed and used even when statistical expertise is not readily available.

4.6.2   Statistical Approach to Determining the Number of PPQ Lots

In contrast to the non-statistical approaches discussed in the previous section, statistical approaches define the number of PPQ lots based on the "expectation" of meeting a desired level of acceptable quality. Determining the number of PPQ lots for a stated objective is not a difficult statistical problem. In many cases, it only requires use of a sample size calculation based on a desired power and statistical hypothesis of interest (Sect. 2.10). The challenging aspect of traditional sample size calculations is that required sample sizes are typically too large to be practical for a PPQ campaign.

Bryder et al. and, more recently, Breen et al. (2016) have offered no less than six statistical procedures that might be considered for determining the number of PPQ lots. These publications are presented as ISPE discussion papers and offer insight from many participants in the pharmaceutical community. The six approaches presented in this set of two papers are briefly described below.

A. Statistical approaches appearing in Bryder et al.

1. *Target process confidence and target process capability*: The number of lots is chosen to ensure with a target process confidence that a target process capability will be maintained. Higher residual risk requires a higher level of confidence to meet the target C_{pk} value. For example, one might determine

the number of lots that if passed ensure an 80% chance (target process confidence) that the process capability (C_{pk}) exceeds 1 (target process capability). This method is based on a confidence interval formula for C_{pk}. Strickland and Altan (2016) have stated that this could lead to larger batch sizes than are practical.

2. *Expected coverage:* This approach relies on order statistics. It is based on the relationship that as more PPQ lots are manufactured, a more informative description of future performance is obtained. As the residual risk increases, the required expectation for coverage increases. As an example, for a passing high risk process using 9 lots, one has 50% confidence that 80% of future lots are within release limits. However, reliance on a nonparametric interval may require more lots than is viewed practical if meaningful confidence and coverage percentages are to be established.

B. Statistical approaches appearing in Breen et al.

1. *Tolerance intervals*: This approach is somewhat similar to the expected coverage approach. Here a tolerance interval is computed by fixing the constant K based on a predefined confidence and coverage. (See Eqs. (2.23) and (2.52) for tolerance interval formulas). PPQ is deemed successful if the tolerance interval falls within the PPQ acceptance criterion. The selected confidence and coverage is based on residual risk.

2. *Probability of batch success*: This approach determines sample size by specifying a probability that a batch will meet all specification limits. The approach can use either a frequentist approach with a given level of confidence, or a Bayesian approach. As before, the selected probability and confidence should be based on the specific residual risk.

3. *Combinatorial approach of analysis of variance with risk assessments*: This approach determines sample size that ensures adequate power to detect an important shift in the process mean. As the criticality of a QA increases, there is a lower tolerance for drift. Drift can be described in terms of the process capability C_{pk}.

4. *Variability-based approach*: Here the sample size is determined using a power calculation for a test that compares the variation of validation lots with the variation of historical lots of a current process.

One other approach of note is a Bayesian approach proposed by Yang (2013). In this approach, quality assurance is defined as the posterior probability that a specified number of future post-PPQ lots will meet specifications. At the end of a successful PPQ campaign, if all lots meet the predefined acceptance criterion, then the posterior probability that a future lot meets the specification exceeds a predetermined level.

In summary, the use of any one approach to compute the number of PPQ lots has inherent advantages and disadvantages. Qualitative approaches force participants to think scientifically about a process, and consider all aspects of the process performance. They also provide information for selecting criteria for statistical approaches. However if used without statistical approaches, it will not most effectively incorporate quantitative knowledge related to the understanding of variability and process reliability. Conversely, statistical approaches give seemingly analytical solutions, but do not entirely consider the holistic process performance provided by a qualitative assessment. Additionally, some of these approaches yield sample sizes that are too large for practical application. Thus, some combination of these two approaches is recommended in order to provide the most appropriate number of PPQ lots.

References

ASTM E2475 (2010) Standard guide for process understanding related to pharmaceutical manufacture and control. ASTM International, West Conshohocken

Breen C, Somayajula D A, Altekar M, Patel P, Lewis R (2016) Determining the number of process performance qualification batches using statistical tools—supplement to prior discussion paper

Bryder M, Etling H, Flemming J, Hu Y, Levy P (2012, updated 2014) Topic 1—stage 2 process validation: determining and justifying the number of process performance qualification batches (Version 2). ISPE discussion paper. (www.ispe.org/discussion-papers/stage-2-process-validation.pdf)

Chao S-B (2005) Risk-based approach for biological process validation: Integrating process validation with product development. J Validation Technol 12(1):54–68

Doymaz F, Ye F, Burdick R (2015) Product homogeneity assessment during validation of biopharmaceutical drug product manufacturing processes. In: Jameel F, Hershenson S, MA K, Martin-Moe S (eds) Appears in quality by design for biopharmaceutical drug product development. Springer Science & Business Media, LLC, New York, pp 649–659

European Medicines Agency (EMA) (2014) Guideline on process validation for finished products—information and data to be provided in regulatory submissions

International Conference on Harmonization (2000) Q7 good manufacturing practice guide for active pharmaceutical ingredients

International Conference on Harmonization (2005) Q9 quality risk management

Levy P (2012) Determining and justifying the number of process validation batches: making initial batch release decisions. ISPE: Lessons from 483s Process Validation Track

O'Donnell K, Greene A (2006a) A risk management solution designed to facilitate risk-based qualification, validation, and change control activities within GMP and pharmaceutical regulatory compliance environments in the EU: part I fundamental principles, design criteria, outline of process. J GXP Compliance 10(4):12–25

O'Donnell K, Greene A (2006b) A risk management solution designed to facilitate risk-based qualification, validation, and change control activities within GMP and pharmaceutical regulatory compliance environments in the EU: part II tool scope, structure, limitations, principle findings, and novel elements. J GXP Compliance 10(4):26–35

Parenteral Drug Association (PDA) (1998) Technical Report No. 29: Points to consider for cleaning validation

Sidor L (2013) PPQ lot tool: determining the number of lots in PPQ. Validation Information Group, PDA Annual Meeting (April)

Sidor L (2014) How many batches for PPQ? AAPS National Biotechnology Conference Short Course (May)

Sidor L, Lewus P (2007) Validation and compliance: using risk analysis in process validation. BioPharm Int 20(2). http://www.biopharminternational.com/validation-compliance-using-risk-analysis-process-validation?id=&sk=&date=&%0A%09%09%09&pageID=2

Strickland H, Altan S (2016) Process validation in the twenty-first century. In: Zhang L (ed) - Chapter 19 of nonclinical statistics for pharmaceutical and biotechnology industries. Springer, Heidelberg, pp 501–531

Yang H (2013) How many batches are needed for process validation under the new FDA guidance? PDA J Pharm Sci Technol 67:53–62

Chapter 5
GMP Monitoring and Continuous Process Verification: Stage 3 of the FDA Process Validation Guidance

Keywords Acceptance quality level (AQL) • Acceptance sampling • Annual product review • Continued process verification • Corrective and preventative action (CAPA) • Critical material attribute (CMaA) • Critical method attribute (CMeA), Critical process parameter (CPP) • Critical quality attribute (CQA) • Lot tolerance percent defective (LTPD) • Operating characteristic (OC) curve • Out of specification (OOS) • Process capability • Statistical control charts

5.1 Introduction

The FDA Process Validation Guidance (2011) advocates a life cycle approach for manufacturing to ensure the process can reliably and consistently provide quality product that meets the therapy's desired efficacy and safety profile. This life cycle approach emphasizes collection and evaluation of appropriate data as evidence to demonstrate that the process is in a controlled state to deliver quality product. It has three stages:

1. Process design,
2. Process qualification, and
3. Continued process verification (CPV).

Chapter 3 discussed the process design stage in which state-of-the-art science and engineering are used to design a process, statistical tools including design of experiment are used to identify sources of variation, and risk assessment is used to establish a control strategy for critical parameters and attributes. A quality by design (QbD) approach is desired to build quality into the process.

Chapter 4 discussed process qualification, which included two substages:

1. Facility, equipment, and systems qualification and
2. Process performance qualification (PPQ).

Substage 1 ensures all facilities, equipment and systems meet cGMP and other regulatory standards and that they are fit for the purpose of a reliable and controlled process to deliver quality product. Substage 2 (PPQ) confirms that the process performs as expected.

© Springer International Publishing AG 2017
R.K. Burdick et al., *Statistical Applications for Chemistry, Manufacturing and Controls (CMC) in the Pharmaceutical Industry*, Statistics for Biology and Health, DOI 10.1007/978-3-319-50186-4_5

The third stage of the FDA process validation considered in this chapter is called continued process verification (CPV). In this stage, data are continuously collected and evaluated to verify the process remains in the desired controlled state. Using the analogy of an orbital spacecraft, process validation Stages 1 and 2 represent the development of an orbital spacecraft and the successful launch into orbit. Stage 3 consists of the work done to ensure the spacecraft remains in orbit. This chapter discusses the key components of CPV and statistical tools that are useful for this purpose.

5.2 Components in Continued Process Verification

It is assumed that Stages 1 and 2 of the validation process provide a good understanding of the manufacturing process and associated analytical methods. Using risk assessment, quality by design, and appropriate quality systems (ICH Q8 (R2), ICH Q9, ICH Q10), an appropriate control strategy will be in place, and the process capability to deliver intended-for-use products will have been validated.

Alsmeyer and Pazhayattil (2014) described a simple case study of CPV for small molecules. The BioPhorum Operations Group (BPOG 2014) provides a position paper on CPV with a case study using a monoclonal antibody manufacturing process. This study is a continuation of a case study in bioprocess development using risk assessment and quality by design (CMC Biotech Working Group 2009). These two case studies provide examples for the following discussion.

Figure 5.1 presents a simplified diagram of key parameters from different sources that potentially need monitoring in Stage 3.

The parameters in Fig. 5.1 are classified into four sets:

1. Critical material attributes,
2. Critical process parameters,
3. Critical quality attributes, and

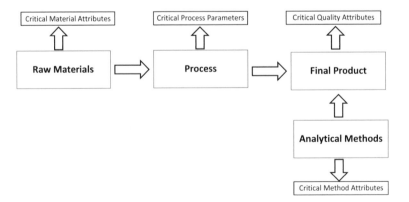

Fig. 5.1 Manufacturing components and structure of CPV monitoring variables

4. Critical method attributes.

Because of the progression shown in Fig. 5.1, critical quality attributes are often called lagging factors, and critical material attributes and critical process parameters are called leading factors (Strickland and Altan 2016).

These categories may not be complete or non-overlapping, but are useful to help determine the quantities that need to be monitored in Stage 3. Modifications of this classification system can be considered for particular circumstances. This structure is useful for understanding the sources of variation as emphasized in the FDA validation guidance.

1. Critical material attributes (CMaAs) are properties or characteristics of input materials used during the manufacturing process. For example, in a typical solid dose manufacturing process, critical characteristics of API and excipients (e.g., water content) can be considered critical material attributes (Alsmeyer and Pazhayattil 2014). In a monoclonal antibody manufacturing process, examples of CMaAs include certain characteristics of working cells, key nutrient levels of the cell culture, and glucose feed levels (BPOG 2014).
2. Critical process parameters (CPPs) are those that relate to the manufacturing process and directly impact product quality. CPPs should be identified in Stages 1 and 2 of process qualification, along with their control range (i.e., design space). These parameters can appear in different unit operations for small molecules (Fig. 2 of Alsmeyer and Pazhayattil 2014) or in different manufacturing steps for monoclonal antibodies (Chap. 10 of BPOG 2014).
3. Critical quality attributes (CQAs) are properties and characteristics of the drug substance or final product. CQAs must meet specifications in order to ensure that the product quality supports its intended safety and efficacy for patients. Typical CQAs relate to product strength, potency, identity, and purity. CQAs for one unit operation may become CMaAs for a subsequent unit operation.
4. Critical method attributes (CMeAs), including critical reagent properties, method parameters, and method accuracy and precision measures, are all candidates for continual monitoring in order to ensure the analytical methods remain fit for purpose. These are often neglected, but much of the data used to assess the CMaAs and CQAs are output from analytical methods. If the analytical methods are not validated and controlled for their intended accuracy and precision, measured CMaAs and CQAs will be compromised. In addition to minimizing the risk of poor measurements, this information is useful in troubleshooting non-conformances of a CQA to determine if the problem is attributable to the manufacturing process or the analytical method.

According to quality by design principles, effective control of CMaAs and CPPs should lead to high confidence that requirements on CQAs will be met. Therefore, the FDA guidance on validation emphasizes sufficient understanding of the process, and states that "Focusing exclusively on qualification efforts without also understanding the manufacturing process and associated variations may not lead to adequate assurance of quality." By monitoring parameters in all four categories, the CPV monitoring program provides evidence that the process is sufficiently understood.

The structure presented in Fig. 5.1 implies the following:

1. If all CQAs are in control, the product quality is in control.
2. If all CMaAs and CPPs are in control, there is a high chance that CQAs will be in control.
3. If all CMeAs are in control, it follows that all data represent the true state of the product and process.

Data for the identified CMaAs, CPPs, CQAs, and CMeAs should be compiled in a format amenable for trending and analysis using available software. The reader is referred to Chap. 10 of BPOG (2014) for a well-structured list of variables to be monitored in each step of a monoclonal antibody drug substance manufacturing process.

5.2.1 Data Collection and Control Limits

Historical data from the four parameter groups discussed above are used for constructing statistical control limits. Once sufficient process understanding is achieved, control limits based on historical process performance are not expected to require revision unless the process has been changed or impacted in some defined manner (e.g., an investigation determines a process shift has occurred). Periodic examination of the appropriateness of the limits may be undertaken based on the frequency of manufacturing.

A sampling plan should be determined for each monitored variable. The plan should include sampling frequency and the type of chart(s) used for trending. An analysis plan should also be created, including the process for constructing control limits, the frequency of analysis, how results will be interpreted, and actions to be taken after a trend or out-of-control event is identified.

New data are best entered into the historical database in a timely manner for trending and analysis so that any potential signals may be investigated in a meaningful manner. Examining data with very low frequency limits the usefulness of the CPV program because it reduces the ability to react to potential factors that may lead to out-of-specification results or excursions from in-process control limits.

5.2.2 Monitoring

Consistent with the 2011 FDA guidance, the goal of CPV is "continual assurance that the process remains in a state of control (the validated state) during commercial manufacture." This goal ensures that high quality products can be consistently supplied to patients. As such, an effective CPV program monitors the chosen parameters for trends and defines actions to be taken if signals are identified.

If the CPV program identifies a signal, any of the critical material, process, quality, or method attributes may require further examination. The investigation type and rigor will depend on the specific attribute that displays the signal, the scientific knowledge about the given parameter and process, an examination of previous investigations, and an analysis of the current data set. One should not assume that the signal is a result of inappropriately set limits and blindly reset control limits so that the process appears in control. However, an examination of the appropriateness of the limits may be considered should the data indicate a need for such an assessment.

In cases where the monitoring program detects a signal, the implications may differ depending on which of the following two cases occurs:

1. One or more CQAs are trending out of control.

   a. One should first examine the analytical method performance. If the method appears out of control, a thorough investigation into the method should be undertaken. Determine if the correct CMeAs are being monitored, if they are being monitored with the appropriate frequency, and if the method is fit for purpose. If the investigation finds that the analytical method is out of control, it should be improved, samples should be re-tested using the updated method, and the quality attribute can then be re-assessed.
   b. If all CMeAs are in control, the out-of-control signal for the CQA is confirmed. The signal is attributed to some portion of the manufacturing process. It is now necessary to examine CMaAs and CPPs.
   c. If one or more of the CMaAs or CPPs are out of control, the process should be re-calibrated. The investigation may indicate the process is not well understood, and more study of the process is warranted. The resulting investigation may lead to a new control and monitoring strategy, possibly including new monitoring variables or control limits.

2. All CQAs are within control, but out-of-control signals occur for other attributes.

   a. If some CMaAs or CPPs are out of control, this implies the upstream parameters may not truly impact the CQA. One should investigate whether the out-of-control variable should continue to be monitored using a risk assessment.
   b. If some CMeAs are out of control, there are two possibilities:

      i. The product quality is consistent, but the method aberration is not large enough to change the method performance or severely affect the quality attribute data.
      ii. The CQAs are within control only because of a serious method aberration. In fact, the CQAs may be out of control, but data distortion due to the method produces a misleading result. In either case, the method and its control strategy should be reviewed. Perhaps an adjustment of the control

strategy solves the problem, or a partial validation or re-validation of the method is warranted. In this situation, product samples may need to be re-tested using a calibrated method.

5.3   Statistical Tools for CPV

Statistical quality tools are used to verify that CQAs are being properly controlled throughout CPV. Statistical control charts, process capability assessment, and acceptance sampling methodology are among the statistical quality applications used to achieve the process monitoring and improvement required in CPV. This section introduces some of these statistical applications applied to CPV. ASTM (2010) provides useful material for further reading.

5.3.1   Acceptance Sampling

Acceptance sampling plans can be incorporated into the overall strategy of CPV for ensuring product quality. The majority of acceptance sampling plans involve attribute sampling, or variables described as qualitative or nominal. However, in many cases quality attributes are physical measurements on a continuous or quantitative scale. In such cases lot acceptance is based on the percentage of individual values in a lot that satisfy a numerical specification.

Acceptance sampling consists of a sampling design and a set of rules for making decisions based on the resulting sample. For situations where only a single sample is selected, the two decisions are

1. Accept the lot or
2. Reject the lot.
3. In a pre-planned multiple sample design, a third decision is to select another sample and then decide to either accept the lot, reject the lot, or continue sampling.

The fundamental tool for selecting a sampling plan is the operating characteristic (OC) curve. An OC curve is a bi-variate graph with probability of passing a lot on the vertical axis and the percentage of units that do not meet the specification limits on the horizontal axis. Figure 5.2 provides an example of an OC curve for an attribute sampling plan in which a sample of 80 items is selected at random from a lot. A lot is "accepted" if there are fewer than two non-conforming (defective) units in the sample. The lot is "rejected" if there are two or more non-conforming units in the sample. The terms "accepted" and "rejected" in this context are used in a generic sense. The action that results from either conclusion depends on the particular application.

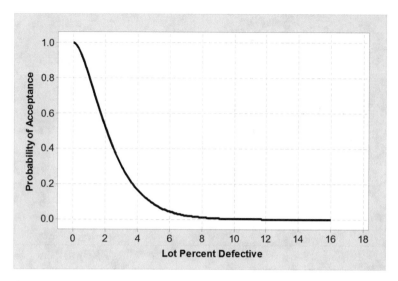

Fig. 5.2 OC curve with sample size = 80 and acceptance number = 1

From Fig. 5.2 is seen that this plan virtually ensures no lot is accepted if the percentage of defective units in the lot exceeds 8%. If the percentage of defective units is 2%, the probability of accepting the lot based on this sampling plan is 52.3%.

When deciding whether to accept or reject a lot, there are two types of errors

1. Rejecting a good lot (Type 1 error) and
2. Accepting a bad lot (Type 2 error).

The risk of committing a Type 1 error is referred to as the producer's risk and is denoted by the Greek letter α. The risk of committing a Type 2 error is called the consumer's risk and is denoted by the Greek letter β. Definitions for "good" and "bad" are typically defined in terms of the percentage of non-conforming (defective) units in the sample. Acceptance quality level (AQL) is the percentage of defective units on the horizontal axis associated with the 95% probability of acceptance on the vertical axis. The lot tolerance percent defective (LTPD) is the percent of defective on the horizontal axis associated with the 10% probability of acceptance. It is also useful to define AQL and LTPD in terms of the proportions $p_1 = \frac{AQL}{100\%}$ and $p_2 = \frac{LTPD}{100\%}$. The AQL and LTPD values for Fig. 5.2 are shown in Fig. 5.3.

Determination of acceptable values for AQL and LTPD require assessment of a variety of criteria including risks, costs, and consumer requirements. The first step in this process is to classify the severity of the defects that might occur. Typical classifications are Critical, Major, and Minor. Defectives of the same category would generally be expected to have the same values for AQL and LTPD.

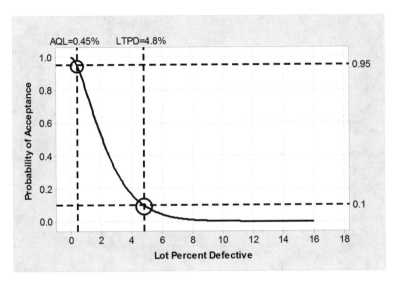

Fig. 5.3 AQL and LTPD for Fig. 5.2

To demonstrate how acceptance sampling can be used in CPV, consider a quality attribute monitored in Stage 3 to ensure it maintains the same quality level attained in Stages 1 and 2. Values will conform to an acceptable quality level if they do not exceed an upper specification limit (*USL*).

A random sample of size n is selected and the quality attribute is measured for each item. The sample mean is then compared to the quantity

$$A = USL - kS \tag{5.1}$$

where S is the sample standard deviation and k is a constant that is a function of AQL and LTPD. The process is considered acceptable if the sample mean of the n items is less than or equal to A. Schilling and Neubauer (2009, p. 186) provide the following approximate formulas for both k and n using what is called the k-method:

$$k = \frac{Z_{1-p_2}Z_{1-\alpha} + Z_{1-p_1}Z_{1-\beta}}{Z_{1-\alpha} + Z_{1-\beta}}$$

$$n = \left(\frac{Z_{1-\alpha}+Z_{1-\beta}}{Z_{1-p_1}-Z_{1-p_2}}\right)^2 \text{ when the variance is known} \tag{5.2}$$

$$n = \left(\frac{Z_{1-\alpha}+Z_{1-\beta}}{Z_{1-p_1}-Z_{1-p_2}}\right)^2 \left(1 + \frac{k^2}{2}\right) \text{ when the variance is unknown}$$

where Z_δ is the percentile of a standard normal distribution with area δ to the left.

To illustrate how acceptance sampling may be applied to Stage 3, consider a power fill process. Suppose that during Stages 1 and 2, the net weight of each vial should be at least 25 g. To determine whether the process maintains this quality level, an acceptance sampling plan requires that the process should be accepted

95% of the time when the proportion of net weight vials below 25 g is AQL $= 0.5\%$, and should be rejected 90% of the time when the proportion of net weight vials less than 25 g is LTPD $= 5\%$. The specification above can be expressed using acceptance sampling terminology as determining a sampling plan with $p_1 = \frac{AQL}{100\%} = 0.005$, $p_2 = \frac{LTPD}{100\%} = 0.05$, $\alpha = 0.05$, and $\beta = 0.10$. Using these values and assuming that the variance of the process is unknown, $Z_{1-p_1} = 2.576$, $Z_{1-p_2} = 1.645$, $Z_{1-\alpha} = 1.645$, and $Z_{1-\beta} = 1.282$, $k = 2.05$ and $n = 31$ (rounding up). If the variance is known, the sample size reduces to $n = 10$ (rounding up).

Schilling and Neubauer (2009) and Burdick and Ye (2016) provide more in-depth discussions of acceptance sampling. Kiermeier (2008) provides R-code for many of the required calculations.

5.3.2   *Statistical Control Charts*

Statistical control charts are useful for continually verifying that a process remains in control. The main goal of statistical control charting is to use probability theory to determine whether an observed deviation is due to a chance cause (also known as a common cause) or to an assignable cause. If a control chart signals the occurrence of an assignable cause, the process is stopped and appropriate actions are taken to eliminate the assignable cause. In addition, preventive actions are put in place to reduce the chance that the assignable cause reappears in the future. One set of rules generally used to determine when an assignable cause occurs is provided by Nelson (1984).

To briefly demonstrate this process, we present results for an individual value chart. An individual value chart is used in Stage 3 to monitor individual values of CQAs for released lots. Suppose that a CQA used for lot disposition is monitored in an individual control chart. A sample of n lots is selected and a single CQA measurement is taken from each lot. The collected sample is represented as Y_1, $Y_2, \ldots, Y_n$. For the procedure that follows, it is assumed that when the process is in control, the sample of n lots behaves as a random sample selected from a normal population with mean μ and standard deviation σ. Based on the probabilities of the normal distribution, the probability that a single observation exceeds the range from $\mu - 3\sigma$ to $\mu + 3\sigma$ is roughly 99.73%. The first rule presented by Nelson (1984) states than an individual value that falls outside this range is a signal that the process is out of control. In practice, μ and σ are unknown and must be estimated from the sample. An unbiased estimator for the unknown process mean μ is the sample average,

$$\bar{Y} = \frac{\sum_{i=1}^{n} Y_i}{n}. \tag{5.3}$$

An estimate of σ using a moving range of two consecutive measurements in the sample is

$$\frac{\overline{MR}}{1.128}$$

$$\overline{MR} = \frac{\sum_{i=1}^{n-1}|Y_{i+1} - Y_i|}{n-1} \tag{5.4}$$

An individual control chart is established by plotting a run chart for the sample values, with horizontal reference lines at $\bar{Y}$ to represent the center line (CL),

$$LCL = \bar{Y} - 3 \times \frac{\overline{MR}}{1.128}$$

$$LCL = \bar{Y} - 2.66 \times \overline{MR} \tag{5.5}$$

to represent the lower control limit (LCL), and

$$UCL = \bar{Y} + 2.66 \times \overline{MR} \tag{5.6}$$

to represent the upper control limit (UCL). Figure 5.4 presents an example of an individual value chart.

A moving range chart as shown in Fig. 5.5 is useful to complement the information provided by the individual control chart. A moving range chart has a horizontal reference line at $\overline{MR}$ and an upper control limit (UCL) of $UCL = 3.267 \times \overline{MR}$.

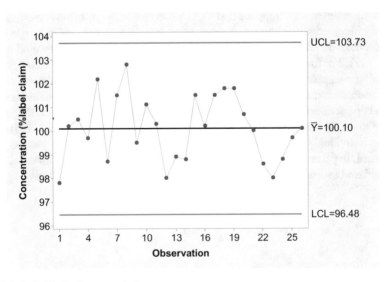

Fig. 5.4 Individual value control chart

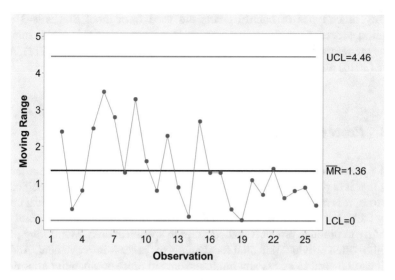

Fig. 5.5 Moving range control chart

Table 5.1 Computations for control charts

Chart	Reference line	Value
Individuals	CL	$\bar{Y} = 100.10$
	LCL	$LCL = \bar{Y} - 2.66 \times \overline{MR} = 96.48$
	UCL	$UCL = \bar{Y} + 2.66 \times \overline{MR} = 103.73$
Moving range	CL	$\overline{MR} = 1.364$
	UCL	$UCL = 3.267 \times \overline{MR} = 4.46$

Figures 5.4 and 5.5 were obtained from a sample of $n = 26$ released lots of a manufacturing process. The measured CQA is concentration expressed as percentage of label claim. A summary of the calculated results are shown in Table 5.1.

A graphical inspection of the plots indicates that none of the individual values or moving ranges are outside their respective control limits. This suggests that the process is in a state of statistical control.

Since decisions based on control charts are based on probability, there is a risk that a future individual value will fall outside the control limits, even though the process is truely in control. Similarly, there is a chance that a future individual value will fall within the control limits, even if an assignable cause is present. The consequences of such errors can be severe, and need to be considered in establishing a risk strategy. These risks can be dramatically increased if one were to use either two or four standard deviation control limits. Two standard deviation limits would result in nuisance signals, whereas four standard deviation limits would fail to detect shifts due to an assignable cause. The three standard deviation control limits described above provide a good balance between these two risks.

Many other types of control charts are used throughout Stages 1–3 of the validation process, and there is a wealth of references on this topic. Interested readers are referred to ASTM (2010), Montgomery (2013), Wheeler (2012), ASTM E2587 (2016) and Altan et al. (2016).

5.3.3 Process Capability and Performance Assessment

When a manufacturing process is in statistical control, this does not necessarily imply that it is producing products that meet predetermined quality specifications. Therefore, it is not only important to evaluate process stability (statistical control) during CPV, but it is also equally important to monitor the process capability (i.e., the ability to produce products that conform to specifications). Monitoring process capability often provides potential focal points for process improvement. Additionally, it can be used to assess any improvements to process capability after process improvements have been implemented. Process capability indices identify the need to reduce common cause variation or to compare processes.

When a process is in statistical control, its quality is predictable. Thus, before assessing process capability, it is necessary that the process be in a state of statistical control. It is common to define process capability in units of standard deviations of the controlled process. We will denote this process standard deviation as σ. In particular, it is common to look at the relationship between the standard deviation and the range between the upper and lower specification limits. This capability index is defined as

$$C_p = \frac{USL - LSL}{6\sigma} \tag{5.7}$$

where LSL is the lower specification limit and USL is the upper specification limit.

When process data are well represented with a normal distribution and the process is centered between the upper and lower specification limits (i.e., the process mean $=(LSL + USL)/2$), the capability index C_p can be expressed as the number of units that are outside of specification. In particular, the proportion of defective product (expressed in parts per million (ppm)) is related to C_p by the equation

$$ppm \text{ defective} = 1,000,000 \times 2\Phi\left(-3C_p\right) \tag{5.8}$$

where Φ is the cumulative standard normal distribution. For example, assume that the process is centered about the specification limits and that $C_p = 1$. Then

$$
\begin{aligned}
ppm \text{ defective} &= 1,000,000 \times 2\Phi(-3 \times 1) \\
&= 1,000,000 \times 2 \times \Phi(-3) \\
&= 1,000,000 \times 2 \times 0.00135 \\
&= 2,700 \text{ ppm.}
\end{aligned}
\tag{5.9}
$$

Thus, a $C_p = 1$ corresponds to a centered process that produces 2700 ppm outside of the specification limits. This relationship between C_p and ppm can be used to establish acceptable values for C_p. Since $C_p < 1$ implies that more than 2700 ppm will be out of the specification limits, and $C_p > 1$ implies less than 2700 ppm out of the specification limits, it is seen that the process improves as C_p increases. In addition to describing the overall capability of a process, C_p can be used to determine where to focus process improvement efforts.

The capability index C_p is not appropriate when a process is operating off-center. In such cases, an alternative capability index is defined as

$$
C_{pk} = \min \left[\frac{USL - \mu}{3\sigma}, \frac{\mu - LSL}{3\sigma} \right]
\tag{5.10}
$$

where μ is the (off-centered) process mean.

Because μ and σ are typically unknown, they must be estimated from a sample. There are several ways to estimate σ, and this unfortunately has created much confusion as to what is the "correct" manner. As in most statistical procedures, the "correct" manner depends on the situation. We demonstrate one approach, but encourage the reader to read more on this topic in the references provided at the end of this section.

Because the control chart in Sect. 5.3.2 indicates the process is stable, we consider the sample to be a random sample of $n = 26$ from a process with mean μ and standard deviation σ. Thus we will use the sample mean $\bar{Y}$ as an estimator for μ and the sample standard deviation S as an estimator for σ. A point estimator and $100(1-\alpha)\%$ lower confidence bound on C_p using these estimators is

$$
\widehat{C}_p = \frac{USL - LSL}{6 \times S}
$$

$$
L = \widehat{C}_p \times \sqrt{\frac{\chi^2_{\alpha:n-1}}{n-1}}
\tag{5.11}
$$

where

$$
S = \sqrt{\frac{\sum_{i=1}^{n}(Y_i - \bar{Y})^2}{n-1}}
$$

$$
\bar{Y} = \sum_{i=1}^{n} Y_i
\tag{5.12}
$$

Table 5.2 Computations for capability indexes

$$\widehat{C}_p = \frac{USL - LSL}{6 \times S} = \frac{105 - 95}{6 \times 1.40} = 1.19$$

$$95\%L = \widehat{C}_p \times \sqrt{\frac{\chi^2_{\alpha;n-1}}{n-1}} = 1.19 \times \sqrt{\frac{14.61}{26-1}} = 0.91$$

$$\widehat{C}_{pk} = \min\left[\frac{USL - \bar{Y}}{3S}, \frac{\bar{Y} - LSL}{3S}\right]$$

$$= \min\left[\frac{105 - 100.10}{3 \times 1.40}, \frac{100.10 - 95}{3 \times 1.40}\right] = 1.17$$

$$95\%L = \widehat{C}_{pk}\left[1 - Z_{1-\alpha}\sqrt{\frac{1}{9n\widehat{C}^2_{pk}} + \frac{1}{2(n-1)}}\right]$$

$$= 1.17\left[1 - 1.645\sqrt{\frac{1}{9 \times 26 \times (1.17)^2} + \frac{1}{2(26-1)}}\right] = 0.88$$

and $\chi^2_{\alpha;n-1}$ is the chi-squared percentile with $n-1$ degrees of freedom and area α to the left. The lower bound on C_p can be used to answer the question, "What is the smallest value of C_p consistent with the uncertainty in the data?" A process is considered capable if L is greater than the desired value for C_p.

A point estimator and $100(1 - \alpha)\%$ lower confidence bound on C_{pk} is

$$\widehat{C}_{pk} = \min\left[\frac{USL - \bar{Y}}{3S}, \frac{\bar{Y} - LSL}{3S}\right]$$

$$L = \widehat{C}_{pk}\left[1 - Z_{1-\alpha}\sqrt{\frac{1}{9n\widehat{C}^2_{pk}} + \frac{1}{2(n-1)}}\right] \tag{5.13}$$

where $Z_{1-\alpha}$ is the percentile of a standard normal distribution with area $1-\alpha$ to the left.

Assume that the specification limits are $LSL = 95\%$ and $USL = 105\%$. Calculations for (5.11) and (5.13) are provided in Table 5.2 using the data from Sect. 5.3.2 where $\bar{Y} = 100.10$ and $S = 1.40$.

As will always be the case, C_{pk} is less than C_p. Some interpret C_p to be the maximum attainable capability that is achieved when the process is centered. Using equation (5.8) and the lower bound of C_p provides the estimated out-of-specification rate of

$$\begin{aligned}
ppm \text{ defective} &= 1,000,000 \times 2\Phi(-3 \times 0.91) \\
&= 1,000,000 \times 2 \times \Phi(-2.73) \\
&= 1,000,000 \times 2 \times 0.0032 \\
&= 6,400 \, ppm.
\end{aligned} \tag{5.14}$$

A process performance index is closely related to a process capability index. These are typically represented as P_p and P_{pk}. A capability index is typically used in a prospective assessment where a process has been demonstrated to be in statistical control. Such an assessement focuses on the ability of the process to meet specifications in the future. A process performance index is used in a retrospective assessment to examine past process behavior and determine how a process will perform in the future if left unchanged. The process under examination may or may not be in statistical control. Some authors differentiate a capability index from a performance index by the manner in which the standard deviation is computed. The capability index employs an estimate of short-term variance, and the performance index employs an estimate of long term variance. More information on the topics considered in this section are provided in Altan et al. (2016), ASTM E2281 (2015) and Montgomery (2013).

5.3.4 Out of Specification and Corrective and Preventative Action (CAPA)

The goal of the CPV program is to detect a process shift before an out-of-specification result is observed. Typically, an out-of-specification result leads to a rigorous investigation, and may ultimately lead to rejection of the batch. Results that do not meet specifications may be observed for unit operations where the CPV program has previously detected signals or where the monitored attribute has displayed less than ideal process capability. However, out-of-specification results may also be obtained for parameters where the CPV program has not previously detected any concerns.

An overall examination of the CPV strategy should be part of any investigation into out-of-specification results. The level of scrutiny given to the CPV monitoring for the given parameter will depend on previous investigations and corrective actions already in place. If the CPV program has previously detected an issue with a given parameter, then the monitoring program is likely functioning properly, and investigative efforts might focus on the effectiveness of previous corrective actions. In contrast, if the CPV program has not previously detected any potential issues, an examination into whether the current monitoring strategy is effective should be undertaken.

When a non-conformity occurs, the following steps are required to investigate and take actions for correction.

1. The magnitude and scope of its risk should be assessed. If there is minimum risk, perhaps no further action is needed. Otherwise, a root cause analysis should be conducted to identify the assignable cause and a solution should be identified.
2. Corrective and preventive actions are taken to eliminate the root cause of the non-comformity and prevent its future occurence.
3. The attribute associated with the non-conformance must be closely monitored to verify that it is now consistently in control and in specification.

This process is demonstrated in the following example. Suppose the potency of a batch of biological product exceeds the upper specification limit. Since potency is a critical quality attribute, the risk of this excursion non-conformity is high. Such a non-conformance could potentially lead to safety problems for patients. Accordingly, a root cause analysis is conducted using the process described in Sect. 5.2.2. The performance of the analytical method for potency is first examined. Assume there is an upward trending in the potency of the negative control. This suggests there was a change in the reference standard. Further investigation leads to the discovery that the shelf-life of the reference standard has been extended twice. To determine if this was the root cause, a new reference standard was qualified and compared to the original reference standard. The comparability analysis showed that the method performance was highly similarly to the method performance when using the previous reference standard before the shelf-life extension. Another few samples from the same batch of the biological produt were tested using the new reference standard and the results were all within specification (corrective action). From this analysis, it was concluded that the excursion was due to method drift. A new process was established to monitor the stability of the reference standard (preventive action). If no aspects of the analytical method had been discovered to be the root cause, a further drill down to the manufacturing process would have been required.

Regulatory agencies expect companies to verify that changes made in response to a CAPA actually work to eliminate the root cause of a non-conformance failure. To do this, it is required to examine data collected after the CAPA and demonstrate that the failure rate intended to be improved by the CAPA satisfies the desired goal. Typically a protocol is drafted that states a post-change sample must satisfy some test criterion related to an upper value for the new defective rate. Burdick and Ye (2016) provide an example of such an application.

5.4 A CPV Protocol and Relation to Annual Product Review

Although CPV is a Stage 3 process validation activity, it should be kept in mind during Stage 1 when collected process knowledge will inform which control points should be monitored and incorporated into the CPV program. The parameters to be monitored under the CPV plan should be largely defined and understood prior to Stage 2 so that important data for generating the Stage 3 CPV limits can be gathered. The CPV protocol should be a living document throughout the first two stages of process validation. When sufficient data are collected to reliably estimate the expected process variability, the CPV protocol should be modified and assessed at some frequency as defined in procedures.

As knowledge of the process accumulates during development and validation, the CPV protocol is updated regarding the variables to be monitored, their sampling plan, monitoring chart type, and control limits.

A good CPV protocol should minimally include the following information:

1. Product information.
2. Personnel, roles, and responsibilities. A designated statistician or someone trained in statistical techniques should be involved throughout the product life cycle.
3. A structured table for all monitored parameters categorized into CMaAs, CPPs, CQAs, CMeAs, and variables corresponding to each attribute. The table should also include the sampling plan, control chart type, and initial limits. Specify which attributes should be monitored with a particular frequency.
4. A description of the process for periodic examination of the appropriateness of the limits and the method for adjusting limits based on updated process knowledge.
5. Identification of the database warehouse and analysis software.
6. All relevant data and knowledge (e.g., design space) accumulated from Stages 1 and 2 should be organized and included for determination of initial control limits.
7. Description of planned analyses, including frequency of analyses, format of documentation, and result evaluation.
8. An appropriate action plan should be established to address aberrant results. Procedures should clearly define what kinds of aberrant results can be handled by designated personnel, and what results require escalation to upper management.
9. A plan for change management should be defined. Over the life cycle of the product, some aspects of the monitoring plan may need to be changed or updated due to an accumulation of experiences and process knowledge, or in response to regulatory requirements.

The CPV protocol should align with the PPQ protocol created in Stage 2. The CPV protocol will specify a frequency for analysis of given parameters, but data should be assessed annually, at a minimum. Since an annual product review (APR) is required for several regulatory jurisdictions, coordinating the annual CPV reporting cycle with the APR cycle is most efficient from an analysis perspective. The CPV protocol should meet the minimum data analysis requirements for the APR. The APR is also a good time to evaluate the performance of the CPV protocol.

5.5   Statistical Support

The FDA guidance on validation defines process validation as "the collection and evaluation of data, from the process design stage through commercial production, which establishes scientific evidence that a process is capable of consistently delivering quality products." This definition characterizes process validation as a joint work between scientists and statisticians, and requires a full integration of statistical involvement throughout the process.

The word "statistical" or "statistics" appears 12 times in the guidance, which highlights the importance of quantitative data analysis methods in the CPV program. Regarding CPV at Stage 3, the guidance specifically emphasizes, "We recommend that a statistician or person with adequate training in statistical process control techniques develop the data collection plan and statistical methods and procedures used in measuring and evaluating process stability and process capability." Additionally, it states "We recommend that the manufacturer use quantitative, statistical methods whenever appropriate and feasible." We strongly recommend that adequate statistical resources are made available for process validation and that statisticians be an integral part of the team throughout all three stages of process validation.

References

Alsmeyer D, Pazhayattil A (2014) A case for stage 3 continued process verification. Pharmaceutical Manufacturing. http://www.pharmamanufacturing.com/articles/2014/stage3-continued-process-verification/?show=all. Accessed 7 June 2016

Altan S, Hare L, Strickland H (2016) Process capability and statistical process control. In: Zhang L (ed) Chapter 21 of nonclinical statistics for pharmaceutical and biotechnology industries. Springer, Heidelberg, pp 549–573

ASTM (2010) Manual on presentation of data and control chart analysis, 8th edn. Neubauer DV (ed), ASTM International, West Conshohocken

ASTM E2281 (2015) Standard practice for process and measurement capability indices. ASTM International, West Conshohocken

ASTM E2587 (2016) Standard practice for use of control charts in statistical process control. ASTM International, West Conshohocken

BioPhorum Operations Group (2014) Continued process verification: an industry position paper with example plan. www.biophorum.com/Page/123/BPOG-CPV-Case-Study.htm. Accessed 7 June 2016

Burdick R, Ye F (2016) Acceptance sampling. In: Zhang L (ed) Chapter 20 of nonclinical statistics for pharmaceutical and biotechnology industries. Springer, Heidelberg, pp 533–548

CMC Biotech Working Group (2009) A-Mab: a case study in bioprocess development, Version 2.1, http://c.ymcdn.com/sites/www.casss.org/resource/resmgr/imported/A-Mab_Case_Study_Version_2-1.pdf. Accessed 7 June 2016

Food and Drug Administration. Center for Drugs Evaluation Research (2011) Process validation: general principles and practices, guidance for industry

International Conference on Harmonization (ICH) (2005) Q9 Quality risk management

International Conference on Harmonization (ICH) (2008) Q10 Pharmaceutical quality system

International Conference on Harmonization (ICH) (2009) Q8 (R2) Pharmaceutical development

Kiermeier A (2008) Visualizing and assessing acceptance sampling plans: the R package acceptancesampling. J Stat Software 26(6):1–20

Montgomery DC (2013) Introduction to statistical quality control, 7th edn. Wiley, New York

Nelson LS (1984) The Shewhart control chart-tests for special causes. J Qual Technol 16:237–239

Schilling RG, Neubauer DV (2009) Acceptance sampling in quality control, 2nd edn. Chapman and Hall, London

Strickland H, Altan S (2016) In: Zhang L (ed) Chapter 19 of nonclinical statistics for pharmaceutical and biotechnology industries. Springer, Heidelberg, pp 501–531

Wheeler D J (2012) Understanding statistical process control (3rd Ed.). SPC Press Knoxville

Chapter 6
Analytical Procedures

Keywords Analytical target profile (ATP) • Bayesian analysis • Method transfer • Prediction intervals • Procedure validation • Procedure qualification • Ruggedness factors • Tolerance intervals

6.1 Introduction

Analytical chemistry is used across the pharmaceutical industry to quantify and identify the components in drug substance, drug product, and raw material to ensure that the final dosage form remains safe and efficacious from lot release throughout the product's shelf life. To understand any potential shifts in the components impacting safety and efficacy, laboratories require analytical procedures which are reliable, fit for purpose, and executed consistently over time. Analytical procedures provide the instructions used by the analyst to ensure consistent use of laboratory equipment, solution preparation, measurement recording, and documentation. As such, analytical procedures form a critical component in any quality system. This chapter considers statistical methods that ensure that these procedures are fit for their intended purpose.

Martin et al. (2013) describe a holistic view of the analytical procedure life cycle. It frames this problem using concepts consistent with Quality by Design (QbD), ICH Q8 (2009), the FDA method validation guidance (2015), and the FDA process validation guidance (2011). The performance requirements of a procedure are defined by the analytical target profile (ATP). The ATP defines the analyte to be measured, the concentration range, procedure performance criteria, and product specifications. The criteria and specifications are established to define the purpose of the analytical procedure.

The analytical procedure life cycle is presented in the following three stages:

1. Stage 1: Procedure development and preparation for Stage 2.
2. Stage 2: Procedure performance validation (Qualification).
3. Stage 3: Procedure performance verification (Transfer and Monitoring).

The detail of each stage is discussed in this chapter. Other references of interest not discussed in this chapter are USP <1030>, <1033>, and <1223>.

R.K. Burdick et al., *Statistical Applications for Chemistry, Manufacturing and Controls (CMC) in the Pharmaceutical Industry*, Statistics for Biology and Health, DOI 10.1007/978-3-319-50186-4_6

6.2   Terminology

6.2.1   Description of an Analytical Procedure

An analytical procedure and relevant terms must be clearly defined in order to design an appropriate analytical study. Descriptors such as "replicates" or "preparations" without further explanation often lead to confusion. Table 6.1 reports terminology used to describe an analytical procedure.

Not all analytical procedures entail all descriptions shown in Table 6.1. For example, liquid laboratory samples that require no further manipulations employ only a test solution. Table 6.2 provides an example of an analytical procedure for a solid dosing form.

Table 6.1 Analytical procedure terminology

Terminology	Description
Laboratory sample	The material received by the laboratory
Analytical sample	Material created by any physical manipulation of the laboratory sample such as crushing or grinding
Test portion	The quantity (aliquot) of material taken from the analytical sample for testing
Test solution	The solution resulting from chemical manipulation of the test portion such as chemical derivatization of the analyte in the test portion or dissolution of the test portion
Reading (individual determination)	The measured numerical value from a single unit of test solution
Reportable value	A summary value of individual readings, such as an average, from one or more units of a test solution. Replication may also occur across any level of the study design

Table 6.2 An analytical procedure for solid dosage coated pills

Terminology	Description			
Laboratory sample	100 coated pills			
Analytical sample	20 pills are removed from the laboratory sample and are crushed together in a mortar and pestle (i.e., composted)			
Test portion	Replicate 1: 1 gram crushed powder aliquot from analytical sample		Replicate 2: 1 gram crushed powder aliquot from analytical sample	
Test solution	Replicate 1: Test portion is dissolved in 1 L solvent		Replicate 2: Test portion is dissolved in 1 L solvent	
Reading (individual determination)	Reading 1 of replicate 1: test solution	Reading 2 of replicate 1: test solution	Reading 1 of replicate 2: test solution	Reading 2 of replicate 2: test solution
Reportable value	Average value of four readings			

6.2.2   Measurement Error Models

In this chapter, we consider the reportable value to be the key output from an analytical procedure and the focus of any investigation. In many cases, a particular analytical procedure may be used for different applications, with a different definition for the reportable value in each application. However, for purposes of discussion in this chapter, the term "reportable value" is used with the understanding that it may not be unique to a particular analytical procedure. A model that is useful for representing a reportable value is

Reportable Value = True Value + Systematic Bias + Random Error   (6.1)

where the true value and the systematic bias are fixed constants and the random error assigns a different error value to each reportable value. This relationship is represented symbolically as

$$Y = \tau + \beta + E \tag{6.2}$$

where Y represents the reportable value, τ (tau) is the true value, β (beta) is the systematic bias, and E is a random error with mean 0 and variance σ^2. (Note that β is not to be confused with its use as a regression slope in Sect. 2.12.) Model (6.2) uses the convention described in Sect. 2.12.7 of representing constants with Greek letters and random effects with upper case Latin letters. In many applications, σ^2 may be further decomposed into components that represent the various causes of variability.

6.2.3   Accuracy

In this text, accuracy concerns the magnitude of the systematic bias, β. The bias is defined as the long-run average of the difference, $Y - \tau$. Note that bias can only be determined if the true value, τ, is known. USP <1225> notes that a reference standard or a well-characterized procedure can be used to assign the value of τ. For relative content procedures used for large molecules, accuracy cannot be defined in this manner. Relative content procedures, sometimes referred to as relative purity procedures, include such procedures as size exclusion and cation exchange chromatography. Generally, minor species observed in purity procedures are product related variants or degradants, and orthogonal procedures are typically not available to provide a value for τ. Thus, the accuracy of the measurement as defined in this context cannot be independently confirmed. In cases where τ is not available, ICH Q2 (2005) states accuracy may be inferred once precision, linearity, and specificity have been established.

6.2.4 Precision

Precision of an analytical procedure is the degree of agreement among reportable values when the procedure is applied repeatedly to multiple samplings (possibly under different conditions) of a homogeneous test solution. The precision of an analytical procedure is quantified by the magnitude of the variance σ^2, or alternatively in terms of the standard deviation, σ. The standard deviation is the preferable measure of precision because it has the same measurement units as Y. The lesser the value of σ, the better the precision. Precision of a test procedure may be influenced by factors that vary during the normal use of the analytical procedure. These are called ruggedness factors, and include factors such as analyst, day, and instrument.

6.3 Stage 1: Procedure Development (Pre-validation)

In order to maximize the likelihood of a successful validation, it is imperative that all aspects of the procedure be well understood prior to the validation. Pre-validation work allows one to best design the experiment employed in the procedure validation. Martin et al. note that pre-validation experiments can be leveraged to support the validation and may reduce work in the validation itself. A lack of pre-validation work will often lead to a failed validation and costly rework.

The following series of questions provided by the USP Statistics Expert Team (2016) should be considered during pre-validation in order to ensure a successful validation experiment.

1. What are the allowable ranges for operational parameters such as temperature and time that impact the performance of the analytical procedure?

   - Robustness of these ranges can be determined using statistical design of experiments (DoE) as described in Chap. 3.

2. Are there ruggedness factors that impact precision?

   - Factors such as analyst, day, and instrument that vary in routine use and impact the precision of a test procedure are called ruggedness factors. When ruggedness factors impact precision, reportable values within the same ruggedness grouping (e.g., analyst) are correlated. Depending on the strength of the correlation, this may necessitate a statistical analysis that appropriately accounts for this dependence. Ruggedness factors can be identified empirically during pre-validation or based on a risk assessment. This topic is addressed in more detail in Sect. 6.4.10.

3. Are statistical assumptions for data analysis reasonably satisfied?

- These assumptions typically include normality, homogeneity of variance, and independence of reportable values. It is useful during pre-validation to employ statistical tests or visual representations to help answer these questions. USP <1010> provides information on this topic as does Sect. 2.12.2.

4. What is the required analytical range for the procedure?
5. Do accepted reference values or results from an established procedure exist for validation of accuracy?

- If not, ICH Q2 states accuracy may be inferred once precision, linearity, and specificity have been established.

6. How many individual readings will be averaged to form a reportable value?

- To answer this question, it is necessary to understand the contributors to the procedure variance and the procedure's ultimate purpose. Estimation of variance components during pre-validation provides useful information for making this decision. A good rule of thumb is to replicate against the source representing the largest component of variance.

7. What are appropriate validation acceptance criteria?

- We provide discussion on this topic throughout Sect. 6.4.

8. How large a validation experiment is necessary?

- Validation experiments should be properly powered to ensure there are sufficient data to conclude accuracy and precision can meet pre-specified acceptance criteria. Computer simulation is a useful tool for performing power calculations as discussed in Sect. 6.4.8.

Based on the answers to these and similar questions, a suitable validation experimental protocol may be designed.

6.4   Stage 2: Procedure Performance Validation (Qualification)

As noted in ICH Q2, the objective of validation of an analytical procedure is to demonstrate that it is suitable for its intended purpose. Suitability for intended purpose can be expressed in several ways. For instance, when a reportable value is used to disposition a product batch, suitability may be expressed in terms of decision error rates (e.g., passing an unacceptable batch, or failing an acceptable batch). In other cases, it may be sufficient to define suitability by placing limits on the quality metrics of the analytical procedure itself (e.g., maximum bias or precision). The life cycle approach suggests that these suitability metrics be documented in an analytical target profile (ATP) statement that guides quality decision making at all stages of the analytical procedure life cycle.

As discussed in the introduction, the term validation aligns with the process described in the USP document <1225>. In the life cycle approach described by Martin et al., this validation process is referred to as qualification.

The validation experiment is the culmination of all the investigational work needed to determine the operational details of the procedure. These details include selected inputs, operating conditions, equipment, limits, ranges, replication strategy, and other factors thought to potentially influence the outcome. The validation experiment is the final check that a newly developed procedure is fit for use.

Traditionally, validation of accuracy and precision provide the essential evidence that a procedure meets the requirements for the intended analytical application. Accordingly, we focus on these two topics in this section. Other factors that are typically characterized in a validation experiment are more descriptive in nature (e.g., range, detection, and quantitation limits), or more internal to the analytical procedure (e.g., linearity). For example, the impact of linearity is captured during DoE, repeatability, and intermediate precision studies because each experiment requires a calibration that includes the impact of the linearity. This is important because the decision rule and ATP provide an overarching criterion for the validation study, and require identification and quantification of all potential uncertainty components. For a full understanding of other characteristics, the reader should consult USP <1225>.

6.4.1 Experimental Design for Validation of Accuracy and Precision

A single experimental design will allow validation of both accuracy and precision. As will be discussed, individual assessment of accuracy and precision is not generally an effective approach. Such an approach is first described, and then better approaches that address bias and precision together are presented.

An example is provided to demonstrate the statistical analysis that follows. This example considers validation of a test procedure using high performance liquid chromatography (HPLC). The measured drug substance (DS) is a USP compendial substance, so information concerning τ is available. Three different quantities of reference standard were weighed to correspond to three different percentages of the test concentrations: 50, 100, and 150%. The unit of measurement for the reportable value is the mass fraction of DS expressed in units of mg/g and $\tau = 1000$ mg/g for all three concentrations. The DS product specification is from 980 to 1020 mg/g (see Weitzel 2012). Similar experiments are often established with levels expressed as a percent of the API label claim for the drug (as opposed to the weight of the entire tablet). Table 6.3 presents the $n = 12$ reportable values and the computed statistics.

Table 6.3 Example data set for procedure validation

Test concentration (%)	Test solution (plate or run)	Reportable value (mg/g)
50	1	1000.57
50	2	996.93
50	3	1002.4
50	4	994.91
100	5	994.16
100	6	992.72
100	7	1000.03
100	8	1004.89
150	9	1002.53
150	10	1004.83
150	11	998.17
150	12	994.15

To begin, the following two assumptions are made:

1. Each row in Table 6.3 is independent. In Sect. 6.4.10 and Sect. 6.4.11, the addition of ruggedness factors that invalidate this assumption is discussed. For example, consider the ruggedness factor "day." Suppose the experiment had been run over four days. Each day a reportable value was obtained from each of the three concentration levels: 50, 100, and 150. If there is variation in the procedure across days, then reportable values made on the same day are correlated and the assumption of independence is violated.
2. The standard deviation of the reportable value is constant across all three concentration levels. Discussion of how to proceed if this assumption is not met is provided in Sect. 6.4.9.

6.4.2 Confidence Intervals for Accuracy and Precision

The model in Eq. (6.2) is used to represent the data in Table 6.3 as

$$Y_{ij} = \tau_i + \beta_i + E_{ij}$$
$$i = 1, \ldots, c \text{ (concentration level)}; \ j = 1, \ldots r; \tag{6.3}$$

where Y_{ij} is the jth reportable value in the ith concentration level, τ_i is the known true value of the ith concentration level, β_i is the procedure bias in the ith concentration level, and E_{ij} is a random error specific to jth reportable value in the ith concentration level. The random error is assumed to have a normal distribution with mean 0 and variance σ^2. For the data in Table 6.3, $c = 3$, $r = 4$, and so the total sample size is $n = c \times r = 12$. We present results for computing confidence intervals on β_i and σ^2 that can be used for validation of accuracy and precision.

6.4.2.1 Case 1: Bias Is Constant Across Concentration Levels

In this case, it is assumed that $\beta_i = \beta$ across all c concentration levels. Note this does not require that τ_i be equal across concentration levels. Since there are an equal number of reportable values for each concentration level, the estimator for β is

$$\hat{\beta} = \frac{\sum_{i=1}^{c} (\bar{Y}_i - \tau_i)}{c}. \tag{6.4}$$

The bounds for a $100(1 - \alpha)\%$ two-sided confidence interval for β are

$$L = \hat{\beta} - t_{1-\alpha/2:n-1}\sqrt{\frac{S^2}{n}}$$

$$U = \hat{\beta} + t_{1-\alpha/2:n-1}\sqrt{\frac{S^2}{n}}$$

$$\bar{Y}_i = \frac{\sum_{j=1}^{r} Y_{ij}}{r} \tag{6.5}$$

$$\bar{Y} = \frac{\sum_{i=1}^{c}\sum_{j=1}^{r} Y_{ij}}{n}$$

$$S^2 = \frac{\sum_{i=1}^{c}\sum_{j=1}^{r} (Y_{ij} - \bar{Y})^2}{n-1}.$$

Validation for precision typically requires only an upper bound, since it is only problematic if the standard deviation is too large. A $100(1 - \alpha)\%$ upper bound on σ is

$$U = \sqrt{\frac{(n-1)S^2}{\chi^2_{\alpha:n-1}}}. \tag{6.6}$$

Note that (6.6) can be calculated with no knowledge of τ_i. Thus, although the true value is required to estimate accuracy, it is not needed to estimate precision.

For the data shown in Table 6.3, $\tau_i = \tau = 1000$ mg/g for each concentration level. The calculated statistics are $\bar{Y}_1 = 998.70$, $\bar{Y}_2 = 997.95$, $\bar{Y}_3 = 999.92$, $\bar{Y} = 998.86$, and $S^2 = 18.55$. Equation (6.5) is now simplified since all τ_i are equal and provides the 90% confidence interval on β

$$L = \bar{Y} - \tau - t_{1-\alpha/2:n-1}\sqrt{\frac{S^2}{n}}$$

$$L = 998.86 - 1000 - 1.796\sqrt{\frac{18.55}{12}} = -3.38 \text{ mg/g}$$

$$U = \bar{Y} - \tau + t_{1-\alpha/2:n-1}\sqrt{\frac{S^2}{n}}$$ (6.7)

$$U = 998.86 - 1000 + 1.796\sqrt{\frac{18.55}{12}} = 1.09 \text{ mg/g}$$

Equation (6.6) provides the upper 95% confidence bound on σ

$$U = \sqrt{\frac{(n-1)S^2}{\chi^2_{\alpha:n-1}}}$$

$$U = \sqrt{\frac{(12-1)18.55}{4.57}} = 6.68 \text{ mg/g}$$ (6.8)

6.4.2.2   Case 2: Bias Changes Across Concentration Levels

In this case, it is necessary to estimate the bias separately for each concentration level. However, since the standard deviation is assumed equal across all concentration levels, it is still possible to use all the data to estimate σ^2. This is referred to as "pooling." In order to pool the variance estimates, an analysis of variance table is constructed as shown in Table 6.4 (see Sect. 2.12.7 for more on the analysis of variance).

where

$$S_C^2 = \frac{r\sum_{i=1}^{c}(\bar{Y}_i - \bar{Y})^2}{c-1}$$

$$S_E^2 = \frac{\sum_{i=1}^{c}\sum_{j=1}^{r}(Y_{ij} - \bar{Y}_i)^2}{c(r-1)}.$$ (6.9)

Table 6.4 Analysis of variance

Source of variation	Degrees of freedom	Mean square
Concentration	$c-1$	S_C^2
Error	$c(r-1)$	S_E^2

Table 6.5 Analysis of variance for example data

Source of variation	Degrees of freedom	Mean square
Concentration	2	$S_C^2 = 3.953$
Error	9	$S_E^2 = 21.796$

Since the bias is different across concentration levels, a separate confidence interval is needed for each concentration level. The bounds for a $100(1 - \alpha)\%$ two-sided confidence interval for β_i are

$$L_i = \bar{Y}_i - \tau_i - t_{1-\alpha/2:c(r-1)}\sqrt{\frac{S_E^2}{r}}$$

$$U_i = \bar{Y}_i - \tau_i + t_{1-\alpha/2:c(r-1)}\sqrt{\frac{S_E^2}{r}} \tag{6.10}$$

where $\bar{Y}_i$ is defined in (6.5) and S_E^2 is defined in (6.9).

A $100(1 - \alpha)\%$ upper bound on σ is

$$U = \sqrt{\frac{c(r-1)S_E^2}{\chi^2_{\alpha:c(r-1)}}}. \tag{6.11}$$

The analysis of variance table for the data in Table 6.3 is shown in Table 6.5.

Using Eq. (6.10) a 90% two-sided confidence interval for the bias in the 100% concentration level is

$$L = \bar{Y}_i - \tau_i - t_{1-\alpha/2:c(r-1)}\sqrt{\frac{S_E^2}{r}}$$

$$L = 997.95 - 1000 - 1.833\sqrt{\frac{21.796}{4}} = -6.33 \text{ mg/g}$$

$$U = \bar{Y}_i - \tau_i + t_{1-\alpha/2:c(r-1)}\sqrt{\frac{S_E^2}{r}} \tag{6.12}$$

$$U = 997.95 - 1000 + 1.833\sqrt{\frac{21.796}{4}} = 2.23 \text{ mg/g}.$$

The confidence intervals for the 50 and 150% concentration levels are made in a similar fashion.

A 95% upper bound on the (pooled) precision from (6.11) is

$$U = \sqrt{\frac{c(r-1)S_E^2}{\chi^2_{\alpha:c(r-1)}}}$$

$$U = \sqrt{\frac{9(21.796)}{3.33}} = 7.68 \text{ mg/g}. \tag{6.13}$$

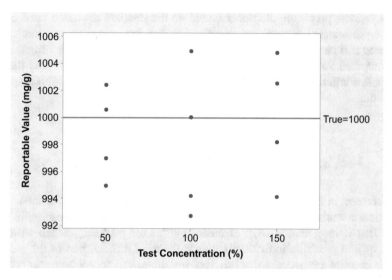

Fig. 6.1   Plot of data in Table 6.3

Note this upper bound is slightly greater than that computed in Case 1 because there are fewer degrees of freedom associated with the pooled estimate of σ (11 in Case 1 and 9 in Case 2). A plot of the data in Table 6.3 is shown in Fig. 6.1.

Note that the reportable values are centered at roughly the same value across levels of concentration. This suggests the bias is constant and that Case 1 is the more appropriate procedure. Also note that the spread appears constant across concentration. This is consistent with the second assumption noted in Sect. 6.4.1.

6.4.3   Using Confidence Intervals to Validate Accuracy and Precision

The confidence intervals provided in the previous section can be used to validate accuracy and precision individually. Because bias can be either positive or negative, it is customary to perform a statistical test of equivalence to validate accuracy. Tests of equivalence are discussed in Sect. 2.11. Assume Case 1 is appropriate and Eq. (6.5) is used to compute a 90% two-sided confidence interval on the bias. A pre-selected value of the equivalence acceptance criterion (EAC) to validate accuracy was selected to be 5 mg/g or 0.5% of the true value. (Section 6.4.4 discusses considerations in selecting an appropriate EAC.) Since the 90% confidence interval from $L = -3.38$ mg/g to $U = 1.09$ mg/g falls completely in the range from −5 to +5 mg/g, the statistical equivalence test is passed, and it can be claimed the procedure is validated for accuracy.

To validate precision, the upper bound on the standard deviation must be less than a pre-selected acceptance criterion. In our example, assume precision is validated if it can be shown that the standard deviation is less than 7 mg/g. Since the computed 95% upper bound in this example, U=6.68 mg/g, is less than the acceptance criterion of 7 mg/g, the procedure has been validated with respect to precision.

6.4.4   Validation Criteria for Accuracy and Precision

As discussed in Sect. 2.11, acceptance criterion should ideally be defined by the analytical scientist and not the statistician. The criterion must be meaningful in the sense that it must define what is meant by "fit for purpose." When validating accuracy and precision individually, this is the most difficult part of the analysis. This is because a procedure that has very small bias can accept a greater standard deviation than a procedure with a large bias. Similarly, a procedure with a relatively small standard deviation can accept a relatively large bias. For this reason, the criteria for these two attributes are linked, making it difficult to get a good assessment of individual criteria. Many companies, as well as industry standards organizations, have default limits that are used for all validations. These may be based on industry benchmarking, but it is arguable whether such an approach truly demonstrates "fit for purpose." For this reason, and to account for the relationship between accuracy and precision as it relates to overall performance, we recommend two other approaches for validation of accuracy and precision. We present these approaches in the next two sections.

6.4.5   Validation of Accuracy and Precision Using Statistical Intervals

Hubert et al. (2004, 2007a, b) proposed validation of both accuracy and precision simultaneously rather than individually as described in the previous section. The reasoning is to take advantage of the natural tradeoffs between these two characteristics. For example, a procedure with a relatively small standard deviation can accept a greater bias than a procedure with a larger standard deviation. Because the intended purpose of an analytical procedure is to provide accurate and precise measurements, one may consider that the procedure is validated if it is shown to provide a high degree of assurance that future measured values will be close to their true values. A criterion that can be used to simultaneously validate accuracy and precision seeks to ensure

$$\Pr(-\lambda < Y - \tau < \lambda) \geq P, \text{ or}$$
$$\Pr(-\lambda + \tau < Y < \lambda + \tau) \geq P \tag{6.14}$$

where $\lambda > 0$ is an acceptable limit defined a priori to be consistent with the purpose of the procedure. The term P is a desired probability value (e.g., $P = 0.90$).

For example, a desired goal for a procedure that reports concentration in mg/mL may be written in the following manner: "The procedure must ensure that at least 90% of the time, measurement error (i.e., the difference between the reported value and the true value) is no greater than 0.5 mg/mL." In terms of Eq. (6.14), this means $\lambda = 0.5$ mg/mL and $P = 0.90$.

Equation (6.14) can be interpreted as either (1) the probability that *the next* reportable value falls in the range from $-\lambda + \tau$ to $\lambda + \tau$ is greater than or equal to P, or (2) the proportion of *all* future reportable values falling between $-\lambda + \tau$ and $\lambda + \tau$ is greater than or equal to P. Accordingly, two statistical intervals have been proposed for validating Eq. (6.14).

1. A prediction interval of reportable values (also referred to as an expectation tolerance interval) and
2. A tolerance interval of reportable values (also referred to as a content tolerance interval).

The prediction interval validates that (6.14) is true for the *next* reportable value, whereas the tolerance interval validates that (6.14) is true for *all future* reportable values with a specific level of confidence. Since the inference associated with the tolerance interval concerns a larger set of values, the tolerance interval is always wider than the prediction interval.

Both intervals can be used in the following manner to validate accuracy and precision simultaneously:

1. Compute the appropriate statistical interval using Eq. (2.21) for the prediction interval and Eq. (2.23) for the tolerance interval.
2. Compute a $100P\%$ prediction interval or a 90% tolerance interval that contains $100P\%$ of the population. A 90% confidence level for the tolerance interval will provide a statistical test with a type I error rate (probability of rejecting the null hypothesis when it is true) of approximately 5%.
3. If the computed interval falls completely in the range from $-\lambda + \tau$ to $\lambda + \tau$, criterion (6.14) is satisfied and the procedure is validated for both accuracy and precision.

When computed by classical statistical methods (as we do below in this section) the interpretation of these intervals is as follows. When the interval estimation methodology is applied repeatedly to many (i.e., an infinite number) of hypothetical future data sets of size n from possibly many different populations, each prediction interval obtained has a $100P\%$ probability of containing the hypothetical next reportable value. Similarly, each $100(1 - \alpha)\%$ tolerance interval has a $100(1 - \alpha)\%$ probability of containing at least $100P\%$ of hypothetical future reportable values.

Our inference about the truth (or not) of relationship (6.14) is thus based on the properties of the statistical procedure. Whether (6.14) is true for any particular sampled population is unknown because it depends on parameters whose values are unknown. In Sect. 6.4.7, Bayesian interval estimation is introduced as a more direct alternative to validation using (6.14).

Huber et al. recommend the testing strategy based on the prediction interval. Yang and Zhang (2015) recommend the tolerance interval. The tolerance interval is the appropriate choice if one desires a statistical test in which the type I error rate is controlled. The tolerance interval is therefore more consistent with the approach described in Sects. 6.4.2 and 6.4.3.

To demonstrate, consider Case 1 (bias constant) and analyze the data in Table 6.3. Suppose (6.14) is defined so that $\lambda = 0.015 \times \tau = 15$ mg/g and $P = 0.90$. Thus, we seek to validate the claim

$$
\begin{aligned}
&\Pr(-\lambda + \tau < Y < \lambda + \tau) \geq P \\
&\Pr(-15 + 1000 < Y < 15 + 1000) \geq 0.90 \\
&\Pr(985 < Y < 1015) \geq 0.90.
\end{aligned}
\tag{6.15}
$$

From Eq. (2.21) the 90% prediction interval is computed as

$$
\begin{aligned}
&L = \bar{Y} - t_{(1+P)/2:n-1} \sqrt{\left(1 + \frac{1}{n}\right) \times S^2} \\
&L = 998.86 - 1.796 \sqrt{\left(1 + \frac{1}{12}\right) \times 18.55} = 990.8 \text{ mg/g} \\
&U = \bar{Y} + t_{(1+P)/2:n-1} \sqrt{\left(1 + \frac{1}{n}\right) \times S^2} \\
&U = 998.86 + 1.796 \sqrt{\left(1 + \frac{1}{12}\right) \times 18.55} = 1006.9 \text{ mg/g}.
\end{aligned}
\tag{6.16}
$$

From Eq. (2.23) using an exact K value of 2.414, the 90% tolerance interval that includes 90% of the future population of reportable values is

$$
\begin{aligned}
&L = \bar{Y} - K\sqrt{S^2} \\
&L = 998.86 - 2.414\sqrt{18.55} = 988.5 \text{ mg/g} \\
&U = \bar{Y} + K\sqrt{S^2} \\
&U = 998.86 + 2.414\sqrt{18.55} = 1009.3 \text{ mg/g}.
\end{aligned}
\tag{6.17}
$$

Since both intervals (6.16) and (6.17) fall within the range from 985 to 1015 defined in (6.15), both intervals validate the procedure. As described, the tolerance interval is wider than the prediction interval, since it makes an inference to a larger set of values. It is also true that the tolerance interval provides a statistical test with a type I error rate near 5%.

One final comment concerning application of this approach. When validated individually, each test has a type I error rate of 5% and the combined error rate can be as high as 10%. Thus, it is not unreasonable to apply a 10% type I error rate with the simultaneous methods described in this section. This means one could use an 80% confidence level for the two-sided tolerance interval. In the present application, the 80% tolerance interval that contains 90% of all future reportable values is from 989.6 to 1008.1 mg/g. This compares to the previously computed 90% tolerance interval of 988.5 mg/g to 1009.3 mg/ml.

6.4.6   Validation of Accuracy and Precision Based on Out-of-Specification Rates

A typical application for an analytical procedure is lot (batch) release. After a lot is manufactured, a reportable value of the product quality is obtained using the analytical procedure. If the reportable value falls within the lower specification limit (LSL) and upper specification limit (USL), it is deemed as satisfying the quality requirement. However, if it falls outside of this range, action must be taken to determine the lot disposition. Thus, an obvious criterion for procedure validation is the probability that a reported value is out-of-specification (OOS). If the process is operating as designed, then a reported OOS alarm in most cases is "false," and can lead to unnecessary time and expense in further examination of the lot. The probability statement in (6.14) can be adapted to consideration of the OOS rate by defining $-\lambda + \tau = LSL$ and $\lambda + \tau = USL$ where LSL and USL are the process lower and upper specifications, respectively, and it is assumed the process is symmetric about τ (i.e., $(LSL + USL)/2 = \tau$). Thus, (6.14) is rewritten as

$$\Pr(-\lambda + \tau < Y < \lambda + \tau) \geq P$$
$$\Pr(LSL < Y < USL) \geq P \qquad (6.18)$$
$$\pi \leq 1 - P$$

where $\pi = 1 - \Pr(LSL < Y < USL)$ is the probability of an OOS signal. A 95% upper bound can be constructed on π using the process described in Sect. 2.6.5. If the upper bound is less than $1-P$, then (6.18) is satisfied and the analytical procedure is validated.

To demonstrate for the present Case 1 example, $LSL = 980$ mg/g, $USL = 1020$ mg/g, $\bar{Y} = 998.86$, $S^2 = 18.55$, and $P = 0.90$. Following the instructions from Eqs. (2.17) to (2.19) with $\alpha = 0.10$ in (2.19), we compute $K_{LSL} = 4.38$ and $K_{USL} = 4.91$. Since both of these values exceed $(n - 1)/\sqrt{n} = (12 - 1)/\sqrt{12} = 3.175$, $K^* = \min(4.38, 4.91) = 4.38$, $\lambda_U = 9.51$, and $U = 0.003$. Since this upper bound is less than $1 - P = 1-0.90 = 0.10$, the procedure is validated against the OOS criterion.

There is an interesting relationship between the upper bound on π and the tolerance interval used to validate (6.14) when $-\lambda + \tau = LSL$ and $\lambda + \tau = USL$. To demonstrate, consider a situation where there is only an upper specification, USL. Let U_1 represent the 95% upper tolerance bound that exceeds $100P\%$ of the population computed with (2.28) and the exact value of K_1. Let U_2 represent the upper 95% confidence bound computed using K_{USL} in Eq. (2.14). When $U_1 = USL$, it must be true that $U_2 = 1 - P$. Thus, the two rules for validation are exactly the same, and have a type I error rate of 0.05. Although the situation with two-sided specifications involves approximations, the two approaches will also generally provide the same result, and the type I error rate is very close to 0.05.

One final adjustment is required for application of this approach. To this point, only the measurement error has been quantified. However, if specifications are used to define "fitness for purpose," it is necessary to also account for process variation. To do this, let σ_P^2 represent the variance of the manufacturing process and σ^2 the variance of the reportable value. (The statistic S^2 is an estimator for σ^2). Now define

$$\rho = \frac{\sigma_P^2}{\sigma_P^2 + \sigma^2}. \tag{6.19}$$

which represents the proportion of the total variance in the reportable value due to the process. An empirical estimate of σ_P^2 might be available from process data, but if not, a subject manner expert can generally provide a reasoned guess for ρ. For example, if the procedure is a bioassay, it is expected that variance due to the analytical procedure is greater than the process variance, and so a value of $\rho = 0.2$ might be reasonable. In contrast, a procedure with relatively little measurement error might employ $\rho = 0.8$.

With a known or well-informed value for ρ, the total of process and measurement variance is written as

$$\sigma_P^2 + \sigma^2 = \sigma^2 \left[\frac{\sigma_P^2}{\sigma^2} + 1 \right] = \sigma^2 \left[\frac{\rho}{1 - \rho} + 1 \right] = \sigma^2 \left[\frac{1}{1 - \rho} \right]. \tag{6.20}$$

The confidence interval for π is now computed as before, but with S^2 replaced with $S^{2*} = S^2 \left[\frac{1}{1 - \rho} \right]$.

In the present example, assume that $\rho = 0.5$. Performing the previous calculations with $S^{2*} = S^2 \left[\frac{1}{1 - \rho} \right] = 18.55 \left[\frac{1}{1 - 0.5} \right] = 37.1$, we obtain $K_{LSL} = 3.096$ and $K_{USL} = 3.471$, $K^* = \min(3.096, 3.471) = 3.096$, $\lambda_U = 6.545$, and $U = 0.029$. Since this upper bound is less than $1 - P = 1 - .90 = 0.10$, the procedure is validated against the OOS criterion.

Burdick et al. (2005) provide an approach for estimating false failure rates and missed fault rates. As noted earlier, the false failure rate is generally very close to the observed OOS rate. However, it is also of interest to know if the missed fault rate is acceptably low. By defining criteria based on misclassification rates, a procedure can be validated using upper confidence bounds on these rates.

Other approaches for establishment of criteria are provided by Chatfield and Borman (2009).

6.4.7   A Bayesian Approach

It is also possible to estimate $\Pr(-\lambda < Y - \tau < \lambda)$ directly using a Bayesian approach (see Sect. 2.13 for a discussion of Bayesian statistics). The validation criterion is thus satisfied if this estimated probability exceeds P. A Bayesian tolerance interval is provided in Wolfinger (1998) and can be computed using the statistical software package WinBUGS (Ntzoufras 2009 or Spiegelhalter et al. 1996).

The WinBUGS code required for Case 2 is shown below:

```
# data
Level[]    Y[]
1      1000.57
1      996.93
1      1002.4
1      994.91
2      994.16
2      992.72
2      1000.03
2      1004.89
3      1002.53
3      1004.83
3      998.17
3      994.15
END
# more data
list(n=12, c=3, tau=1000)
model{
   # Priors
   for(i in 1:c){ beta[i] ~ dnorm(0,0.000001) }
   sigma ~ dunif(0, 100)
   precision <- pow(sigma,-2)
   # Likelihood
```

(continued)

```
  for(obs in 1:n){
    Dif[obs] <- Y[obs] - tau
    Dif[obs] ~ dnorm (beta[ Level[obs] ],precision)
  }
}
```

Boxplots that compare the posterior distributions for beta[i] are shown in Fig. 6.2. The distributions are labeled with the concentration level index (i.e., [1] = 50, [2] = 100, [3] = 150). These boxplots are created by WinBUGS and are different than those described in Sect. 2.4. Boxes represent inter-quartile ranges and the solid black line at the (approximate) center of each box is the mean. The arms of each box extend to cover the central 95% of the distribution. The horizontal line behind the boxes is the overall mean of the posterior means.

The two-sided 90% credible interval for beta2 is -5.885 to $+3.307$ which may be compared to the classical frequentist two-sided 90% confidence interval computed in (6.12) of -6.33 to $+2.23$. The upper 95% credible bound for sigma is 8.49 which may be compared to the classical frequentist upper 95% confidence bound of 7.68 shown in (6.13). Figure 6.2 supports the conclusions from Fig. 6.1 that Case 1 (constant bias) provides a more appropriate model for these data.

The WinBUGS code required for Case 1 is shown below:

Fig. 6.2 Comparison of posterior distributions

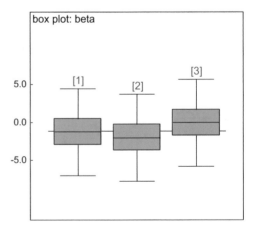

```
# data
Y[]
1000.57
...
END
# more data
list(n=12, tau=1000)
# Let WinBUGS pick initials
model{
  # Priors
  beta ~ dnorm(0,0.000001)
  sigma ~ dunif(0, 100)
  precision <- pow(sigma,-2)
  # Likelihood
  for(obs in 1:n){
    Dif[obs] <- Y[obs] - tau
    Dif[obs] ~ dnorm (beta,precision)
  }
}
```

The posterior sample obtained from executing this code was imported into R for further analysis. This sample consists of 300,000 pairs of values for β and σ drawn from their joint posterior. A two-sided 90% central credible interval for β can be obtained using the R function

```
quantile(beta,c(0.05,0.95))
```

The interval is -3.50 to $+1.22$ mg/g which is comparable to the corresponding classical frequentist interval of -3.38 to $+1.09$ that was computed in (6.7).

A one-sided upper 95 credible bound for sigma can be obtained using the following R function:

```
quantile(sigma,c(0.95))
```

The resulting upper bound is 7.204 mg/g which is comparable to the classical frequentist upper bound of 6.68 mg/g computed in (6.8).

It is of interest to estimate the posterior probability that 90% of future values will be within $\tau \pm \lambda$ where $\tau = 1000$ and $\lambda = 15$. This can be obtained using the following two lines of R code:

```
Content <- pnorm(tau + lambda, tau + beta, sigma) - pnorm(tau
- lambda, tau + beta, sigma)
mean(Content >= 0.9)
```

The result is 0.989. Since this value is greater than $0.90 = 90\%$, the method is validated. The frequentist approach computes an upper bound on $\Pr(-\lambda < Y - \tau < \lambda)$ using results from Sect. 2.6.5. An example of this approach was given in (6.18) where $\lambda = 20$.

A sample of 100,000 draws from the posterior predictive distribution of future reportable results may be obtained using the following random normal R function:

```
Y.fut <- rnorm(100000,beta+tau,sigma)
```

The central two-sided 90% credible posterior predictive interval of future values may be obtained using the R function

```
quantile(Y.fut,c(0.05,0.95))
```

The resulting interval is from 990.3 to 1007.5 which may be compared to the classical prediction interval of 990.8–1006.9 computed in (6.16).

It is of interest to estimate the posterior predictive probability that future values will be within $\tau \pm \lambda$. This can be obtained using the following two lines of R code:

```
Y.fut.in <- (tau - lambda <= Y.fut) & (Y.fut <= tau + lambda)
mean(Y.fut.in)
```

The resulting probability is 0.989. There is no comparable classical frequentist estimate available.

It is of interest to estimate the posterior predictive probability that future values will be OOS (outside 980 to 1020). The following two lines of R code will provide this estimate:

```
OOS <- (Y.fut <= 980) & (1020 >= 1020)
mean(OOS)
```

The resulting probability is 0.00137 which can be compared to the result obtained previously using classical frequentist approaches of 0.003 shown in Sect. 6.4.6.

A Bayesian content tolerance interval to contain 90% of future values with 90% credibility can be obtained using Algorithm 11.2 of Krishnamoorthy and Mathew (2009). The R code is a bit more involved because it requires an iterative search. The implementation of this algorithm in R is given below.

```
P<-0.90; C<-0.90
L<- tau + beta + qnorm((1-P)/2,0,sigma)
U<- tau + beta + qnorm((1+P)/2,0,sigma)
ndraws<-length(L)
mu.bar <- mean(c(L,U))
intervals<-1000 # 10000 takes too long
mu.2.range <- seq(min(U),max(U), ( max(U)-min(U) )/intervals
)
PP <- matrix (rep(NA,3*length(mu.2.range)),ncol=3)
i <-0
for(mu.2 in mu.2.range){
  i<-i+1
  mu.1 <- -mu.2 + 2*mu.bar
     PP[i,]<-c(mu.1,mu.2,mean((U<=rep(mu.2,ndraws))&(L>=rep
(mu.1,ndraws))))
}
# select values in PP for which the proportion covered is
near C
# may need to adjust the factors of C to select a narrow range
of values close to C
PP.out<- PP[ (PP[,3]<=1.02*C) & (PP[,3]>=0.98*C),]
TI <- c(mean(PP.out[,1]),mean(PP.out[,2]))
TI
```

The resulting Bayesian content tolerance interval is 986.5 to 1011.3 which is comparable to the 988.5–1009.3 as computed in (6.17).

While the Bayesian results are quite close to those obtained using classical statistical methods, the interpretation of these intervals and probabilities are different from the frequentist interpretations. The Bayesian results were obtained using relatively uninformative prior distributions for β and σ. Had more informative distributions been available, the estimates could be much different, and arguably more informative. The other difference is that the Bayesian methodology replaces analytical solutions (some of which are necessarily approximate) with computer algorithms for which there are no approximations. However, these computer algorithms have their own challenges related to Markov Chain Monte-Carlo (MCMC) convergence verification and the requirement for large MCMC samples to minimize simulation error. An additional advantage of the Bayesian approach is that it is easily extended to more complex models for which frequentist analytical approaches are intractable.

6.4.8 Power Considerations

As noted in Sect. 6.3, it is important to conduct pre-validation work to gain understanding of the procedure. Part of this work should include a power analysis to determine the probability of passing validation under given scenarios. Computer simulation is extremely useful for this purpose. Statistical power is defined as the probability that one meets the acceptance criterion given a true value of the parameter of interest. To demonstrate, a simulation program was written to determine the power of a validation test based on the requirement that the probability of an OOS (π) is less than $1 - P = 0.10$ with a type I error rate of 5%. The specifications are $LSL = 980$ and $USL = 1020$. The test is to be conducted as described in Sect. 6.4.6. The simulation was conducted with 100,000 iterations. Table 6.6 presents the results for validation designs with $n = 6$, $n = 12$, and $n = 20$ using a 90% confidence coefficient (type I error rate of 0.05). Table 6.7 reports the same design with an 80% confidence coefficient (type I error rate of 0.10).

Note that the power in the last row where $\pi = 0.10$ is the estimated type I error rate. Since these values are all less than the desired rate (0.05 in Table 6.6 and 0.10 in Table 6.7), this provides an additional argument for applying an 80% confidence coefficient on the confidence interval for π. The simulation results also show that with the typical sample sizes used in a validation study, the criterion ($1 - P = 0.10$ in this case) must be much greater than the true value of π in order to provide a reasonable chance of passing the validation.

Table 6.6 Power for several designs with 90% confidence

True value of π	$n = 6$	$n = 12$	$n = 20$
0.001	0.566	0.875	0.997
0.005	0.375	0.644	0.941
0.010	0.270	0.455	0.808
0.10	0.041	0.028	0.039

Table 6.7 Power for several designs with 80% confidence

True value of π	$n = 6$	$n = 12$	$n = 20$
0.001	0.775	0.968	0.999
0.005	0.586	0.842	0.983
0.010	0.455	0.684	0.919
0.10	0.087	0.068	0.086

6.4.9   Violation of Homogeneity Across Concentration Levels

Procedures based on chemical or biological principles will sometimes demonstrate different variances as concentrations vary. This violates one of the assumptions made during discussions to this point. In such a situation, it is sometimes possible to transform the data so that the standard deviations can be assumed equal across the concentration range. The analyses that combine the data from all concentration levels as described above can then be performed using the transformed data, with appropriately transformed validation criteria.

It is extremely important that pre-validation work be used to determine necessary transformations that will allow the pooling of data across concentration levels. Failure to do so could lead to either unnecessary experimentation, or an underpowered validation experiment. Section 4.3 of USP chapter <1032> presents an excellent review of this topic. The normality transformations described in Sect. 2.6.10 also often stabilize the variance.

6.4.10   Experimental Designs to Incorporate Ruggedness Factors

In order to validate a procedure across the total environment in which it is expected to operate, it is sometimes necessary to manipulate ruggedness factors in the experimental design. Examples of ruggedness factors include analysts, equipment, and days. Table 6.8 reports the same data shown in Table 6.3, but with information concerning the analyst that performed the preparation work for the assay.

We again assume that the reportable value has a constant variance across all three concentration levels, and that bias is constant (Case 1) so that we may

Table 6.8 Example data set with ruggedness factors

Test concentration (%)	Analyst	Test solution (plate or run)	Reportable value (mg/g)
50	1	1	1000.57
50	1	2	996.93
50	2	3	1002.4
50	2	4	994.91
100	3	5	994.16
100	3	6	992.72
100	4	7	1000.03
100	4	8	1004.89
150	5	9	1002.53
150	5	10	1004.83
150	6	11	998.17
150	6	12	994.15

combine all concentration levels into a single data set. Unlike the previous analysis with Table 6.3, the 12 rows in Table 6.8 are not independent unless analysts do not impact the reportable value. For example, rows 1 and 2 were both prepped by analyst 1. If analyst impacts the reportable value, the values in rows 1 and 2 are more similar with each other than with the values in other rows of the table. In such a situation, we state the responses made by the same analyst are correlated. We described such a situation previously in Sects. 2.7 and 2.12.7.

In the present example, represent the reportable value with the statistical model

$$Y_{ij} = \tau + A_i + E_{ij}$$
$$i = 1, \ldots, a \text{ (analyst)}; \ j = 1, \ldots r; \tag{6.21}$$

where Y_{ij} is the reportable value for the jth replicate of the ith analyst. The number of analysts in this example is $a = 6$, and each analysts performs $r = 2$ independent repetitions. The random error A_i represents between analyst variability. It is assumed to have a mean of zero and a variance σ_A^2. The random error E_{ij} is the within analyst variability which has an assumed mean of zero and variance σ_E^2. The parameter τ represents the true value (1000 mg/g for all concentration levels in our example).

The ANOVA table for model (6.21) is shown in Table 6.9. Formulas for S_A^2 and S_E^2 are provided in Table 2.16. The numerical values for the data in Table 6.8 are shown in Table 6.10.

The model shown in (6.21) assumes that analyst is a random effect. That is, a sample of six analysts used in the experiment is viewed as a random sample from a population of analysts that will perform the procedure in the future. In some situations, it may be reasonable to treat analyst as a fixed effect. This case is considered in the next section.

The total variance associated with the procedure is the sum of the variance components, $\sigma_{Total}^2 = \sigma_A^2 + \sigma_E^2 = (\theta_1 - \theta_2)/r + \theta_2 = \theta_1/r + (1 - 1/r)\theta_2$. This sum is called the intermediate precision. The estimator for the intermediate precision is

$$S_{Total}^2 = \frac{S_A^2}{r} + \frac{(r-1)S_E^2}{r}. \tag{6.22}$$

Table 6.9 ANOVA for model (6.21)

Source of variation	Degrees of freedom	Mean square	Expected mean square
Between analysts	$n_1 = a - 1$	S_A^2	$\theta_1 = \sigma_E^2 + r\sigma_A^2$
Within analysts	$n_2 = a(r - 1)$	S_E^2	$\theta_2 = \sigma_E^2$

Table 6.10 ANOVA, for example

Source of variation	Degrees of freedom	Mean square
Between analysts	$n_1 = 5$	$S_A^2 = 29.165$
Within analysts	$n_2 = 6$	$S_E^2 = 9.708$

Using the Satterthwaite approximation given in (2.120), the degrees of freedom associated with this estimator is

$$m = \frac{S_{Total}^4}{\dfrac{S_A^4}{r^2 \times n_1} + \dfrac{(r-1)^2 \times S_E^4}{r^2 \times n_2}} \tag{6.23}$$

Using the data in Table 6.10,

$$S_{Total}^2 = \frac{29.165}{2} + \left(1 - \frac{1}{2}\right)9.708 = 19.437 \tag{6.24}$$

and

$$m = \frac{(19.437)^2}{\dfrac{(29.165)^2}{2^2 \times 5} + \dfrac{(2-1)^2 \times (9.708)^2}{2^2 \times 6}} = 8.13 = 8 \text{ (rounded)} \tag{6.25}$$

Notice that $S_{Total}^2 = 19.437$ is greater than the estimate of the measurement variability obtained when assuming all 12 rows of the data table are independent $(S^2 = 18.55)$. This demonstrates the problem of not properly modeling the ruggedness effects to account for correlation. Namely, one will underestimate the true procedure variance, and possibly validate a procedure that is not truly fit for purpose.

The prediction and tolerance intervals used for validation in the previous section can be easily modified for this correlated condition. In particular, simply replace S^2 with S_{Total}^2 and the error degrees of freedom $(n-1)$ with m (rounded to the nearest integer). The formula for $\bar{Y}$ remains unchanged. Thus, the 90% prediction interval is computed as

$$L = \bar{Y} - t_{(1+P)/2:m}\sqrt{\left(1 + \frac{1}{a \times r}\right) \times S_{Total}^2}$$

$$L = 998.86 - 1.86\sqrt{\left(1 + \frac{1}{12}\right) \times 19.437} = 990.3 \text{ mg/g}$$

$$\tag{6.26}$$

$$U = \bar{Y} + t_{(1+P)/2:m}\sqrt{\left(1 + \frac{1}{a \times r}\right) \times S_{Total}^2}$$

$$U = 998.86 + 1.86\sqrt{\left(1 + \frac{1}{12}\right) \times 19.437} = 1007.4 \text{ mg/g}$$

The 90% tolerance interval that contains 90% of the population is computed as

$$L = \bar{Y} - K\sqrt{S_{Total}^2}$$

$$L = 998.86 - 2.59\sqrt{19.437} = 987.4 \text{ mg/g}$$

$$U = \bar{Y} + K\sqrt{S_{Total}^2}$$ (6.27)

$$U = 998.86 + 2.59\sqrt{19.437} = 1010.3 \text{ mg/g}$$

where

$$K = \sqrt{\frac{\left(1 + \dfrac{1}{a \times r}\right)Z_{(1+P)/2}^2 \times m}{\chi_{0.1:m}^2}}$$ (6.28)

$$K = \sqrt{\frac{\left(1 + \dfrac{1}{6 \times 2}\right)(1.64)^2 \times 8}{3.49}} = 2.59.$$

Note that both of the computed intervals are wider than their counterparts computed earlier (990.8 to 1006.9 for the prediction interval and 988.5 to 1009.3 for the tolerance interval). This difference occurs for two reasons:

1. S_{Total}^2 is generally greater than S^2 and
2. The error degrees of freedom, m, is generally less than $n-1$.

Thus, incorporation of ruggedness effects requires more experimental runs to obtain the same power as a completely independent design. If during pre-validation work a ruggedness factor has been discovered to not impact the intermediate precision, do not include it in the analysis. This will needlessly decrease statistical power.

The intervals in (6.26) and (6.27) can be recommended for validation as described in this chapter. The same substitutions can be applied to the formulas described in Sect. 6.4.6 to estimate OOS.

To finish the example, we now account for the process variation assuming $\rho = 0.5$ and compute the tolerance interval using $S_{Total}^{2*} = S_{Total}^2/(1 - \rho) = 38.873$. The resulting tolerance interval from 986.1 to 1011.6 falls within the range from $LSL = 980$ to $USL = 1020$, and the procedure is validated as fit for purpose.

We have considered the case where there is only a single ruggedness factor. If more ruggedness factors are included, more power is lost for a fixed number of experimental runs due to additional partitioning of σ_{Total}^2. Again, one is reminded to not employ ruggedness factors unless they have a demonstrable impact on the intermediate precision.

If a ruggedness factor can be more properly considered as a fixed effect rather than a random effect, power will not be as dramatically impacted. This topic is discussed in the next section.

6.4.11 Incorporating Fixed Effect Ruggedness Factors

In some situations, ruggedness factors are more properly treated as fixed effects. For example, suppose that a major contributor to the intermediate precision of an analytical procedure is the instrument used in the procedure. Suppose there are four instruments in the laboratory, and these will be the only instruments used to perform the procedure in the foreseeable future. Since these are the only four instruments that will be used when performing the procedure, instrument is a fixed effect. Even though it is a fixed effect, differences among the instruments will contribute to the total variation, since only one instrument will be selected for a given application. Thus, it is still necessary to account for this component of variance in the intermediate precision.

As another example, Schwenke and O'Connor (2008) argue that in many cases, analysts can be considered a fixed effect. They argue an analyst is a trained professional proficient on the procedure through a lab-sponsored training program. As such, they are viewed as fixed effects since the training program has made them interchangeable. In many labs, only a small set of analysts perform a given procedure. If they are all used in the validation process, then assuming analyst to be fixed effect is a reasonable assumption.

Analysis of the data in Table 6.8 is now performed assuming analyst to be a fixed effect. The statistical model used to describe the fixed design is

$$Y_{ij} = \tau + \alpha_i + E_{ij}$$
$$i = 1, \ldots, a \text{ (analyst)}; \ j = 1, \ldots r;$$

(6.29)

where Y_{ij} is the reportable value for the jth replicate of the ith analyst. The term α_i is a fixed unknown constant that represents the ith analyst and replaces the random variable A_i shown in the random model (6.21). The variance of the a values of α_i is defined as

$$\sigma_\alpha^2 = \frac{\sum_{i=1}^{a} \alpha_i^2}{a - 1}.$$

(6.30)

The random error E_{ij} has an assumed mean of zero and variance σ_E^2. The parameter τ represents the true value (1000 mg/g for all concentration levels in our example). The total variance associated with the procedure is the sum of the variance components, $\sigma_{Total}^2 = \sigma_\alpha^2 + \sigma_E^2$.

Using an approximation described in Dolezal et al. (1998), all the formulas described in the previous section can be used with a fixed effect by simply replacing n_1 with n_1^* where

$$n_1^* = \frac{[n_1 + 2\lambda]^2}{n_1 + 4\lambda}$$

$$\lambda = \frac{n_1}{2}\left[\frac{S_A^2}{S_E^2}\left(\frac{n_2 - 2}{n_2}\right) - 1\right].$$

(6.31)

In the present example,

$$\lambda = \frac{5}{2}\left[\frac{29.165}{9.708}\left(\frac{6 - 2}{6}\right) - 1\right] = 2.507$$

$$n_1^* = \frac{[5 + 2 \times 2.507]^2}{5 + 4 \times 2.507} = 6.673$$

(6.32)

$$m = \frac{(19.437)^2}{\dfrac{(29.165)^2}{2^2 \times 6.673} + \dfrac{(2 - 1)^2 \times (9.708)^2}{2^2 \times 6}} = 10.55 = 11 \text{ (rounded)}.$$

Thus, the degrees of freedom used in the prediction and tolerance intervals has increased from $m = 8$ to $m = 11$, and length of the intervals will be properly reduced. In this problem, the 90% tolerance interval that contains 90% of the population computed in (6.27) assuming random analysts is from 987.4 to 1010.3 mg/g. If analysts are treated as fixed, then the interval that results replacing $m = 8$ with $m^* = 11$ is from 988.3 to 1009.5 mg/g. Although the difference is relatively modest in this example, this adjustment will have a major impact on results when $a = 2$ or 3.

This example demonstrates the importance of identifying whether ruggedness factors are fixed or random in the validation experiment. If the validation experiment includes all levels of a ruggedness factor that will be employed in the future, then properly treating it as a fixed effect will increase the likelihood of a successful validation.

6.5 Stage 3: Procedure Performance Verification and Analytical Procedure Transfer

Once the validation is done, it is important to continually monitor the performance of the analytical procedure. A useful statistical tool for this purpose is a control chart of measurements made with the reference standard (refer to Chap. 5 for information on control charts). It is also good practice to perform a system suitability test before every application. USP, ICH, and FDA all provide recommendations as to the need for system suitability tests. Procedures used for this purpose will vary by the procedure and the company.

The purpose of an analytical procedure transfer is to ensure that the receiving laboratory can perform an analytical procedure with the same ability as the transferring laboratory.

6.5.1 Objectives and Regulatory Guidance for Transfers

Some guidance for procedure transfers is provided in General USP Chapter <1224>. The purpose of <1224> is to summarize the types of transfers that may occur, including the possibility of waiver of any transfer, and to outline the potential components of a transfer protocol. However, the chapter does not provide any statistical methods.

A procedure transfer study requires a preapproved transfer protocol that includes details pertaining to the procedure, the sample types being tested, and predetermined acceptance criteria. The acceptance criteria often consider both bias and variability. The acceptance criteria must be satisfied in order to successfully demonstrate the receiving lab is qualified to perform the analytical procedure.

6.5.2 Experimental Designs for Transfers

USP <1224> refers to three types of studies employed in procedure transfer:

1. Comparative testing,
2. Covalidation, and
3. Re-validation.

As described in <1224>, comparative testing requires the analysis of a predetermined number of samples of the same lot by both the transferring and the receiving labs. (More than one lot can be employed if the measurements of the two labs are properly matched.) Covalidation occurs when more than one lab is involved in the initial procedure validation. Re-validation occurs when the receiving lab performs its own independent validation of the procedure.

Statistical designs are not provided in USP <1224>, and so the procedure transfer design is typically determined by the individual company. Consider the summary data in Table 6.11 where each lab is assigned $n = 10$ independent samples from the same lot of material. The response variable is the amount of active ingredient measured in mg.

Table 6.11 Summary of procedure transfer

Parameters	Point estimator	Computed estimate
μ_1—Mean of transferring lab	$\bar{Y}_1$	247.7
μ_2—Mean of receiving lab	$\bar{Y}_2$	249.4
σ_1^2—Variance of transferring lab	S_1^2	10.2
σ_2^2—Variance of receiving lab	S_2^2	27.1

6.5.3 An Equivalence Test for Bias

Bias between the labs is defined as the difference in the lab averages, $\mu_1 - \mu_2$. The equivalence test described in Sect. 2.11 based on the 90% confidence interval on the difference in means can be used for this purpose. Since the two samples are independent, the appropriate confidence interval on the mean difference is provided in Eq. (2.58) where we assume variances are not equal. Based on historic reference sample measurements in the transferring lab, the EAC is taken to be 10 mg. Thus, the 90% confidence interval on the difference $\mu_1 - \mu_2$ must fall entirely within the range from -10 to $+10$ mg.

We begin by computing the degrees of freedom for the confidence interval.

$$
df = \frac{\left(\frac{S_1^2}{n_1} + \frac{S_2^2}{n_2}\right)^2}{\frac{S_1^4}{n_1^2(n_1 - 1)} + \frac{S_2^4}{n_2^2(n_2 - 1)}}
$$

$$
= \frac{\left(\frac{10.2}{10} + \frac{27.1}{10}\right)^2}{\frac{(10.2)^2}{(10)^2(10 - 1)} + \frac{(27.1)^2}{(10)^2(10 - 1)}} = 14.9 = 15 \text{ (rounded).}
$$

(6.33)

The lower and upper bounds of the confidence interval are now computed as

$$
L = \bar{Y}_1 - \bar{Y}_2 - t_{1-\alpha/2:df}\sqrt{\frac{S_1^2}{n_1} + \frac{S_2^2}{n_2}}
$$

$$
= 247.7 - 249.4 - 1.75\sqrt{\frac{10.2}{10} + \frac{27.1}{10}} = -5.1 \text{ mg}
$$

(6.34)

$$
U = \bar{Y}_1 - \bar{Y}_2 + t_{1-\alpha/2:df}\sqrt{\frac{S_1^2}{n_1} + \frac{S_2^2}{n_2}}
$$

$$
= 247.7 - 249.4 - 1.75\sqrt{\frac{10.2}{10} + \frac{27.1}{10}} = 1.7 \text{ mg.}
$$

Since the 90% confidence interval falls entirely within the range from -10 to $+10$ mg, equivalence of means between the laboratories has been demonstrated.

Rugaiganisa (2016) has proposed an approach for setting the EAC for the equivalence test. As an alternative to this procedure, one may wish to establish transfer criteria using an ATP as described by Martin et al. (2013).

6.5.4 Tests for Precision

Precision of a procedure is described by the magnitude of the variance (or standard deviation). As discussed in Schwenke and O'Connor (2008), the necessity and form of an equivalence test for precision is not obviously apparent. If the receiving lab provides better precision, that is a good thing, even if the two procedure precisions are not equivalent. Thus, a difference testing approach as opposed to equivalence testing might be considered.

In particular, one might test the null hypothesis $\sigma_1^2 \geq \sigma_2^2$ versus the alternative hypothesis $\sigma_1^2 < \sigma_2^2$ and conclude the receiving lab is no worse than the transferring lab if one does not reject the null hypothesis. Such a test could be performed by computing an upper bound on the ratio σ_1^2/σ_2^2 as described in Sect. 2.8.3. If this upper bound is greater than 1, then one is unable to reject the null hypothesis, and will conclude the procedure transfer is successful. However, one would need to ensure that the power associated with the test is sufficiently high to discover a situation where $\sigma_1^2 < \sigma_2^2$.

Alternatively, one might compute a range of expected variances in the receiving lab based on the variance in the transferring lab. The receiving lab passes the transfer if the computed variance falls in this range. Such an approach is similar to using a control chart or a system suitability test. A 95% upper prediction bound based on n_1 observations to contain the variance of a future sample of size n_2 from the same normal population is

$$U = S_1^2 \times F_{0.95, n_2-1, n_1-1} \tag{6.35}$$

(see page 64 of Hahn and Meeker (1991)).

That is, if the receiving lab is performing the procedure in the same manner as the transferring lab, the transfer criteria are satisfied if $S_2^2 \leq U$ where U is defined in (6.35). To demonstrate, using the data in Table 6.11,

$$\begin{aligned} U &= S_1^2 \times F_{0.95, n_2-1, n_1-1} \\ &= 10.2 \times 3.18 = 32.4. \end{aligned} \tag{6.36}$$

Since $S_2^2 = 27.1$ is less than 32.4, the procedure transfer satisfies the precision requirement.

References

Burdick RK, Borror CM, Montgomery DC (2005) Design and analysis of gauge R&R experiments: making decisions with confidence intervals in random and mixed ANOVA models. ASA-SIAM Series on Statistics and Applied Probability, SIAM, Philadelphia

Chatfield MJ, Borman PJ (2009) Acceptance criteria for method equivalency assessments. Anal Chem 81:9841–9848

Dolezal KK, Burdick RK, Birch NJ (1998) Analysis of a two-factor R&R study with fixed operators. J Qual Technol 30:163–170

Food and Drug Administration. Center for Drugs Evaluation Research (2011) Process validation: general principles and practices, guidance for industry

Food and Drug Administration. Center for Drugs Evaluation Research (2015) Analytical procedures and methods validation for drugs and biologics, guidance for industry

Hahn GJ, Meeker WQ (1991) Statistical intervals: a guide for practitioners. Wiley, New York

Hubert Ph, Nguyen-Huu J-J, Boulanger B, Chapuzet E, Chiap P, Cohen N, Compagnon P-A, Dewé W, Feinberg M, Lallier M, Laurentie M, Mercier N, Muzard G, Nivet C, Valat L (2004) Harmonization of strategies for the validation of quantitative analytical procedures: a SFSTP proposal—part I. J Pharm Biomed Anal 36:579–586

Hubert Ph, Nguyen-Huu J-J, Boulanger B, Chapuzet E, Chiap P, Cohen N, Compagnon P-A, Dewé W, Feinberg M, Lallier M, Laurentie M, Mercier N, Muzard G, Nivet C, Valat L, Rozet E (2007a) Harmonization of strategies for the validation of quantitative analytical procedures: a SFSTP proposal—part II. J Pharm Biomed Anal 45:70–81

Hubert Ph, Nguyen-Huu J-J, Boulanger B, Chapuzet E, Cohen N, Compagnon P-A, Dewé W, Feinberg M, Laurentie M, Mercier N, Muzard G, Valat L, Rozet E (2007b) Harmonization of strategies for the validation of quantitative analytical procedures: a SFSTP proposal—part III. J Pharm Biomed Anal 45:82–96

International Conference on Harmonization (2005) Q2 (R1) Validation of analytical procedures: text and methodology

International Conference on Harmonization (2009) Q8 (R2) Pharmaceutical development

Krishnamoorthy K, Mathew T (2009) Statistical tolerance regions. Wiley, Hoboken

Martin GP, Barnett KL, Burgess C, Curry PD, Ermer J, Gratzl GS, Hammond JP, Herrmann J, Kovacs E, LeBlond DJ, LoBrutto R, McCasland-Keller AK, McGregor PL, Nethercote P, Templeton AC, Thomas DP, Weitzel J (2013) Stimuli to the revision process: lifecycle management of analytical procedures: method development, procedure performance qualification, and procedure performance verification. Pharm Forum 39(5). http://www.usp.org/uspnf/notices/stimuli-article-lifecyclemanagement-analytical-proceduresposted-comment. Accessed 11 Mar 2014

Ntzoufras I (2009) Bayesian modeling in WinBUGS. Wiley, New York

Rugaiganisa A (2016) Comparative analytical method transfer. Presentation at the 2016 ASA Biopharmaceutical Section Regulatory-Industry Statistics Workshop, September

Schwenke JR, O'Connor DK (2008) Design and analysis of analytical method transfer studies. J Biopharm Stat 18:1013–1033

Spiegelhalter D, Thomas A, Best N, Gilks W (1996) BUGS 0.5 Examples Volume 1(version i). Accessed 21 April 2014. http://www.mrc-bsu.cam.ac.uk/bugs/

USP Statistics Expert Team (2016) In-process revision: <1210> Statistical tools for procedure validation. Pharm Forum 42(5). http://www.usppf.com/pf/pub/index.html

USP 39-NF 34 (2016) General Chapter <1010> analytical data—interpretation and treatment. US Pharmacopeial Convention, Rockville

USP 39-NF 34 (2016) General Chapter <1030> biological assay chapters—overview and glossary. US Pharmacopeial Convention, Rockville

USP 39-NF 34 (2016) General Chapter <1032> design and development of biological assays. US Pharmacopeial Convention, Rockville

USP 39-NF 34 (2016) General Chapter <1033> biological assay validation. US Pharmacopeial Convention, Rockville

USP 39-NF 34 (2016) General Chapter <1223> validation of alternative microbiological methods. US Pharmacopeial Convention, Rockville

USP 39-NF 34 (2016) General Chapter <1224> transfer of analytical procedures. US Pharmacopeial Convention, Rockville

USP 39-NF 34 (2016) General Chapter <1225> validation of compendial procedures. US Pharmacopeial Convention, Rockville

Weitzel MLJ (2012) The estimation and use of measurement uncertainty for a drug substance test procedure validated according to USP<1225> Accred Qual Assur 17:139–146

Wolfinger RD (1998) Tolerance intervals for variance component models using Bayesian simulation. J Qual Technol 30:18–32

Yang H, Zhang J (2015) Validation based on total error: a generalized pivotal quantity approach to analytical method. PDA J Pharm Sci Technol 69:725–735

Chapter 7
Specifications

Keywords Acceptance limits • Blend uniformity • Compendial tests • Composite assay • Dissolution • Percentiles • Process capability • Protein concentration • Release and shelf life limits • Similarity factor (f_2) • Simulation • Three sigma limits • Tolerance intervals • Uniformity of dosage units

7.1 Introduction

Setting specifications for drug substances and drug products is a complex process. Many factors must be considered such as patient requirements, clinical and development experience, and global regulatory expectations. The specification setting process starts with determining the critical attributes and parameters of the product and manufacturing process that need to be controlled to ensure a high quality, safe, and efficacious product for the patient. During the development process, risk assessments, prior knowledge, and experimentation should yield which attributes and parameters need to be controlled and the range over which they can vary and still produce a high quality product. The next step is to understand the global regulatory expectations for setting specifications including compendial requirements. There must be an understanding of the markets the product will be sold in since regulatory requirements vary from market to market. International Conference on Harmonization Q6A (1999a) and International Conference on Harmonization Q6B (1999b) represent efforts to harmonize expectations around specification setting. A review of regulatory and compendial expectations will likely add attributes to the list that require specifications. It will also be important to understand which drug product attributes must meet specifications throughout the shelf life of the product.

7.1.1 *Definition and Regulatory Expectations*

Two key global guidance documents for specification setting are ICH Q6A that covers specifications for chemical substances and ICH Q6B that covers

© Springer International Publishing AG 2017
R.K. Burdick et al., *Statistical Applications for Chemistry, Manufacturing and Controls (CMC) in the Pharmaceutical Industry*, Statistics for Biology and Health, DOI 10.1007/978-3-319-50186-4_7

specifications for biotechnological/biological products. In ICH Q6A and Q6B, a specification is defined as:

> a list of tests, references to analytical procedures, and appropriate acceptance criteria, which are numerical limits, ranges, or other criteria for the tests described. It establishes the set of criteria to which a drug substance or drug product should conform to be considered acceptable for its intended use. "Conformance to specifications" means that the drug substance and/or drug product, when tested according to the listed analytical procedures, will meet the listed acceptance criteria. Specifications are critical quality standards that are proposed and justified by the manufacturer and approved by regulatory authorities as conditions of approval.

One key concept in this definition is that the specification is more than just the acceptance criterion (e.g., a range the result most be within or equal to). This definition also includes the analytical procedure that must be used to obtain the result. Another key concept is that the manufacturer must propose and justify the specification to the appropriate regulatory authority. Even if the acceptance criteria are already known via regulatory or compendial expectations, it is incumbent on the manufacturer to justify that the product can meet the specifications.

There are many guidance and requirements documents covering specifications for drug substances and drug products from regulatory organizations and compendia such as the Food and Drug Administration (FDA), European Medicines Agency (EMEA), World Health Organization (WHO), International Council on Harmonization (ICH), United States Pharmacopeia (USP), European Pharmacopoeia (EP), and Japanese Pharmacopoeia (JP). Current versions can be found at the organizations' websites. This chapter will introduce some of the key guidance documents as they relate to the application of statistics.

7.1.2 Conformance to a Specification

It is important to understand how to determine if a test result meets the acceptance limits. As described in Sect. 2.3, rounding of data is critical to making such decisions. The USP General Notices and Requirements Section 7.20 (2016) provides clear guidance on how to round test results to determine if the results conform to the specification. The USP states that:

> The observed or calculated values shall be rounded off to the number of decimal places that is in agreement with the limit expression. . ... When rounding is required, consider only one digit in the decimal place to the right of the last place in the limit expression. If this digit is smaller than 5, it is eliminated and the preceding digit is unchanged. If this digit is equal to or greater than 5, it is eliminated and the preceding digit is increased by 1.

For example, assume the attribute of interest is composite assay and the acceptance criterion (limit expression) is that the result as measured by the defined analytical procedure must be within or equal to 95.0–105.0%LC (percent of label claim). If the unrounded test result obtained is 94.95%LC, the result would be rounded using the hundredth's place to 95.0%LC which would conform to the specification. However, if the test result was 94.94%LC, the result would round to 94.9%LC and would not conform. The appropriately rounded result is often referred to as the "reportable result."

7.1.3 Reportable Versus Recordable Results

The rounding to create the reportable result as shown in Sect. 7.1.2 leads to an unfortunate situation that is often encountered in the pharmaceutical industry where data are stored either in paper files or in electronic databases to the number of decimal places in the acceptance limit. The number of decimal places in the acceptance limit does not necessarily equal the number of significant digits in the result. Excessive rounding removes information from the data that could be helpful for future statistical assessments. Figure 7.1 provides an example of the effect of rounding to the acceptance limit. Assume that the %RSD (percent relative standard deviation) measurement has an acceptance limit of maximum 6%RSD. The ten batches that are recorded with one decimal place show an increase in the %RSD. If the data are recorded to the number of places in the acceptance limit, it is not possible to see the increasing trend. Since the acceptance limit is maximum 6% RSD, the manufacturer would be interested to learn the data are trending toward the maximum acceptable limit. However, such movement is only apparent when the data are recorded and trended with the extra decimal point. As the example makes clear, it is important to separate reportable values (rounded to the number of decimal places in the acceptance limit) and recordable values (rounded to the number of significant digits) for use in future statistical analyses.

Batch	Unrounded %RSD	Rounded %RSD
1	4.6	5
2	4.5	5
3	4.5	5
4	4.7	5
5	4.6	5
6	4.8	5
7	5.0	5
8	5.2	5
9	5.1	5
10	5.4	5

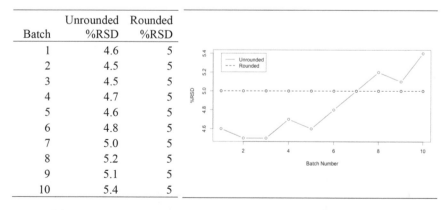

Fig. 7.1 Example of the effect of rounding

7.2 Considerations in Setting Specifications

Ideally, specifications would not be based on statistical calculations. Rather, they should be derived from patient requirements, clinical experience, and scientific understanding to ensure a safe and efficacious product whenever possible. However, there are times when these inputs are not easily linked to a particular critical quality attribute or process parameter and do not lead to quantifiable requirements for the attribute or parameter in question. For these situations, manufacturers and regulators often turn to statistical process capability as the basis for setting specifications. Although necessity may require this approach, it is fraught with potential dangers that must be considered when using this strategy.

7.2.1 Frame of Reference

When determining acceptance limits for a specification, it is essential to understand what the limits refer to. To apply statistical methods appropriately, it is important to define the sampling population and the population parameter or parameters required to make the inference of interest (e.g., mean or standard deviation). In this chapter, the definition of the population and parameter is referred to as the "Frame of Reference." First the population must be defined. This determines how the product should be sampled from batch and how the results should be obtained from the analytical method. The result from the analytical method might come from measuring an individual dosage unit, a composite of several dosage units, or the average of several composite results. The parameter of interest should also be defined as well. For example, the intent of the specification might be to estimate the distribution of individual values for dosage units from a batch such as uniformity of dosage units. In contrast, the intent of another specification might be to estimate the overall batch mean such as composite assay or protein concentration.

If the acceptance limits apply to individual dosage units, data used to set the limits must be capable of estimating the variation of individual dosage units. If averages are being evaluated, the data must be suitable for estimating variation expected in averages. The statistical approach used for setting the acceptance limits must take this in account. For example, if statistical intervals are to be used to determine the acceptance limits, it must be determined if the interval should be constructed to reflect the location of the batch mean or if the interval should contain individual dosage unit values from the batch. A statistical interval designed for individual dosage unit values will naturally be wider since individual values will have more variation than values that represent batch averages such as a composite sample value. As a result, it is not reasonable to require that individual dosage unit values satisfy limits derived to assess batch means. If the frame of reference for a specification is not clear, it is very hard to provide acceptance limits that make sense.

7.2.2   Incorporation of All Sources of Variation

When using statistical process capability to set acceptance limits, the data used in the analysis should represent the full variability expected for the product and process. Many textbooks recommend at least 20–30 results to assess process capability if it can be assumed that all the factors that create variation in the process are "active" for those 20–30 results. This is often not the case in pharmaceutical manufacturing situations. For example, raw materials can be a major contributor to process variation, and batches of raw materials are often used for several product batches in a row. Thus, it is possible that in a campaign of 30 product batches, a raw material batch may have changed only one or two times. As a result, the full impact of variation due to raw material batches will be underrepresented in the 30 product batches.

It is nearly impossible to know if all the sources of variation have been captured in a set of data used for specification setting. Thus, the best that can be done is to ensure that the approach used for setting the acceptance limits accounts for our lack of knowledge about the sources of variation. Tolerance intervals (see Sect. 2.6.7 in Chap. 2 and Sect. 7.4.3 for more information) and simulation (see Sect. 7.4.4) are statistical tools that can be employed. It is important to note that tolerance intervals (and most other statistical approaches) cannot represent variation that has not been observed in the data. However, tolerance intervals are adjusted to the amount of data available as discussed in the next section.

7.2.3   Small Data Sets for Specification Setting

The use of small data sets for specification setting not only leads to an underrepresentation of all the sources of variation, but also leads to additional problems that must be addressed. Specifications are usually set when a product is in the approval stage with limited data from full scale production runs. With such limited data, appropriate statistical approaches will often yield wide acceptance ranges since there is uncertainty about the distribution of the data from the production process. These wide intervals may not provide useful acceptance limits as it is often claimed "one can drive a truck through them." At this point, it is typical for negotiation between regulatory authorities and manufacturers to ensue in order to determine limits that seem "practical." There is a danger that the "practical" range will be unsuitable as more sources of variability manifest themselves, and this creates increased probabilities of out-of-specification results. Since neither the "practical" nor statistically based acceptance limits are based on direct links to safety and efficacy, no one really knows if results outside the acceptance limits are truly reflective of unacceptable quality. This issue can cause manufacturers significant challenges in supplying the market.

7.2.4 Disincentive to Improve Process Variation

If specifications are set based on process capability, there is little incentive for the manufacturer to improve the process. If process variation is reduced, regulatory agencies may request that acceptance limits be tightened. If the manufacturer knows that process capability will be needed to set the limits, it might lead developers to make sure as much variation as possible is included in the development data for the specification setting process rather than a focus on what knowledge is needed to optimize the process and formulation.

7.3 Compendial Standards and Tests with Commonly Expected Acceptance Limits

Compendial tests are found in the national/regional pharmacopeia such as USP, EP, and JP. In general, there are more compendial tests for small molecule products than biologics and vaccines. It is expected that a product will meet the standards whenever tested during the entire shelf life of the product. During the specification setting phase for a product, it is useful to understand the performance needed to meet the standards and in some cases to help determine product specific parameters that must be specified in the regulatory filing (e.g., the value of Q for dissolution testing).

7.3.1 Compendial Standards and Specifications

There has been much discussion of the similarities and differences between compendial standards and specifications. It is not uncommon for regulatory filings to include wording such as "when tested will comply with USP $<$XXX$>$" when providing the specification for a specific attribute. The USP made changes to the General Notices and Requirements Section 3.10 in USP 39-NF34 (2016b) on "Applicability of Standards" to clarify that the standards provided are not be considered specifications or statistical sampling plans:

> The standards in the relevant monograph, general chapter(s), and General Notices apply at all times in the life of the article from production to expiration. It is also noted that the manufacturer's specifications, and manufacturing practices (e.g., Quality by Design, Process Analytical Technology, and Real Time Release Testing initiatives), generally are followed to ensure that the article will comply with compendial standards until its expiration date, when stored as directed. Every compendial article in commerce shall be so constituted that when examined in accordance with these assays and test procedures, it meets all applicable pharmacopeial requirements (General Notices, monographs, and general chapters). Thus, any official article is expected to meet the compendial standards if tested, and any official article actually tested as directed in the relevant monograph must meet such standards to demonstrate compliance.

Some tests, such as those for Dissolution and Uniformity of Dosage Units, require multiple dosage units in conjunction with a decision scheme. These tests, albeit using a number of dosage units, are in fact one determination. These procedures should not be confused with statistical sampling plans. The similarity to statistical procedures may seem to suggest an intent to make inference to some larger group of units, but in all cases, statements about whether the compendial standard is met apply only to the units tested. Repeats, replicates, statistical rejection of outliers, or extrapolations of results to larger populations, as well as the necessity and appropriate frequency of batch testing, are neither specified nor proscribed by the compendia; such decisions are based on the objectives of the testing. Frequency of testing and sampling are left to the preferences or direction of those performing compliance testing, and other users of USP–NF, including manufacturers, buyers, or regulatory authorities."

To address the issue that product must meet the compendial standards whenever tested, several approaches have been proposed. It is logical to think that the manufacturer would need to apply a tighter "standard" at product release to ensure that the product will meet the compendial standard whenever tested. Since it is impossible to create 100% assurance unless all dosage units in the batch are tested at release, a level of assurance that is less than 100% is selected using statistically-based sampling methodologies. One of these approaches will be discussed in Sect. 7.3.5. The next sections discuss compendial standards for several dosage forms and critical quality attributes that have commonly expected specifications.

7.3.2   Uniformity of Dosage Units

Uniformity of dosage units is a measurement intended to ensure that every dosage unit contains the amount of drug substance intended with little variation among dosage units within a batch. In the sample taken for testing, each dosage unit is tested individually and the acceptance criteria are intended to limit the amount of variation among the individual values.

7.3.2.1   Uniformity of Dosage Units for Immediate Release Products Covered by USP <905>, EP 2.9.40, and JP 6.02

The UDU test for these dosage forms can be conducted by either of two approaches: Content Uniformity (CU) or Weight Variation. Content Uniformity is based on an assay for the content of the drug substance in the individual dosage units. Weight Variation is applicable to certain dosage forms such as granules/powders and solutions in single unit dose containers. Weight Variation is also possible for capsules and tablets where the drug substance comprises a large portion of the total weight of the dosage unit. Specifically the dosage unit must contain at least 25 mg of the drug substance and comprise more than 25% of the total weight.

The UDU compendial standard was harmonized among the USP, EP, and JP pharmacopeia. However, a few differences between the pharmacopeia exist. The EP

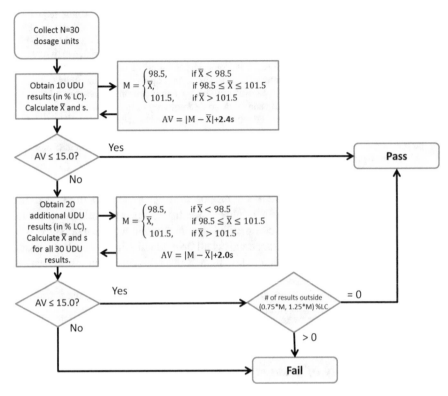

Fig. 7.2 Flow Diagram of Harmonized UDU test assuming T = 100%LC, L1 = 15%, and L2 = 25%

and JP use the term Mass Variation rather than Weight Variation. The EP and JP also allow the use of mass variation when the relative standard deviation of drug substance is not more than 2% even if the dosage units do not meet the 25 mg/25% condition.

The harmonized UDU test has two stages: Stage 1 (n = 10 dosage units assayed) and Stage 2 (n = 30, an additional 20 dosage units assayed). Figure 7.2 describes the test where the target for the product is 100%LC (T = 100%), the maximum acceptance limit is 15% (L1 = 15%) and the maximum allowed range for each dosage unit relative to the reference value (M) is 25% (L2 = 25%). It is important to note that the AV (Acceptance Value) calculation changes in Stage 1 and 2. The multiplier in front of the standard deviation is 2.4 for Stage 1 and 2.0 for Stage 2. T, L1, and L2 are specified in the monograph for a drug product but typically they are set to values defined in Fig. 7.2. In some situations, it is possible to have different values for T, L1, and L2. This would likely be negotiated with the regulatory agencies at the time of initial application. In those cases, the use of simulation techniques discussed in Sect. 7.4.4 is helpful in determining the appropriate values to propose.

Table 7.1 shows two sets of example calculations for the harmonized UDU test. The first example in the table shows a Stage 1 passing result. In this example, the

AV value is 8.7% which is less than the L1 value of 15% so the test yields a "pass" and stops at Stage 1. The second example shows a passing result that went to Stage 2. The AV value for Stage 1 is 18.7% which is greater than the L1 value of 15% so an additional 20 UDU results are obtained and combined with the 10 results from Stage 1. The AV value using all thirty results is 12.2% and passes. In Stage 2, all

Table 7.1 Two examples of UDU test calculations

Stage 1	(%LC)	Calculations	(%)
1	101.4	T	100.0
2	98.7	L1	15.0
3	105.5	L2	25.0
4	99.4	k	2.4
5	104.8		
6	94.3	Average	101.3
7	100.4	s	3.62
8	105.9	M	101.3
9	103.2	AV	8.7
10	99.6		
Pass Stage 1, AV ≤ 15.0%			

Stage 1	(%LC)	Calculations	(%)
1	94.4	T	100.0
2	95.2	L1	15.0
3	85.1	L2	25.0
4	104.4	k	2.4
5	98.4		
6	92.3	Average	97.6
7	97.9	s	7.45
8	105.9	M	98.5
9	92.6	AV	18.7
10	110.2		
Go to Stage 2, AV > 15.0%			

Stage 2:	(%LC)		(%LC)	Calculations:	(%)
11	95.8	21	99.1	T	100.0
12	95.3	22	100.7	L1	15.0
13	97.6	23	98.2	L2	25.0
14	103.6	24	98.3	k	2.0
15	105.4	25	106.4		
16	105.1	26	100.8	Average	96.5
17	103.1	27	97.0	s	5.11
18	99.5	28	100.1	M	98.5
19	101.8	29	102.3	AV	12.2
20	102.8	30	102.5	0.75M	73.9
				1.25M	123.1
Pass Stage 2, AV ≤ 15.0%, All UDU values within limits					

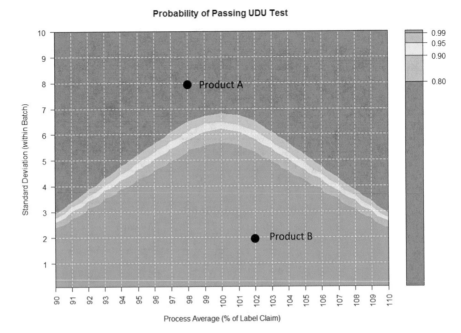

Fig. 7.3 Heatmap showing the probability to Pass the harmonized UDU compendial test for various process averages and within batch standard deviations (T $=$ 100%LC, L1 $=$ 15%, L2 $=$ 25%)

individual results must be checked to determine that all results are within 0.75 M and 1.25 M%LC to allow for a Stage 2 pass (this assumes L2 $=$ 25%).

When evaluating UDU performance for a product, it is important to understand if the product will meet the UDU standard reliably. Figure 7.3 provides a heatmap that illustrates the probability to pass the UDU test for a given process average and within batch standard deviation (when T $=$ 100%, L1 $=$ 15 and L2 $=$ 25%). The heatmaps in this chapter are created by use of computer simulations as described in Sect. 7.4.4. Access to example programs is listed in Sect. 7.8. A chart like this can be used with the development or historical UDU data for a product to determine if the product is expected to consistently pass the UDU test. For example, Product A in Fig. 7.3 typically has a process average of 98%LC and a within batch standard deviation for UDU results of 8%LC. This product has less than an 80% chance of passing the UDU test. Product B typically has a process average of 102%LC and a within batch standard deviation of 2%LC. This product has a greater than 99% chance of passing the UDU test. The heatmap does not take into account batch to batch variation. This variation can be visualized by plotting development and/or historical batches on the chart. If the data from the batches are clustered tightly together and within the greater than 95–99% probability to pass region, the process is likely performing adequately and the batch to batch variation is not of concern. If the batches are widely scattered and include results that fall in the red portion of the heatmap, this process most likely needs improvement.

7.3.3 Blend Uniformity

The blending of drug substance(s) and excipients is a challenging process operation in pharmaceutical manufacturing. For powders, it is possible to blend too little or too much with both situations potentially creating a mixture that is not homogeneous. As a result, blend uniformity testing is usually required during process performance qualification (PPQ) and may be required in routine production especially for complex and generic products. Blend uniformity testing is complicated due to the difficulties in obtaining representative samples from the blend. It is usually expected that the samples of the blend are no more than three times the dosage unit size. Thief sampling is typically employed to take the samples and with the physical characteristics of the blend and the design of thief, sampling errors are sometimes significant enough to falsely indicate the blend is not uniform.

The acceptance criteria for this testing have been a hotly debated topic for many years and are still in discussion between the industry and health authorities. The FDA has withdrawn two draft guidance documents that had provided some standards to follow. The first is the ANDAs: Blend Uniformity Analysis which was released in 1999 and withdrawn in 2002. A Product Quality Research Institute (PQRI) project was underway in that timeframe and the guidance was withdrawn to allow for the research to be completed. The PQRI project resulted in a second FDA draft guidance in 2003, Powder Blends, and Finished Dosage Units—Stratified In-process Dosage Unit Sampling and Assessment. This guidance was withdrawn in 2013 because it was no longer consistent with the current agency thinking. The criteria for blend uniformity in the ANDA guidance were the following:

1. Analyze 6–10 samples from appropriate locations in the blender or drum; sample should be no more than three times the weight of the dosage unit.
2. Average of the sample results should be within or equal to 90.0–110%LC.
3. %RSD should be less than or equal to 5.0%.

The 2003 guidance document attempted to link the blend uniformity results to the UDU results which were to be conducted using a specific type of sampling called stratified sampling. The guidance recommended that key development and PPQ batches (or current production batches for an existing product) include testing of blends and dosage units in such a way that the results could be used to develop a correlation between these two types of testing. The logic of this proposal was that the UDU testing of dosage units sampled using an appropriate stratified approach can only be acceptable if the blend is sufficiently uniform. As a result once both the blend and stratified UDU testing are demonstrated successfully, routine testing of the blend should not be necessary. For the stratified sampling plan for UDU, dosage units should be sampled at a least 20 locations during the compression/filling process. The locations should be spaced appropriately throughout the process and should include significant events during the process such as the filling or empting of intermediate bulk containers and the start and end of the run. The blend uniformity results were expected to meet the following criteria:

1. Collect three replicate samples for at least 10 locations in the blender. Assay one sample per location.
2. RSD should be less than or equal to 5.0%.
3. All results should be within 10.0% (absolute) of the mean of the results.

If the results on the blend did not meet the criteria above, it was recommended to analyze the second and third samples from each location and compare that to the testing of the UDU testing of the dosage units at the 20 locations during the compression/filling run (a detailed plan with criteria for the UDU results was provided in the withdrawn draft guidance).

After the FDA draft guidance was withdrawn in 2013, International Society for Pharmaceutical Engineering (ISPE) sponsored a group consisting of representatives from the FDA, industry, and academia to develop and evaluate alternative approaches to assessing Blend and Content Uniformity. To date, the group has published three papers that provide a proposal for assessing Blend and Content Uniformity: Bergum et al. (2014), Garcia et al. (2015), and Bergum et al. (2015). The plan is similar to the approach in the withdrawn stratified sampling guidance but with a few key changes in order to align with the FDA's current thinking such as more explicit instructions on when to test the second and third replicate blend samples and to provide criteria that offer more statistical confidence that future UDU samples will comply with the USP $\langle 905 \rangle$ standard.

At this time appropriate specifications and testing plans for blend uniformity are not clear. Each manufacturer must determine the best approach for their product(s). The ISPE work may provide an approach that can be considered but this work is still evolving. Manufacturers will need to justify their approach including the number and size of the samples and acceptance criteria. Manufacturers may also consider other approaches such as the use of process analytical technologies (PAT) and near infrared (NIR) to assess blend uniformity.

7.3.4 Dissolution

The dissolution test attempts to mimic how the active drug substance in a dosage form is absorbed into the body and is a required release test for many dosage forms. The dosage unit is placed into a vessel with a known volume of media. Some form of gentle stirring or flow of the media occurs in the vessel. The media is sampled at a specific time (or a series of times) and the amount of active drug substance in the sample is determined via an analytical technique such as high performance liquid chromatography (HPLC).

Table 7.2 Acceptance criteria for the compendial dissolution test for immediate release products

Stage	Number of samples	Acceptance criterion
S_1	6	Each unit is not less than Q +5%
S_2	6	Average of 12 units (S_1 + S_2) is equal to or greater than Q and no unit is less than Q -15%
S_3	12	Average of 24 units (S_1 +S_2 + S_3) is equal to or greater than Q, not more than 2 units are less than Q -15% and no unit is less than Q -25%

The dissolution compendial standard was harmonized across the USP, EP, and JP pharmacopeia in USP $\langle 711 \rangle$, Ph. Eur. 2.9.3, and JP 6.10. There are some differences between the pharmacopeia in areas such as apparatus definition and the handling of extended release and delayed release products.

7.3.4.1   Immediate Release Products

The immediate release dosage form compendial test is multi-stage and is described in Table 7.2. The criteria depend on the parameter Q, which is an amount of dissolved active ingredient specified in the monograph. Q is typically set at 70, 75, or 80% dissolved and this is determined on a product specific basis at the time of specification proposal. The other important parameter is the time of sampling for the test. This is often in the range of 15–60 min. The manufacturer must propose the value of Q and the time of sampling for the release test in the filing. Development data and simulation can be used to help determine the appropriate Q and time of sampling. While not specifically called out in Table 7.2, if a result is less than Q -25% at any stage, the data fail the dissolution test.

The pivotal clinical batches, primary registration (stability), and key development batches are usually tested using the profile format for dissolution. The profile format is when the sample is tested at several time points such as 15, 20, 25, 30, 45, and 60 min. Figure 7.4 shows some typical development data from an immediate release solid oral dosage product. The figure shows two possible choices for Q for a 30 min sampling time. Figure 7.5 shows a histogram and boxplot of the 30 min dissolution data. At 30 min, all the individual dissolution values are well above Q = 75% and just above Q = 80%. The choice of Q can be further refined by using simulation techniques as shown in Sect. 7.4.4.

Table 7.3 contains two examples of the compendial test with a Q value of 75% LC. The example on the left illustrates a Stage 1 pass. All six dissolution results are greater than or equal to Q +5%. The example on the right shows a Stage 2 pass. For Stage 1, one of the six results is less than Q +5% (Unit #2 = 78%LC). The test passes Stage 2 since the average dissolution value of all 12 results is 84%LC and all results are greater and or equal to Q -15%.

Figure 7.6 provides a heatmap that illustrates the probability to pass the immediate release dissolution test. The x-axis shows the distance of the process average from the Q value. The y-axis is the within batch standard deviation. This chart can

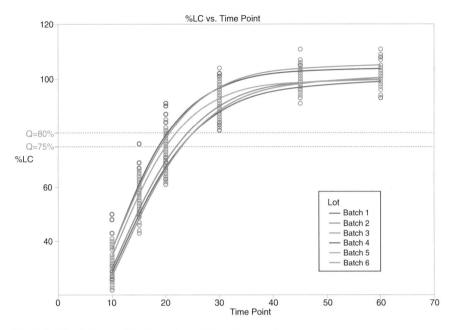

Fig. 7.4 Dissolution profiles for an immediate release product

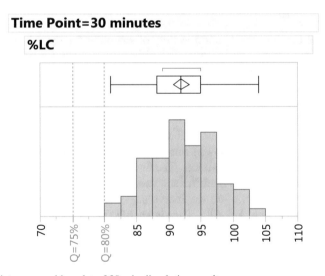

Fig. 7.5 Histogram and boxplot of 30 min dissolution results

be used with the development or historical dissolution data to determine if the product is expected to consistently pass the compendial dissolution test. For example, Product A in Fig. 7.6 typically has a process average that is 1.2%LC above the Q value and a within batch standard deviation for UDU results of 7%LC.

Table 7.3   Two examples of dissolution test calculations

Q Sample time	75%LC 30 min	Q Sample time	75%LC 30 min		
Unit	Stage 1 (%LC)	Unit	Stage 1 (% LC)	Unit	Stage 2 (%LC)
1	90	1	88	7	77
2	83	2	78	8	81
3	89	3	82	9	86
4	85	4	90	10	85
5	81	5	85	11	82
6	97	6	83	12	87
Stage 1 pass. All results greater than or equal to Q + 5%		Stage 1. All results are not greater than or equal to Q + 5%. Go to stage 2		Average S1 + S2	84
				Stage 2 pass. Average of S1 + S2 is greater than or equal to Q and all results are greater than or equal to Q = 15%	

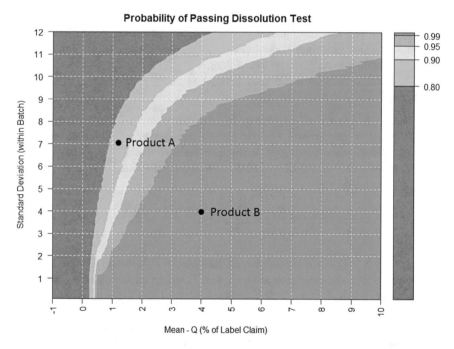

Fig. 7.6   Heatmap showing the probability to pass the harmonized dissolution compendial test for various process averages and within batch standard deviations

This product has between 80 and 90% chance of passing the dissolution test. Product B typically has a process average that is 4%LC above the Q value and a within batch standard deviation of 4%LC. This product has a greater than 99% chance of passing the dissolution test. Just like the heatmap in Fig. 7.3, this graph does not take into account batch to batch variation. This variation can be visualized by plotting development and/or historical batches on the chart. If the data from the batches are clustered tightly together and within the greater than 95–99% probability to pass region, the process is likely performing adequately and the batch to batch variation is not of concern. If the batches are widely scattered and include results that fall in the red portion of the heatmap, this process most likely needs improvement.

7.3.4.2 Dissolution Profile Comparisons, Similarity Factor (f_2)

While not a direct part of dissolution specification setting, an important aspect of dissolution testing is the ability to compare profiles. Profile comparisons are used for many purposes such as:

1. Demonstrating bioequivalence to obtain biowaivers
2. Comparing formulations in development
3. Supporting process, formulation, equipment, and site changes as discussed in two key FDA guidance documents: Immediate Release Solid Oral Dosage Forms: Scale-up and Postapproval Changes (SUPAC-IR) (1995) and SUPAC-MR Modified Release Solid Oral Dosage Forms: Scale-up and Postapproval Changes (SUPAC-MR) (1997a).

Equation (7.1) is a similarity factor (f_2) for comparing dissolution profiles that was proposed in Moore and Flanner (1996) and further discussed in the FDA Guidance Dissolution Testing of Immediate Release Solid Oral Dosage Forms (1997b) as well as SUPAC-IR and SUPAC-MR guidance documents.

$$f_2 = 50 \times \log \left\{ \left[1 + \frac{1}{n} \sum_{t=1}^{n} (R_t - T_t)^2 \right]^{-0.5} \times 100 \right\}. \tag{7.1}$$

where log is log10, n is the number of time points, R_t is the dissolution value of the reference (pre-change) batch at time t, and T_t is the dissolution value of the test (post-change) batch at time t. Values between 50 and 100 are evidence of similarity between the reference and test.

The f_2 calculation was not developed from a statistical perspective and thus it has many undesirable properties such as increasing the number of time points in the calculation makes it easier to show similarity. Also the equation does not weight time point data relative to the amount of variation at the time points even though the differing variability at the time points is almost always evident. These issues and many more have been detailed in several publications such as Liu et al. (1997) and LeBlond et al. (2016).

There are several guidance documents that discuss some of the issues that arise when using the f_2 calculation with some variations on the requirements provided. The following requirements (as listed in the FDA guidance for dissolution testing, August 1997) must be met before using the f_2 to compare dissolution profiles:

1. Test 12 units for both sets of data to be compared: test and reference.
2. There should be data from at least 3–4 dissolution time points and the time points should be the same for both the test and reference.
3. Average dissolution values should be used in the calculation of f_2 for the test and reference data.
4. Only one time point after reaching 85% dissolved for both products should be used in the calculation.
5. The coefficient of variation (or %RSD) for early time points (e.g., 15 min.) should not be more than 20% and should not exceed 10% for all other time points.

If the conditions described above are not met, the FDA guidance indicates that other methods may be used to compare the curves. Alternative methods may also be used if the f_2 value does not accurately reflect the similarity of the test and reference. Alternative approaches are reviewed in LeBlond et al. (2016).

Table 7.4 shows two examples of f_2 calculations and Fig. 7.7 shows the dissolution profiles for the two examples. Example 1 shows a pair of profiles that have an f_2 value of 65.4 which is greater than 50 and indicates similarity. Example 2 has an f_2 value of 46.0 which is less than 50 and does not support similarity using the f_2 approach.

Table 7.4 Two examples of f2 calculations

Example 1			
Mean dissolution (%LC)			
Time point (min)	Reference	Test	Difference squared
10	29	34	25
15	53	58	25
20	70	75	25
30	84	89	25
45	95	99	16
		Sum of squared difference = 116	
		f2 = 65.4	
Example 2			
Mean dissolution (%LC)			
Time point (min)	Reference	Test	Difference squared
10	25	40	225
15	45	60	225
20	67	78	121
30	82	90	64
45	90	99	81
		Sum of squared difference = 716	
		f2 = 46.0	

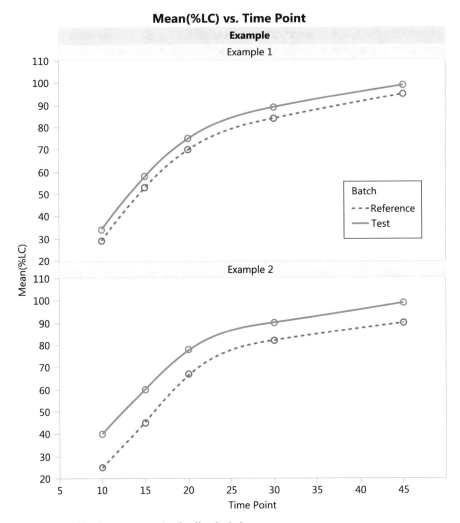

Fig. 7.7 Profiles for two examples for f2 calculations

7.3.4.3 Extended Release Products

Extended release products are formulated so that the active ingredient is released slowly over time. Dissolution specifications for these products require multiple time points with limits for each sampling time. Typically three time points are selected for dissolution testing for release:

1. Early time point to provide assurance against premature release of drug.
2. Intermediate time point to define the release profile.
3. Late time point to provide assurance of full release. It is common to select a time point where the dosage form is at least 80% dissolved.

Table 7.5 Acceptance criteria for the compendial dissolution test for extended release products

Stage	Number of samples	Acceptance criteria
S_1	6	No individual unit lies outside each of the stated ranges and no individual is less than the stated amount at the final test time
S_2	6	The average value of the 12 units ($S_1 + S_2$) lies within each of the stated ranges and is not less than the stated amount at the final test time. No unit is more than 10% of labeled content outside of each of the stated ranges and no unit is more than 10% below the stated amount at the final test time
S_3	12	The average value of the 24 units ($S_1 + S_2 + S_3$) lies with each of the stated ranges and is not less than the stated amount at the final test time. Not more than 2 of the 24 units are more than 10% of the labeled content outside of the stated ranges and note more than 2 of the 24 units are more than 10% of labeled content below the stated amount at the final test time. None of the units are more than 20% of labeled content outside of the each of the stated ranges or more than 20% of labeled content below the stated amount at the final test time

Table 7.5 provides the compendial acceptance criteria. The test has three stages similar to the immediate release test. For each time point except the final time point, a range is provided. For the final time point, a lower limit is provided. In Stage 1, 6 units are tested. All individual values must fall within or equal to the ranges for each time point to achieve a passing result. If not, the test proceeds to Stage 2. In Stage 2, an additional 6 units are tested for a total of 12 values. The average value for the 12 units must be within the stated ranges and no unit can be more than 10% outside of the stated ranges for a passing result. If the data do not meet the Stage 2 criteria, the test proceeds to Stage 3. In Stage 3, an additional 12 units are tested for a total of 24 units. The average value for the 24 units must be within the stated ranges and no more than two units can be more than 10% outside of the stated ranges and no unit can be more than 20% outside the stated ranges.

Figure 7.8 shows dissolution profile data for an extended release product. The manufacturer must propose the sampling times and ranges for the dissolution release test. The shaded purple area is a potential choice for the range (25–45% LC) when the early time point is 0.5 h. The gray shaded area contains a possible range (62–82%LC) when the intermediate time point is 4 h. The red line at 85%LC shows that several later time points are potential choices for the final time point such as 10 or 12 h. The stated ranges in this example are 20% wide for the early and intermediate time points. This is a typical range but it is possible to propose wider ranges if additional justification is provided such as clinical data and scientific rationale.

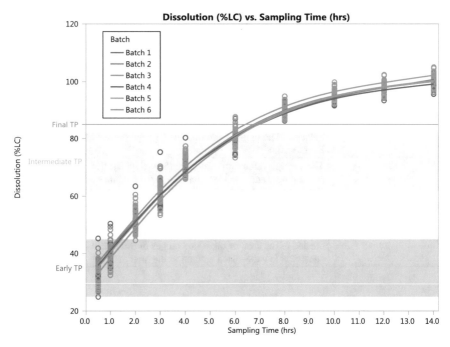

Fig. 7.8 Dissolution profiles for an extended release product

7.3.5 ASTM 2709/2810 (Bergum/CUDAL Method)

As discussed in Sect. 7.3.1, the USP has revised its general notices section to clarify that USP tests are not required to be used as release tests and they are not intended to make inferences about the performance of a batch tested. However, the text makes it clear that products must pass the compendial requirement whenever tested from production to the end of shelf life. As a result, there is much activity in the industry to develop release tests (sampling plans and acceptance criteria) that ensure a high probability of passing compendial tests with strong focus on the UDU test. The requirement to pass USP tests whenever tested can present a compliance issue if a batch of product tested at release just meets the acceptance criteria of the USP test and is released to the market. The batch may not pass a future USP test in this situation. ASTM E2709 (2014) is a methodology that can be used to assure future samples will have a high probability of passing the compendial criteria with a defined statistical confidence. ASTM E2810 (2011) is the application of the methodology described in ASTM 2709 to the harmonized UDU test.

The ASTM 2709/2810 methodology is also known as Content Uniformity and Dissolution Acceptance Limits (CUDAL) or the Bergum Method. It is a well-known procedure that is used in the pharmaceutical industry for estimating the probability that a batch will pass a multi-stage test, such as the compendial tests for

UDU and dissolution. Historically, this methodology was applied by many manufacturers to establish criteria for PPQ studies.

The confidence provided by the ASTM 2810 method is achieved by calculating a C% confidence region on the average and standard deviation estimated from the UDU sample. The extremes of this confidence region are compared to a lower bound, LB%, on the probability of passing the harmonized UDU test. If the confidence region is within the lower bound, LB, then the sample, and hence the lot, has demonstrated the necessary confidence for the market. The interpretation of passing an ASTM 2810 plan conducted at C% confidence with a lower probability bound of LB% is the following: "with C% confidence there is at least LB% probability that a future sample taken from the batch will meet the UDU test." The ASTM standard contains tables to implement plans for the following confidence and lower bound probabilities (C%/LB%): 95/90, 95/95, 95/99, and 90/95. A verified SAS program is described by Bergum and Li (2007). The actual code can be downloaded at the ISPE website http://www.ispe.org/blend-content-uniformity-initiative/tools.

The statistical statement provided by an ASTM 2810 plan is achieved by determining the worst-case mean and standard deviation as defined by the joint confidence region at the selected confidence level (C%). An example of this region is shown in Fig. 7.9 as the area inside the blue triangle. Figure 7.9 shows an example where the sample average is equal to 100.8 % LC and the standard deviation equals 4.91%LC for 100 dosage units tested. This plan was conducted at 95% confidence with a 95% lower bound. The upper right corner (worst-case mean and standard

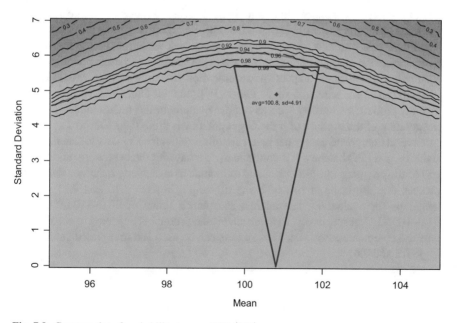

Fig. 7.9 Contour plot of probability to pass USP ⟨905⟩

deviation for the batch) of the joint confidence region is just touching the estimated 95% LB. It is important to note that nearly all the other possible means and standard deviations in the confidence region will provide much more than 95% chance of passing a future USP test. In fact, most of the area of triangular region falls in the contour indicating a 99% to greater than 99.99% chance of passing the USP test.

While the ASTM method is a statistically valid approach, it is conservative. Here "conservative" means the ASTM approach will not pass many inherently acceptable lots with sample sizes used for UDU testing. For example, when taking a sample of size 30, a process with a true mean of 100%LC, a standard deviation of 5% will pass the UDU test approximately 100% of the time, but would only pass a 95%/95% ASTM 2810 plan 7.7% of the time. If the ASTM approach is used, a much larger sample must be tested so that the confidence region size is not overly large. There are other statistical approaches that provide reasonable assurance and are efficient in terms testing resources such as tolerance intervals. The ASTM standard explicitly states that other methodologies can be utilized. Several alternative approaches are discussed in De los Santos et al. (2015) along with greater detail on the conservative nature of the ASTM method.

7.3.6   Composite Assay

Composite assay is used to determine if a batch of product contains the appropriate amount of drug substance/active ingredient. This assay involves compositing dosage units (typically in the range of 5–20 dosage units) into a sample preparation. For small molecule products, the preparation is assessed via chromatography methods to determine the amount of drug substance in the sample as a percent of label claim. Regulatory authorities expect a specification of either 90.0–110.0%LC or 95.0–105%LC for standard small molecule products. Other ranges are possible given a product's situation such as unique manufacturing technologies, expected degradation of the product or if the target potency is more than 100%LC at release.

Several composite assays per batch are often tested for key development, formal stability, and PPQ batches. It is common to do assays for the beginning, middle, and end of the compression/filling run and to include assays at any locations during the process that might be at risk of being off-target. Figures 7.10 and 7.11 provide heatmaps for evaluating data against acceptance limits of 90.0–110.0%LC and 95.0–105%LC, respectively, to determine if the product is expected to consistently pass the composite assay test. These charts can be used and interpreted similarly to Figs. 7.3 and 7.6.

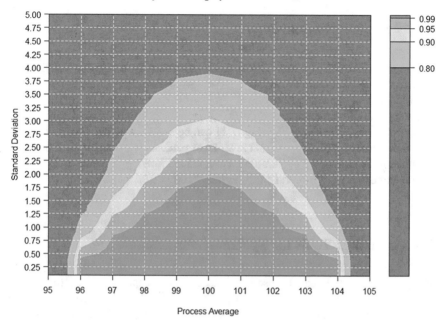

Fig. 7.10 Heatmap showing the probability to pass composite assay for various process averages and within batch standard deviations with acceptance limits of 95.0–105%LC

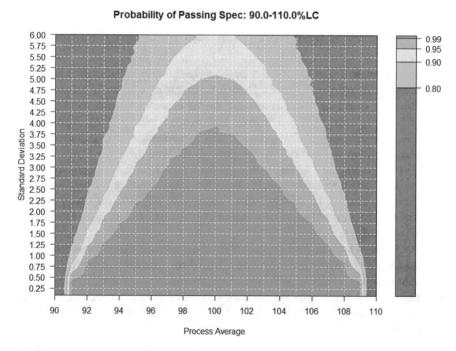

Fig. 7.11 Heatmap showing the probability to pass composite assay for various process averages and within batch standard deviations with acceptance limits of 90.0–110%LC

7.3.7   Protein Concentration

Protein concentration is a key critical quality attribute for a biologic which measures the quantity of protein in the product. The units of this attribute are usually in
mg/mL. Regulatory authorities typically expect a specification of $\pm 10\%$ of the
target concentration. Statistically, protein concentration can be viewed similarly as
composite assay. The heatmap in Fig. 7.11 can be used to evaluate protein concentration data after expressing the data in terms of percent of target concentration. For
example, if the target concentration for a product is 25 mg/mL, a value of 23 mg/mL
can be expressed as (23 mg/mL)/(25 mg/mL)*100 = 92% of target.

7.4   Statistical Tools for Specifications

Many statistical tools can be used to help set the acceptance limits for critical
quality attributes and parameters of a product. While there is a large volume of
literature on this subject, there is no agreement on a best practice for using data to
determine appropriate acceptance limits. This section will cover the key statistical
tools/approaches that have been successfully used.

7.4.1   Amount of Data Needed and How to Collect the Data

A common concern when using statistical approaches is how much data is needed
to set reasonable acceptance limits. There is no simple answer to this question since
it depends on many factors such as what are the sources of variation in the data,
have all the sources of variation been incorporated into the data available for
analysis, and which statistical approach will be utilized. As noted previously in
this chapter, there is often no way to know if the data available contain all the
sources of variation that can be expected in the future. In fact, it is probably best to
assume that all sources of variation have not been experienced yet. At times,
specifications need to be updated since sources of variation may only become
evident once the product is in routine manufacturing and/or when analytical method
technology changes.

How the data are obtained is probably more important than how much data is
needed since many statistical tools that are used for specification setting adjust to
the amount of data available. Well-designed experiments are essential to make sure
the data contains as many of the expected sources of variation as possible and that
these sources can be identified. Two key areas to consider in designing experiments
for specification setting are producing the product and measuring the product:

1. Producing batches of drug substance or drug product: Were the batches made within a wide or narrow range of operating parameters? Were different batches of raw materials used in manufacture of the drug substance or drug product?
2. The analytical methods for measuring the product: What is the method variation? Do factors such as day/time of testing, analyst, instrumentation, sample preparation, and assay run contribute to the method variation? How will the analytical method be used for batch release? Will there be a single value from the method or will multiple runs of the method be used and an average or geomean of the runs be used for batch release? Many of these questions are addressed in Chapter 6.

7.4.2 Data Distribution

Most of the statistical approaches used in specification setting require the assumption that the data follow a normal distribution. If the data are not well represented by the normal distribution, three approaches can be considered. The first two are to consider a transformation of the data or to use of another data distribution as discussed in Sect. 2.6.10. A third option is to use statistical methods that do not make assumptions (or limited assumptions) about the data distribution. These approaches are called nonparametric. Many of the statistical methods discussed in this chapter have nonparametric versions. Section 7.4.5 discusses the use of percentiles to set acceptance ranges which can be a reasonable approach when the data do not follow a normal distribution although percentiles require a fairly large set of data to be effective $(n > 100)$.

7.4.3 Tolerance Intervals and Three Sigma Limits

Statistical intervals are often used in specification setting activities and follow the general equation:

$$\bar{Y} \pm K\sqrt{S^2} \qquad (7.2)$$

This formula requires the sample mean and variance of the data set. The critical piece in Eq. (7.2) is the choice of K. Note that this equation assumes that the acceptance limits will be two-sided due to the use of the "$\pm$" sign. One-sided acceptance limits are also possible and can be constructed using the same approaches described here.

When the data are normally distributed, a simple approach for the choice of K is to use the theoretical normal distribution to pick appropriate K values. This approach has a long history in the pharmaceutical industry and comes from statistical process control (SPC) theory. Typically K is set equal to three which yields a range that represents "three sigma limits," the usual limits on an SPC chart. For data that follow a normal distribution, this range would encompass or cover 99.73% of the data. Only 0.27% should fall outside these limits. This is why three sigma limits seem a natural approach to setting specifications.

The pros of using three sigma limits (or any value of K such as 2 or 4) as the acceptance range are that the calculations are simple and easy to interpret. The cons of using these limits is that they do not provide an adequate estimate when the sample size is small (n < 100). In small sample situations, the three sigma range will not reliably provide the expected coverage of 99.73% as discussed in Dong et al. (2015).

Tolerance intervals provide an alternative to three sigma limits (see Sect. 2.6.7 for discussion and computational details). Since tolerance intervals are specified with both a confidence level $(1 - \alpha)$ and proportion (P) of the data covered, these intervals provide excellent flexibility for many specification setting scenarios. Tolerance intervals also adjust to the amount of data available and will cover the proportion of data specified. The downside of using tolerance intervals is that the interval may become overly wide in small data situations especially when the confidence and proportion of the future outcomes are set to high values. Figure 7.12 shows the K values for a variety of confidence levels and coverage. When the sample size is less than 10, the K values can be quite large.

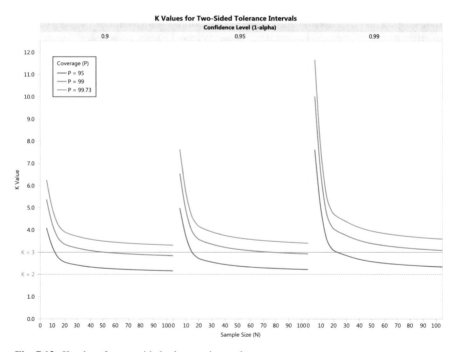

Fig. 7.12 K values for two-sided tolerance intervals

When using either three sigma limits or equivalent tolerance intervals it should be recognized that the specifications are set so that the process will at best achieve a process capability index of 1 (see Sect. 5.3.3 for more information on process capability). This means the process is just capable of meeting the specifications routinely. Any additional variation or shifts in the process mean will lead to a process that has a capability index of less than 1. If process data are the only source of information being used to set specifications, it may make sense to use the equivalent of a four sigma limits to allow for small increases in the process variation and/or shifts in the process mean. This would also allow for a process to achieve a capability index of greater than 1.

The challenge in using tolerance intervals is determining the appropriate confidence and coverage for the situation at hand. There is no accepted standard for confidence and coverage in the industry or among the various health authorities. Practical choices are described below and shown in Fig. 7.13.

- 95% confidence and 99% coverage. K is usually between 3 and 5 for most sample sizes of interest. K quickly approaches a value of three as the sample size approaches 50 ($K = 3.13$) but drops slightly below three after 80 results with $K = 2.92$ for 200 results.

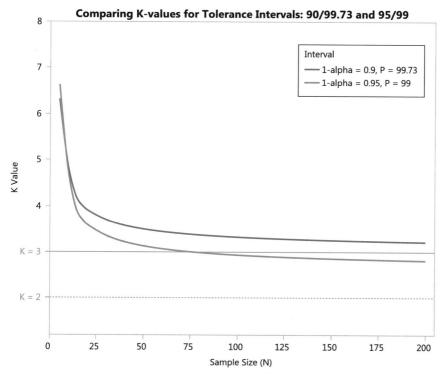

Fig. 7.13 Comparison of K values for 90% confidence/99.73% coverage and 95% confidence/99% coverage

- 90% confidence and 99.73% coverage. K is also between 3 and 5 for most sample sizes of interest. After 200 results, the value of K is still slightly above three at 200 results ($K = 3.22$). This choice would leave some "room" for slight process shifts or increases in process variation which are often experienced once a product goes into full scale production.

There are other types of tolerance intervals that can also be considered for establishing acceptance limits. These are a two-sided tolerance interval to control the tails (Hahn and Meeker 1991) of the distribution and two one-sided tolerance intervals. Both of these intervals focus on ensuring that no more than a specified proportion of the population is contained in the tails of the distribution. The tolerance interval discussed so far in this chapter only ensures that interval covers the specified proportion and is sometimes referred to as a tolerance interval to control the center. It is possible that one tail could have a greater percentage of the population in the tail than the other. Both the two-sided interval to control the tails and the two one-sided tolerance intervals provide a means to control the proportion of data in each of the tails separately. The tolerance interval formula for controlling the tails becomes

$$L = \bar{Y} - K_L\sqrt{S^2}$$
$$U = \bar{Y} - K_U\sqrt{S^2}. \tag{7.3}$$

where K_L and K_U are chosen so that $P(Y < L) < p_L$ and $P(Y > U) < p_U$ where p_L and p_U are desired maximum proportion of the population in the lower and upper tails of the population, respectively.

For the two-sided tolerance interval to control the tails, appropriate K values are provided in Hahn and Meeker (1991). The two one-sided tolerance interval approach is constructed just like it sounds. A lower one-sided tolerance interval is calculated for lower limit with the appropriate proportion to be allowed in the lower tail. The upper limit is calculated using an upper one-sided tolerance interval with the appropriate proportion to be allowed in the upper tail.

Two one-sided tolerance intervals have also played a role in recent work to establish a statistical test for release testing of inhalation products for delivered dose uniformity (DDU). In this situation, limits would already need to be specified and the two one-sided tolerance intervals would be calculated from the release data to ensure that the tails of the distribution contain no more than a specified proportion outside the limits with a specified confidence. This test is referred to as a Parametric Tolerance Interval Two One-Sided Test or (PTI-TOST) and has been extensively discussed in the literature (see Novick et al. 2009 and Tsong et al. 2015). Similar tests have also been proposed for UDU (Tsong and Shen 2007) and dissolution (Hauck et al. 2005).

7.4.4   *Simulation*

Simulation can be used to help set specifications if a reasonable model of the process and the sources of variability for the attribute or parameter can be developed. Usually Monte Carlo techniques are used to simulate data for the attribute. In this type of simulation, a model is developed to link together all the operations that occur related to the attribute. A distribution is assumed for each source of variation in the process that affects the attribute's final value. In a simulation run, a determination for the attribute is created by generating values for each of the sources of variation and then the values are combined according to the model to produce a final value for the attribute. To generate a distribution of final attribute values, several thousand simulation runs are conducted. This final attribute distribution can then be used to set specifications and to explore the effect of the various sources of variation on the attribute.

For example, a vaccine product typically experiences several temperatures during its distribution and shelf life. Assume it is known that exposure to higher temperatures can cause a measurable loss in potency. In order to understand the distribution of potency values at time of use and also set specifications for expiry, simulation can be used. Figure 7.14 shows the storage conditions that the product experiences during its distribution and storage. Potency for vaccine products is often log normally distributed so the potency is transformed using the natural log to allow for the use of the normal distribution in the simulation. The simulation starts by generating a release potency from the expected distribution of release values.

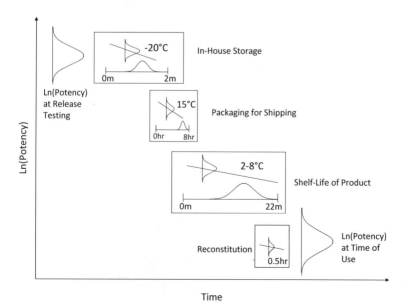

Fig. 7.14 Vaccine potency simulation model

This might be known from data taking during development and/or from the process targets during the dilution of the drug substance and lyophilization (freeze drying) steps that occur before the release testing is conducted. Assume this product has the potential to experience loss at several conditions:

- warehouse storage at $-20\,°C$ for up to 2 months
- packaging for shipping at $15\,°C$ for up to 8 h
- time in distribution at 2–$8\,°C$ for up to 22 months
- reconstitution for administration to the patient for 0.5 h.

The amount of time at each storage condition is obtained by generating a time for each storage condition using an assumed mean time of storage and the variability of the time at that storage condition. A slope representing the loss of potency for each storage condition is also obtained from an appropriate distribution. The loss is considered to vary in the simulation since the loss is estimated from limited studies conducted during development. Varying the amount of loss experienced for each simulation run allows for understanding of the impact of not knowing exactly the loss expected at that storage condition. The time and slope for each storage condition are combined across all of the storage temperatures to calculate the total potency loss for the lot. Subtracting this amount from the release potency yields the potency at time of use for the simulated lot. Figure 7.15 shows a histogram of the potency results at time of use after running the simulation for a total of 10,000 lots. This distribution can then be used to help determine specifications.

Simulation can also be used to assess probability to pass compendial tests such as uniformity of dosage units and dissolution. R code is provided at the book website to simulate several of the compendial tests discussion in Sect. 7.3 and these are listed in Sect. 7.8. The dissolution simulation can be used to help determine the appropriate sampling time and Q value for the release test. The simulation requires the user to input the mean and standard deviation expected for the dissolution values and a proposed Q value. The process generates dissolution

Fig. 7.15 Simulation results: ln(potency) at time of use

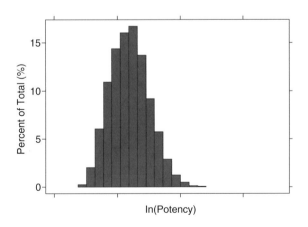

Table 7.6 R statements needed to run dissolution simulation and display key results

```
> # simulation run with 10,000 batches with the mean = 82% and

> # sd = 4% and the Q value set to 80% dissolved/%LC
> simRun1 <- simDissoTest(reps=10000, mean=82, sd=4, Q=80)

> # percent pass Stage 1
> mean(simRun1[,1])*100
[1] 0

> # percent pass at Stage 1 or 2
> mean(simRun1[,1]==1 | simRun1[,2]==1)*100
[1] 96.17

> # percent pass overall
> mean(simRun1[,12])*100
[1] 99.69
```

data from a normal distribution with the user selected mean and standard deviation. The program simulates data for the required number of batches which is also selected by the user. Table 7.6 shows the R statements needed to run the dissolution simulation and evaluate the data generated (note the # symbol means that line is a comment and the ">" is the prompt from the R Console Window). The simulation in Table 7.6 was run with a mean of 82%LC, standard deviation of 4%LC, and the Q value was set to 80%LC. 10,000 batches were simulated (reps = 10,000). The simulation indicates that setting the Q value at 80% for this process will mean that almost all batches will require Stage 2 testing since 0% passed at Stage 1. Approximately 96% of all tests will pass at Stage 2 and the rest will require Stage 3 testing. We can expect a 0.31% (100–99.69%) failure rate overall in routine manufacturing with these assumptions. Simulations such as this can help the manufacturer understand the risk of failure and testing requirements under various assumptions.

Design of Experiments (DOE) models can also be leveraged in combination with simulation techniques to help set and justify acceptance limits. If a DOE was conducted using the key factors in the process that affects the attribute of interest, the model developed from the DOE can be utilized to understand how variation in the processing factors will cause the attribute to vary. Figure 7.16 shows such a simulation for a downstream purification step in a biologic product. The factors studied in the DOE were Load Factor (g/L), Feed pH, and Elution pH. The attribute of interest is the Yield (%). The DOE data was first analyzed to develop a model relating Yield to the DOE factors. This model has been called a "transfer function" in the literature since it predicts how values for the factors transfer their impact to the attribute of interest (Little 2016). In this example, the model fit from the DOE study is

$$\text{Yield}(\%) = 66.7\text{-}9.4 \times \text{Load}$$
$$\text{Factor} + 6.5 \times \text{Elution pH-}4.0 \times \text{Feed pH} \qquad (7.4)$$
$$\text{Factor*Feed pH.}$$

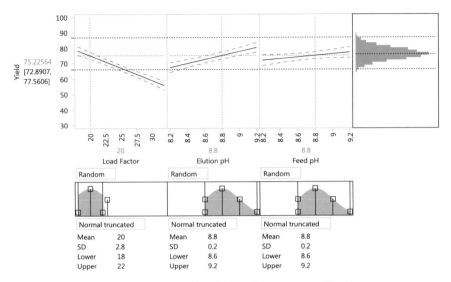

Fig. 7.16 Simulation using DOE model for yield for downstream purification step

Once the model is developed, the factors are set to the values where the process is expected to operate. The variation expected in the factors is also specified. In Fig. 7.16, the Load Factor was set to 20 g/L and the variation expected was defined with a truncated normal distribution with a standard deviation of 2.8. The Elution and Feed pH factors were similarly specified and variation in measuring the Yield was also added to the simulation. The histogram of simulated Yield values on the top right of Fig. 7.16 shows Yield can be expected to vary between 66 and 87% with the settings and assumptions used for the simulation. The settings of the factors and the expected variation can be changed to investigate how changing these will impact the expected Yield. Different distributions for the expected variation can also be specified. This approach can support a deeper understanding of the impact of set points for the factors and the impact of control strategies that reduce the expected variation in the factors and the attributes of interest. It can also help justify acceptance limits that are wider than the results observed during development. One simple approach to setting limits using simulated data is to use percentile methods discussed in the next section.

7.4.5   Percentiles

A percentile is the value in a set of data below which a certain percentage of the data fall. For example, the 99th percentile is the value in a set of data below which 99% of the data reside. Percentiles can be used in a similar fashion as the three sigma limits and tolerance intervals. If acceptance limits for a parameter were set so that 95% of the data were contained with the range, the 2.5th percentile and the 97.5th

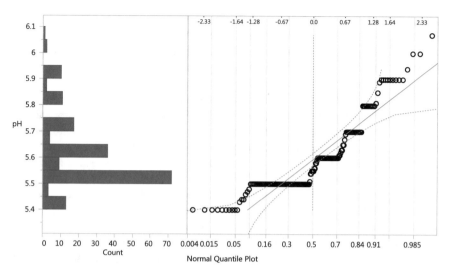

Fig. 7.17   Histogram and normal quantile plot of pH data

percentile could be used. Using percentiles requires a large data set to be effective. If percentiles near zero or 100 are needed, at least 100 results are needed with even more data needed as the percentiles approach zero and 100. The advantage of using percentiles is that the data do not need to follow any particular distribution. A disadvantage, in addition to requiring large data sets, is that percentiles will not provide any "buffer" for the acceptance limits. The limits will always be an actual data point or very close to an actual data point. There will be no gap between the data observed so far and the proposed acceptance range.

Figure 7.17 shows a histogram and normal quantile plot of pH data comprising 180 results. It is clear from both graphs that this data set is not normally distributed. If it was desired to set the acceptance range to contain 99% of data in the distribution of pH, the 0.5th and 99.5th percentiles are needed. The 0.5th and 99.5th percentiles of the pH data are 5.4 and 6.07 which are the minimum and maximum of the data. In contrast, a tolerance interval with 95% confidence and 99% coverage is (5.2–6.0). The tolerance interval assumes the data are normal distributed that does not reflect a reasonable acceptance range for the pH data.

7.5   Release and Stability Specifications

Chapter 8 provides detailed information on how to analyze data from stability studies. A key focus of the chapter is estimating the shelf life of a product using the methodologies described in ICH Q1E (2003) and several alternative approaches. The approaches in Chapter 8 assume that the acceptance limits are known. For some attributes, typical acceptance limits are known such as composite assay (90.0–110.0 or 95.0–105.0%LC) and protein concentration (target concentration ±10%). However, for some quality attributes such as degradates and charged variants, stability studies can be leveraged to help determine acceptance limits in conjunction with scientific understanding and patient safety. In this situation, one must first pick a reasonable shelf life (e.g., 24 or 36 months) and then determine acceptance limits that will support that shelf life given the results from the stability studies. Scientific understanding, patient safety, regulatory acceptance, and manufacturability must be taken into account in these situations. This analysis often requires an iterative approach where different shelf lives are assumed and the data is evaluated to see what acceptance limits would be needed to achieve the assumed shelf life.

A short summary of the ICH Q1E approach is provided to illustrate how stability analysis can be incorporated into acceptance limit determination. In ICH Q1E, a stability study for shelf life estimation is usually assumed to include at least 3 batches tested at regular intervals (typically at 0, 3, 6, 9, 12, 18, 24, and 36 months). The analysis described in ICH Q1E uses analysis of covariance (ANCOVA) to determine if the data from the batches in the study can be combined (or pooled) to estimate the shelf life of the product (or expiry). There are usually three "models" as shown in Fig. 7.18 for a simple stability study with batch and time as factors. The analysis determines which of these three models is appropriate for the data in the study.

Once the model is selected, a 95% confidence interval for the mean is constructed for the data allowing for pooling of the intercepts or slopes as appropriate. Pooling usually leads to longer shelf life estimates. The shelf life estimate is

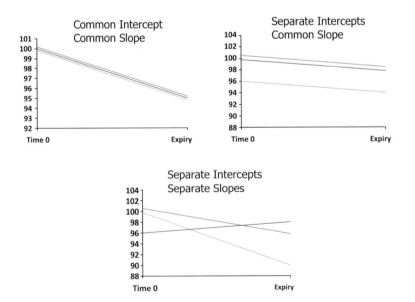

Fig. 7.18 Stability models for ICH Q1E ANCOVA analysis

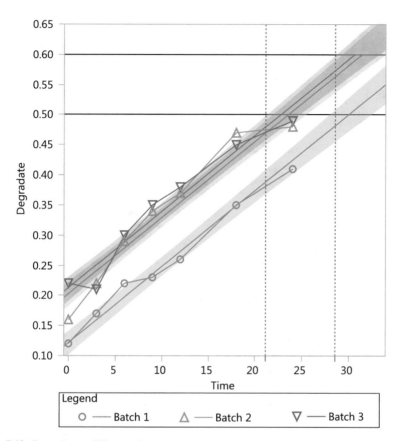

Fig. 7.19 Degradate stability results

the time where the confidence interval intersects the acceptance limit. Figure 7.19 shows the analysis of a degradate from a three batch stability study. In this example, the slopes can be pooled but the intercepts are statistically different so the Separate Intercept, Common Slope model is the appropriate model. The graph shows the shelf life is estimated to be 21 months if the acceptance limit for this degradate is set to 0.5%. If it is possible to increase the acceptance limit to 0.6%, the shelf life estimate is 28 months. Using a higher acceptance limit for a degradate would require understanding if the limit being proposed was acceptable to regulatory authorities and safe for the patient. Toxicology studies and clinical experience are often used to support the choice of the acceptance limit along with the stability analysis.

   It is possible to have separate release and shelf life acceptance limits for an attribute. In these cases, it has usually been demonstrated that the attribute changes on stability and this change needs to be reflected in acceptance limits. The degradate example shown in Fig. 7.19 could be such an attribute. If the acceptance

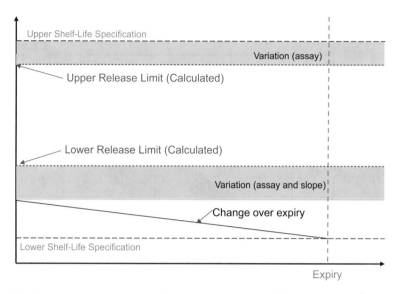

Fig. 7.20 Release limits—assuming attribute decreases over shelf life

limit for shelf life is set to 0.6%, a release limit below the 0.6% limit would need to be determined to ensure that the 0.6% limit will be met at the expiry. Figure 7.20 graphically shows how one derives a release limit. A "buffer" is added or subtracted from the shelf life limits to determine the release limits. This buffer is composed of estimates for the change expected over the shelf life (slope), assay variation, and variation of the slope estimate.

The calculations discussed in Chap. 8 provide a way to determine release limits that will support shelf life limits at expiry. For the example provided in Fig. 7.19, an upper release limit of around 0.25% would support the upper shelf life limit of 0.6% with 95% confidence. Depending on the specific situation, manufacturers will include both the shelf life limit and release limit in the regulatory filing or include only shelf life limit in the specification and designate the release limit as an "internal" limit. Some manufacturers also use release limits for attributes without significant change over shelf life to protect against releasing a product too close to the shelf life limits. If a batch of product is very close to one of the shelf life limits, it is possible that assay variation alone could cause an out-of-specification (OOS) result. Release limits in this case provide a measure of protection that such a situation will not occur.

7.6   Real Time Release/Large Sample Sizes

With the implementation of process analytical technology (PAT) in the pharmaceutical industry, there is an increased interest in developing acceptance criteria when a large number of measurements can be examined for a production batch. For example, dosage units of sample size 10–30 are typically assessed for uniformity of dosage units (UDU) for batch release testing using traditional measurement systems (e.g., HPLC). With the PAT technology, this sample size can be increased substantially to hundreds or thousands of dosage units per batch. The acceptance criteria for the traditional small sample testing are not appropriate for much larger sample sizes that are possible using PAT methods. In the last 10 years, there have been numerous approaches discussed in industry meetings and journals for developing appropriate large sample size criteria.

Uniformity of dosage units has been the most common application area for the development of large sample size specifications. By using PAT methods such as near infrared (NIR) spectroscopy, it is possible to measure the active ingredient content of a large number of dosage units during the compression or filling process. The data collection is complete when the batch run is finished and the batch can be "released" based on the data collected. This scenario is called real time release.

One of the first approaches proposed for large sample size UDU testing is a nonparametric counting test (Sandell et al. (2006); Bergum and Vukovinsky (2010)). This test focused on the meeting or exceeding the performance of a compendial UDU test (see Sect. 7.3.1) to detect unacceptable dose uniformity for various ranges of the process performance. The compendial test requires the acceptance value to be less than or equal to 15%LC. This requirement essentially means that a high proportion of individual content values should be within 85–115%LC. The counting test simply counts the number of individual values that are outside the 85–115%LC range. Equations are provided in both papers that define the number of individual content values that can be outside the 85–115% range for a given sample size. The Sandell and Bergum papers provide different limits with the Bergum paper having tighter limits. The approach in the Sandell paper is usually referred to as the Large-N test and the approach in the Bergum paper is the Modified Large-N test. The second and third columns of Table 7.7 show a comparison of the number of results that can be outside the 85–115%LC by sample size. The Modified Large-N approach was proposed to address regulatory feedback that the test should provide quality equivalent to or better than the compendial UDU test across the entire test range.

The Large-N and Modified Large-N counting tests have some desirable properties. One key property is that they are nonparametric tests. This means the test is not based on assuming a particular distribution such as the normal distribution for the UDU data. These tests are also quite simple to implement in a manufacturing environment. A table of sample sizes and acceptance values is all that is needed to assess the data. These tests allow differing sample sizes that are likely to occur when using PAT technologies for real time release testing. It is common for the PAT sampling strategies to be based on time where dosage units are sampled evenly over the production run. It is possible for the small differences in sample sizes to occur for each production run. This flexibility is very easy to achieve with counting tests.

Two more large sample size criteria for dose uniformity are presented in Chapter 2.9.47 of European Pharmacopoeia (EP) 8.8 (2016b). The chapter

Table 7.7 Acceptance values for Large-N UDU counting tests

Sample size	Large-N acceptance value (Sandell)	Modified Large-N acceptance value (Bergum)	EP 2.9.47
100	4	3	3
250	11	7	7
500	23	15	13
1000	47	30	25

discusses two alternatives: a parametric and nonparametric approach. The parametric version is a modification of the acceptance value ($AV = |M - \bar{X}| + ks$) formula used in the compendial test and discussed in Sect. 7.3.2. For this approach, the k value of the formula is adjusted based on the sample size. As the sample size increases, the k value increases. The nonparametric option is a version of the counting rules of Bergum and Sandell and is slightly tighter than the Modified Large-N Test as shown in the fourth column of Table 7.7. The EP chapter also limits the number of results outside of $(1 \pm L2 \times 0.01)M$ based on sample size. When L2 = 25% and M = 100%, this amounts to a range of 75–125%LC. This requirement addresses the concern that large deviations in dose uniformity should not be permitted. As the sample size increases, a larger number of results are allowed outside of the range.

As on-line measurement systems progress, there will certainly be more applications for large sample specifications beyond uniformity of dosage units. One concern as small sample size tests are translated to large sample sizes is the imposition of "zero tolerance" criteria. This concept was brought to the forefront in a paper by Murphy and Griffiths (2006). A zero tolerance criterion is one that allows no test results from a batch outside a specified range. This type of criterion gives the impression of ensuring that large deviations in the batch will not be allowed. The reality is that this is not the case and there are significant downsides to such criterion. For example, this creates a reluctance to test large numbers of results per batch since each test increases the risk of being outside the specified range. This becomes especially important as PAT technologies provide manufacturers an opportunity to learn more about their manufacturing process and products by allowing the nondestructive measurement of a large number of dosage units.

7.7 Incorporation of Clinical Experience

Clinically relevant specifications have long been a desire of the industry. It would be ideal if acceptance limits were determined based on safety and efficacy rather than process capability determinations. Much work has been done to find ways to link clinical performance and safety requirements to appropriate acceptance limits. One of most discussed areas is dissolution specifications for solid oral dosage forms

Table 7.8 Computer programs

Section	Program
7.3.2	Simulation of harmonized UDU test in R
7.3.2	Heatmaps for UDU test in R
7.3.4.1 and 7.4.4	Simulation of compendial dissolution test for immediate release products in R
7.3.4.1	Heatmaps for dissolution test for immediate release products in R
7.3.4.2	f_2 calculation in R

(Dickinson et al. 2008; Sunkara and Chilukuri 2003). One approach is to create formulations that vary key parameters such as particle size and tablet hardness. These formulations are then tested using the dissolution analytical method and the formulations are tested in the clinic. The dissolution data from the formulations deemed to have acceptable clinical performance are then used to determine the appropriate Q value and sampling time for the release tests (see Sect. 7.3.4).

Clinical experience can also be incorporated by combining both process capability and clinical performance. Capen et al. (2007) used statistically based limits for potency and then verified that product within these limits produced acceptable clinical performance in a potency ranging study.

Even though there are numerous presentations and articles on clinically relevant specifications, regulatory expectations are still evolving with regard to approaches for developing clinically relevant specifications.

7.8 Computer Programs

We have written several computer programs in the R language for performing some of the calculations and simulations discussed in this chapter. These are located at the book website. Table 7.8 describes the programs available at the website.

References

ASTM E2709 (2014) Standard practice for demonstrating capability to comply with an acceptance procedure. ASTM International, West Conshohocken

ASTM E2810 (2011) Standard practice for demonstrating capability to comply with the test for uniformity of dosage units. ASTM International, West Conshohocken

Bergum JS, Li H (2007) Acceptance limits for the new ICH USP 29 content uniformity test. Pharm Technol 30(10):90–100

Bergum J, Vukovinsky K (2010) A proposed content-uniformity test for large sample sizes. Pharm Technol 34(11):72

Bergum JS, Prescott JK, Tejwani RW, Garcia TP, Clark J, Brown W (2014) Current events in blend uniformity and content uniformity. Pharm Eng 34(2):1–10

Bergum J, Parks T, Prescott J, Tejwani R, Clark J, Brown W, Muzzio F, Patel S, Hoiberg C (2015) Assessment of blend and content uniformity. Technical discussion of sampling plans and application of ASTM E2709/E2810. J Pharm Innovation 10:84–97

Capen R, Shank-Retzlaff ML, Sings HL, Esser M, Sattler C, Washabaugh M, Sitrin R (2007) Establishing potency specifications for antigen vaccines: Clinical validation of statistically derived release and stability specification. BioProcess Int 5(5):30–43

De los Santos P A, Pfahler L B, Vukovinsky K E, Liu J, Harrington B (2015) Performance characteristics and alternative approaches for the ASTM E2709/2810 (CUDAL) method for ensuring that a product meets USP <905> uniformity of dosage units. Pharm Eng 35(5): 44–57

Dickinson PA, Lee WW, Stott PW, Townsend AI, Smart JP, Ghahramani P, Hammett T, Billett L, Behn S, Gibb RC, Abrahamsson B (2008) Clinical relevance of dissolution testing in quality by design. AAPS J 10(2):280–290

Dong X, Tsong Y, Shen M, Zhong J (2015) Using tolerance intervals for assessment of pharmaceutical quality. J Biopharm Stat 25(2):317–327

European Pharmacopoeia (EP) 8.8 (2016a) 2.9.40 Uniformity of dosage units

European Pharmacopoeia (EP) 8.8 (2016b) 2.9.47 Demonstration of uniformity of dosage units using large sample sizes

Food and Drug Administration. Center for Drugs Evaluation Research (1995) Immediate release solid oral dosage forms, scale-up and postapproval changes: chemistry, manufacturing and controls, in vitro dissolution testing, and in vivo bioequivalence documentation, guidance for industry

Food and Drug Administration. Center for Drugs Evaluation Research (1997a) SUPAC-MR modified release solid oral dosage forms, scale-up and postapproval changes: Chemistry, manufacturing and controls, in vitro dissolution testing, and in vivo bioequivalence documentation, guidance for industry

Food and Drug Administration. Center for Drugs Evaluation Research (1997b) Dissolution testing of immediate release solid oral dosage forms, guidance for industry

Food and Drug Administration. Center for Drugs Evaluation Research (1999, withdrawn 2002) ANDAs: Blend uniformity analysis, guidance for industry

Food and Drug Administration. Center for Drugs Evaluation Research (2003, withdrawn 2013) Powder blend and finished dosage units—Stratified in-process dosage unit sampling and assessment, guidance for industry.

Garcia T, Bergum J, Prescott J, Tejwani R, Parks T, Clark J, Brown W, Muzzio F, Patel S, Hoiberg C (2015) Recommendations for the assessment of blend and content uniformity: modifications to the withdrawn FDA draft stratified sampling guidance. J Pharm Innovation 10:76–83

Hahn GJ, Meeker WQ (1991) Statistical intervals: A guide for practitioners. Wiley, New York

Hauck WW, Foster T, Sheinin E, Cecil T, Brown W, Marques M, Williams RL (2005) Oral dosage form performance tests: new dissolution approaches. Pharm Res 22(2):182–187

International Conference on Harmonization (1999a) Q6A Specifications: test procedures and acceptance criteria for new drug substances and new drug products: chemical substances

International Conference on Harmonization (1999b) Q6B Specifications: test procedures and acceptance criteria for biotechnological/biological products

International Conference on Harmonization (2003) Q1E Evaluation for stability data

Japanese Pharmacopoeia (JP) (2016) 17th edition 6.02 Uniformity of dosage units

LeBlond D, Altan S, Novick S, Peterson J, Shen Y, Yang H (2016) In vitro dissolution curve comparisons: A critique of current practice. Dissolution Technol:14–23

Little TA (2016) Essentials in tolerance design and setting specification limits. BioPharm Int:41–45

Liu J, Ma M, Chow S (1997) Statistical evaluation of similarity factor f_2 as a criterion for assessment of similarity between dissolution profiles. Drug Inf J 31:1255–1271

Moore JW, Flanner HH (1996) Mathematical comparison of dissolution profiles. Pharm Technol 20(6):64–74

Murphy JR, Griffiths KL (2006) Zero-tolerance criteria do not assure product quality. Pharm Technol 30(1):52–60

Novick S, Christopher D, Dey M, Lyapustina S, Golden M, Leiner S, Wyka B, Delzeit H, Novak C, Larner G (2009) A two one-side parametric tolerance interval test for control of delivered dose uniformity: part 1–characterization of FDA proposed test. AAPS PharmSciTech 10(3):820–828

Sandell D, Vukovinsky K, Diener M, Hofer J, Pazden J, Timmermans J (2006) Development of a content uniformity test suitable for large sample sizes. Drug Inf J 40:337–344

Sunkara G, Chilukuri DM (2003) IVIVC: An important tool in the development of drug delivery systems. Drug Delivery Technol 3(4). http://www.drug-dev.com/Main/Back-Issues/IVIVC-An-Important-Tool-in-the-Development-of-Drug-256.aspx

Tsong Y, Shen M (2007) Parametric sequential quality assurance test of dose content uniformity. J Biopharm Stat 17(1):143–157

Tsong Y, Dong X, Shen M, Lostritto R (2015) Quality assurance test of delivered dose uniformity of multiple-dose inhaler and dry powder inhaler drug products. J Biopharm Stat 25(2):328–338

USP 39-NF 34 (2016a) General chapter <905> uniformity of dosage units. US Pharmacopeial Convention, Rockville

USP 39-NF 34 (2016b) General notices 3.10: conformance to standards, applicability of standards. US Pharmacopeial Convention, Rockville

USP 39-NF 34 (2016c) General notices 7.20: rounding rules. US Pharmacopeial Convention, Rockville

Chapter 8
Stability

Keywords Accelerated stability • Fixed batch ANCOVA • Matrix and bracketing designs • Pharmaceutical shelf life • Predictive stability model • Random batch hierarchical model • Regulatory guidance • Release limit estimation • Risk assessment • Stability design and analysis • Probability of achieving desired shelf life

8.1 Introduction

The stability of a drug product is defined by the rate of change over time of key quality attributes on storage under specific conditions of temperature and humidity. Understanding the stability of a pharmaceutical product (or any of its components) is important for proper quality design at many stages of the product life cycle. Table 8.1 lists some examples of pharmaceutical stability studies and their objectives.

Stability is intimately connected to many other key quality aspects of a drug product. For instance, interpretation of the rate of change of a key measure requires knowing the associated product release levels, recommended storage conditions, packaging, and stability acceptance limits. Proper interpretation of stability study data requires an understanding of the chemical-kinetic processes, the accuracy and precision of the associated test method, and the statistical limitations of the stability study experimental design. The statistical methodology used to achieve a given objective often depends on the quantity and quality of the data available and on the life cycle stage of the product.

A stability study should always be regarded as a scientific experiment designed to test certain hypotheses (such as equality of degradation rates among lots) or estimate certain parameters (such as shelf life). Similar to any other scientific process, the outcome of a stability study should lead to knowledge that permits the pharmaceutical manufacturer to better understand and predict product behavior.

© Springer International Publishing AG 2017
R.K. Burdick et al., *Statistical Applications for Chemistry, Manufacturing and Controls (CMC) in the Pharmaceutical Industry*, Statistics for Biology and Health, DOI 10.1007/978-3-319-50186-4_8

Table 8.1 Applications of stability studies in pharmaceutical development

Product development stage	Objective
Chemical characterization (Pharmaceutics)	Accelerated studies to define degradation pathways
Formulation development	Establish retest period for active ingredient. Excipient/packaging selection and compatibility studies
Clinical studies	Verify stability of clinical supplies. Assess risk of temperature excursions
Product registration	Shelf life estimation. Release limit estimation. Determine process capability with respect to release or acceptance limits. Comparison of stability of clinical, development, registration batches
Postapproval commitment	Shelf life confirmation/extension with long term studies annual stability monitoring
Life cycle management of marketed product	Determination of predictive model from historical data. Shelf life extension. Assess risk of temperature excursions. Routine trending. Justification of scale-up, process, formulation, dosage strength, manufacturing site, packaging or other postapproval changes. Establish equivalency with new formulations/packages. Annual stability reports
Trending of reference material	Estimate stability or retest date for analytical standards or calibrators

8.2   Regulatory Guidance

As with most drug development activities, a stability study is not merely a regulatory requirement. Rather, it is a key component in acquiring scientific knowledge that supports continued quality, safety, and efficacy. Regulatory guidance is provided to ensure that the required knowledge base is available to support a pharmaceutical product throughout its shelf life.

The regulatory aspects of drug product stability are governed by a number of interrelated regulatory guidance documents. One must use caution in interpreting and using this guidance. Often, their individual scopes are somewhat narrow and may not always include the specific objectives of interest to the developer. There is no guarantee that following such guidance will lead to a manufacturing process that is approvable and will yield high quality product. Each guidance is established by a separate committee and despite harmonization efforts, inconsistencies are inevitable. In addition, current guidance often lags behind advancements in scientific and statistical best practices. Blind adherence to guidance may lead to suboptimal decision making. A developer must always consider the specific study objectives and the approach taken should always be scientifically and statistically justified.

In the USA, the European Union, and Japan, these guidance documents are provided through the International Conference on Harmonization of Technical Requirements for Registration of Pharmaceuticals for Human Use (ICH).

1. ICH Q1A (R2, 2003a), ICH Q1B (1996), ICH Q1C (1997), ICH Q1D (2002), and ICH Q1E (2003b) govern stability testing of drug substances and products,
2. ICH Q2(R1) (2005) and ICH Q2B (1996) govern validation of analytical methods for (among other things) stability testing,
3. ICH Q3A (2003c) and ICH Q3B (Revised) (2003d) govern impurity levels in drug substances and products,
4. ICH Q6A (1999a) and ICH Q6B (1999b) govern acceptance criteria, and
5. ICH Q5C (1995) governs stability testing of biotechnological/biological products.

8.3 Modeling Instability

No pharmaceutical is perfectly stable. In other words, all pharmaceuticals exhibit some degree of instability. This instability is quantified as a finite rate of loss in potency, a rate of increase in products of chemical degradation, or a rate of change in a key quality attribute over time. Understanding the magnitudes of these rates is a critical aspect of any pharmaceutical development program. In this section we discuss predictive kinetic models for such instability.

We will start with shopping lists of key response, experimental, and controlled variables that often appear in drug product stability studies. The lists are by no means exhaustive. Then we will discuss how each is typically incorporated into the kinetic model.

8.3.1 Stability Study Variables

8.3.1.1 Response Variables

The response variable that is monitored over time should include, as appropriate, results from the physical, chemical, biological, microbiological, and/or key performance indicators of the dosage form. The potency level of the active ingredient(s) and the levels of related degradants or impurities are always considered. In the case of instability due to degradation, mass balance should be accounted for. Only quantitative variables will generally be amenable to statistical analyses as described below.

Other critical or key quality attributes also need to be considered if they have acceptance limits. Dissolution may be especially critical for modified release products or products of poorly soluble compounds where bioavailability may change over time when dissolution is the rate limiting step for drug absorption. Other responses that may be included in stability studies are color, moisture or solvent level, preservative, antioxidant, or pH.

ICH Q2 (R1) governs the validation of analytical methods used for stability testing. Quantitative, precise, unbiased, stability indicating test methods are essential components of stability experimentation. All information obtained from a stability study ultimately comes from the test methods used in the study. No amount of statistical sophistication can compensate for a poor measurement system. The following recommendations promote the use of high quality test methods and data management approaches.

1. Use quantitative test methods which produce a result on a continuous measurement scale. These have higher information content than those that produce binary (yes/ no), discrete (1, 2, 3,...), ordinal (low, medium, high), or categorical (A, B, C,...) responses.
2. The validity of information obtained from a stability study ultimately depends on the selected analytical test methods. Where possible, these test methods should be thoroughly characterized and validated prior to use. Partial validation may be acceptable for early development studies, but the uncertainties of the test method need to be accounted for in data evaluation. The validation should include determination of bias and precision as a function of true concentration level. The sources (components) of variance (replication, instrument, operator, day, lab, calibration run, etc.) should be identified and quantified. Such understanding is important in both study design (e.g., assuring adequate sample size or study power) and analysis (e.g., deciding whether to average replicated test values). See Chap. 6 for more information on analytical methods.
3. Excessive rounding and binning (e.g., conversion of measured values to below detection or below quantitation) of continuous measured values should be avoided. Such practices are common when reporting individual analytical results. However, when values such as stability testing results are used as input for further data analyses, over-rounding and binning of test results lowers the data information content and distorts the measurement error structure. These abuses may limit the statistical procedures available for data analysis and lead to biased estimates or incorrect conclusions. All information contained in the original measurements should be used for efficient, sound decision making. More discussion on this topic is provided in Sect. 2.3.
4. Design stability databases with data analysis in mind. Statistical analyses may lead to approval of a longer shelf life, or provide information needed to better manage a product throughout its life cycle. Often statistical analyses are not performed or included in submissions because hand re-entry of data is required, excessive reformatting must be performed, or there may be a delay in obtaining analytical results.
5. As recommended by Tsong (2003), include a trained statistician on the study design and analysis team.

Analytical data are obtained at great expense. The information contained within these data represents a proprietary advantage to the product developer. Thus the analytical methods used to produce the data, and the computing/statistical methods used to extract information from them should be of the highest possible quality.

8.3.1.2 Experimental Fixed Factors

Factors such as storage time, dosage strength, and packaging type are called fixed factors (fixed effects) because the levels (e.g., 3 months of storage, 300 mg) have specific meaning that is the same whenever that level is used in the study.

Storage time is the primary experimental fixed factor always included in a stability study. ICH Q1A recommends testing every 3, 6, and 12 months during the first, second, and subsequent years, respectively. Thus for a typical 3 year study, testing would occur at 0, 3, 6, 9, 12, 18, 24, and 36 months.

Other experimental fixed factors that may be included as part of more complex studies include container type or size, closure type, fill size, desiccant type or amount, manufacturing site, batch size, or other covariates.

8.3.1.3 Experimental Random Factors

Factors such as replicate number or batch number are called random factors (random effects) because the levels used (say replicate 1, 2, or 3) are not the same whenever they are applied. Rather, these levels are assumed to represent experimental conditions drawn at random from a hypothetical infinite population of all possible replicates.

The ICH Q1A guidance recommends that stability studies for new product registration include at least three batches of drug product (say batches 1, 2, and 3). If batch is considered as a fixed factor, then inference concerning the shelf life pertains only to the three batches in the study. If, on the other hand, batch is considered to be a random factor, then the conclusions of the study are valid for the hypothetical infinite population of all (future) batches. To properly treat batch as a random factor requires use of a hierarchical or multi-level statistical model. Such models can often be expressed as "mixed models" as described in Sect. 2.12.7.

It has been our experience that a random effects model often leads to a longer shelf life estimate because the fixed model estimate is based on the "worst case" lot. Since the shelf life specification is meant to apply to all future batches, the random effects model may be a more pertinent model. However in most new product registrations governed by ICH Q1E, where the minimum of three batches are available, batch is treated as a fixed variable and any inferences are strictly limited to the three batches on hand. This is done, quite simply, because data from only three batches, without some additional prior knowledge, are insufficient to project to the larger population of all future batches. This caveat must be born in mind when interpreting the results of stability analyses.

Table 8.2 Temperature and humidity control of stability studies

Label storage	Long term condition	Intermediate condition	Accelerated condition
Controlled room temperature	$25 \pm 2°C/$ $60\% \pm 5\%$ relative humidity	$30 \pm 2°C/65\% \pm 5\%$ RH	$40 \pm 2°C/75\% \pm 5\%$ RH
Refrigerated	$5 \pm 3°C$		$25 \pm 2°C/60\% \pm 5\%$ RH
Frozen	$-15 \pm 5°C$		$5 \pm 3°C/$ambient RH
Minimum time period covered by data at submission	12 months	6 months	6 months

8.3.1.4   Controlled Factors

Temperature and humidity are usually controlled during a stability study. The product label storage condition will dictate the conditions used in studies conducted to estimate shelf life. Table 8.2, taken from ICH Q1A, describes typical conditions used in a stability study.

For controlled room temperature products susceptible to moisture loss, lower relative humidity (RH) storage conditions may be appropriate (40% for long term and intermediate and 15% for accelerated conditions). For controlled room temperature products stored in water impermeable containers, ambient RH might be appropriate for different temperature conditions.

While storage temperature and RH are quantitative variables that are often included in drug product stability studies, they are not generally included in a stability model. As specified by ICH Q1E, each storage condition (i.e., long term, intermediate, accelerated) is evaluated separately. An exception to this is with analysis of accelerated stability studies as briefly discussed in Sect. 8.6.3. Inclusion of storage condition variables as predictors in the stability model can be useful in judging the risk of temperature or humidity excursions which can occur during storage of finished drug products in warehouses or in a patient's environment. Such evaluations are briefly discussed in Sect. 8.6.4.

8.3.2   Predictive Stability Models

The design and analysis of a stability study requires the specification of a kinetic, predictive model for the instability of each key attribute. A model should include mechanistic, experimental, and statistical aspects. That is, the model must take into account the physical or chemical mechanisms that result in changes of the average measure over time. It must account for the fixed effects of factors whose levels are systematically varied as part of the stability study. Further, it must account for the statistical variation introduced by random factors whose levels are not specifically controlled, but vary in the study (e.g., product lot, analyst, delay between sample collection and analysis, and reagent or analytical standard lot).

An understanding of the physicochemical mechanisms of instability of a drug product is an essential component of drug product life cycle support. Knowledge of these mechanisms allows a developer to anticipate flaws, design an appropriate formulation and packaging system, and to troubleshoot and support the product throughout its life cycle. The reader is directed to Carstensen (1995) for a discussion of kinetic mechanism.

When the rate of change in the response measure is nearly constant over time and the change is monotonic (strictly increasing or decreasing), linear regression (LR) as discussed in Sect. 2.12 may be used to describe the stability profile. LR provides a simple description that is easily understood by non-statisticians, requires few assumptions, and can be executed in widely available software. While the assumptions of LR are few, they must be satisfied for any estimates, inferences, or predictions made with the regression model to be valid. The assumptions of LR were presented in Sect. 2.12.2. In the context of stability testing, these assumptions are:

1. The response (Y) is linearly related to the storage time.
2. There is no uncertainty in the storage time value.
3. Errors in the response measurements are normally distributed.
4. Errors in the response measurements are mutually independent.
5. The error variance in the response measurement is the same at all points in time.

Section 2.12.2 provides methods for verifying these assumptions.

If the kinetic processes of instability are more complex, the profile may exhibit curvature. If a theoretical kinetic model is available, then nonlinear regression approaches may be the best way to draw inferences, estimate shelf life, and make predictions from the stability data. In some cases, a nonlinear model can be linearized by transformation. For instance, the first-order kinetic model $Y = Ae^{-kt}$ can be rewritten $log(Y) = log(A) - kt$. Thus a log transformation of the response, Y, may improve conformance to assumption 1 above.

When the stability profile is not well represented by a straight line and ignorance of the underlying kinetic process prevents use of a nonlinear theoretical model, various transformations of the response or time scales (or both) may be tried in an effort to allow LR to be used for analysis. It is important that the transformation be valid for all possible response or time values. For instance, transformations such as $log(Y/(A - Y))$ or $log(t)$ are undefined for $Y = A$ or $t = 0$, respectively. Section 2.12.6 provides more information on fitting nonlinear models using variable transformation.

Transformation of the time scale can have subtle effects on estimation efficiency but generally presents few statistical issues. However, transformation of the response scale may fundamentally change the error structure of the response measurements (assumptions 2 and 5). In favorable cases, a transformation may be found that will improve conformance to all five basic LR assumptions. However, if the transformation does not have a theoretical basis, extrapolations beyond the scope of the study must be made with caution.

As indicated in ICH Q1A, a linearizing transformation or use of an appropriate nonlinear model for the effect of storage time should be considered if the stability profile cannot be represented by a straight line. A thorough discussion of the important topic of physicochemical mechanisms is beyond the scope of this chapter. The chosen model for the effect of storage time should be scientifically and statistically justified.

A zero order kinetic (straight line) model is often sufficient to describe the relationship between the average of the response and storage time. This is particularly true when the response changes only a small amount from its initial or potential value (say less than 15%).

8.4 Shelf Life Estimation

8.4.1 Definition of Shelf Life

A coherent discussion of shelf life estimation requires an absolutely clear definition of shelf life. To date, regulatory guidance does not provide such a definition. Instead, guidance suggests estimation methodology and leaves the definition largely to the imagination of the product sponsor. This has led to a fascinating diversity of opinions about what shelf life really means and how best to estimate it.

Let's see what definition can be gleaned from the following ICH Q1E excerpt (emphasis added).

> The purpose of a stability study is to establish, based on testing a minimum of **three batches** of the drug substance or product, a retest period or shelf life and label storage instructions applicable to **all future batches** manufactured and packaged under similar circumstances. The **degree of variability of individual batches** affects the confidence that **a future production** batch will remain within acceptance criteria throughout its retest period or shelf life... An appropriate approach to retest period or shelf life estimation is to analyze a quantitative attribute (e.g., assay, degradation products) by determining the earliest time at which the 95 percent confidence limit for the **mean** intersects the proposed acceptance criterion.

Amazingly, the guidance implies that it is advisable to estimate a shelf life for all future batches based on testing results from only three batches while acknowledging batch to batch variability. Ignoring the obvious naivety here and looking further we notice the emphasis on the **mean**. Is this the mean of each individual batch or the mean of the process that makes batches (or something else)? The guidance is not clear here, but the spirit of the passage above suggests, and we assume in this chapter, that it is the individual mean of each future batch that must conform. Not all accept this definition of shelf life. Some suggest that shelf life should be set to control the spread of individual results using prediction or tolerance bounds (see, for instance, Kiermeier et al. 2004). Unfortunately, stability results are generally obtained from tests on composited units so such proposals cannot claim to control unit dose uniformity.

ICH Q1E further encourages the consistent use of appropriate statistical methods to analyze long term primary stability data to establish, with a high degree of confidence (e.g., 95%), a shelf life during which the *mean* of each stability indicating attribute will remain within acceptance criteria for all future batches produced by a given process. Ideally, the statistical method will be based on a model that accounts for known kinetic processes and measurement variability. However, when kinetic mechanisms are not understood, an approximate straight line, or zero order kinetic model, is often used. Use of a straight line model to approximate (say) an exponential kinetic process may be adequate as long as the overall loss in potency is less than 15%. In such cases, the use of a simple linear regression method is appropriate. In other cases, transformations or the use of nonlinear models may be required. Other commonly made assumptions are normality and independence. When such assumptions are not appropriate, statistical support is recommended.

8.4.2   Single Batch

Because batch to batch variation in initial levels and rates of change can be expected, drawing inferences about a process from only a single batch is clearly risky. Regulators generally expect shelf life estimation for the population of batches to be based on data from at least three batches. It is best to regard a shelf life estimate based on data from a single batch as a preliminary estimate that strictly applies to that batch only.

An appropriate approach, using batch number 2 from the potency data set of Shuirmann (1989), provided at the book website, is illustrated below. The unit of measure of this potency response is percent of label claim (%LC) and the unit of measure for time is months. If the data for batch 2 are in an R data frame called stab1, a simple linear regression is obtained using the following R statements:

```
fit<-lm(Potency~Months,data=stab1)
summary(fit,correlation=TRUE)
anova(fit)
```

This produces the following summary and ANOVA tables:

```
Coefficients:
            Estimate Std. Error t value Pr(>|t|)
(Intercept) 100.24914    0.42208  237.51  < 2e-16 ***
Months       -0.18013    0.03411   -5.28 0.000746 ***


Residual standard error: 0.9044 on 8 degrees of freedom
Multiple R-squared: 0.7771,      Adjusted R-squared: 0.7492
F-statistic: 27.88 on 1 and 8 DF,  p-value: 0.0007459


Correlation of Coefficients:
        (Intercept)
Months -0.74

Analysis of Variance Table
Response: Potency
          Df  Sum Sq Mean Sq F value     Pr(>F)
Months     1 22.8057 22.8057  27.883 0.0007459 ***
Residuals  8  6.5433  0.8179
---
Signif. codes:   0 '***' 0.001 '**' 0.01 '*' 0.05 '.' 0.1 ' ' 1
```

The time at which the 95% one-sided lower bound on the average response intersects the stability acceptance limit provides an estimate for the batch shelf life. The formula for this 95% lower bound as presented in (2.81) for the given storage time X_P is

$$L = b_0 + b_1 X_P - t_{0.95:n-2} \times \sqrt{\frac{1}{n} + \frac{(X_P - \bar{X})^2}{\displaystyle\sum_{i=1}^{n} (X_i - \bar{X})^2}} \times S \qquad (8.1)$$

For the data shown above, $b_0 = 100.249\%LC$ (shown as (Intercept) Estimate in the output), $b_1 = -0.18013\%LC/month$ (shown as the "Months" estimate in the output), $n = 10$, $S = 0.9044\%LC$ (labeled as "Residual standard error" in the output), $\sum_{i=1}^{n} (X_i - \bar{X})^2 = 702.9$, $\bar{X} = 9.1$, and $t_{0.95:n-2} = t_{0.95:8} = 1.860$. For this example, assume the lower one-sided stability acceptance limit is 90%LC, where "%LC" is shorthand for "% of label claim." To determine shelf life, one sets L equal to 90%LC in (8.1) and then solves for X_P. That is,

$$L = b_0 + b_1 X_P - t_{0.95:n-2} \times \sqrt{\frac{1}{n} + \frac{(X_P - \bar{X})^2}{\sum\limits_{i=1}^{n}(X_i - \bar{X})^2}} \times S$$

$$90 = b_0 + b_1 X_P - t_{0.95:n-2} \times \sqrt{\frac{1}{n} + \frac{(X_P - \bar{X})^2}{\sum\limits_{i=1}^{n}(X_i - \bar{X})^2}} \times S \tag{8.2}$$

$$0 = -90 + b_0 + b_1 X_P - t_{0.95:n-2} \times \sqrt{\frac{1}{n} + \frac{(X_P - \bar{X})^2}{\sum\limits_{i=1}^{n}(X_i - \bar{X})^2}} \times S$$

The R function uniroot () can be used to solve this quadratic equation for the shelf life, X_P, as follows:

```
Lower.Limit<-90
intercept<-coef(fit)[1]
slope<-coef(fit)[2]
n<-length(stab1$Potency)
s<-sqrt(anova(fit)[2,3])
Sxx<-var(stab1$Months)*(n-2)
xbar<-mean(stab1$Months)
# Ref Chow and Liu page 362
# Note that the interval has to be chosen carefully to include only
one of the roots or will get an error
Shelf.Life<-uniroot(f=function(x,L,a,b,n,s,Sxx,xbar){
        L-a-b*x+qt(0.95,n-2)*s*sqrt(1/n+(x-xbar)^2/Sxx)
        },
interval=c(0,100),tol=0.00000000001,L=Lower.Limit,
   a=intercept,b=slope,n=n,s=s,Sxx=Sxx,xbar=xbar)$root
Shelf.Life
```

For this example where the stability limit is 90%LC, the resulting shelf life is $X_P = 43.6$ months. This simple linear regression and the two-sided confidence interval on the average are illustrated in Fig. 8.1 which is produced by the following R statements:

```
new<-data.frame(Months=seq(0,60,0.1))
pred.w.clim<-predict(fit, new, interval = "confidence",level = 0.90)
pred.w.plim<-predict(fit, new, interval = "prediction",level = 0.90)
matplot(new$Month,cbind(pred.w.clim, pred.w.plim[,-1]),
        col=c(1,2,2,3,3),lty=c(1,2,2,3,3), type="l", ylab="Potency",
xlab="Month")
```

Fig. 8.1 Shelf life
estimation for a single batch

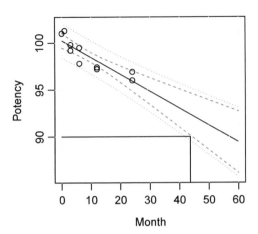

```
points(stab1$Months,stab1$Potency)
grid()
lines(c(0,Shelf.Life),rep(Lower.Limit,2))
lines(rep(Shelf.Life,2),c(Lower.Limit,0))
```

The individual observed potencies are plotted as circles and the predicted batch mean level, obtained by simple linear regression, is indicated by the solid black line through the data. The red dashed lines give the two-sided 90% confidence intervals for the average batch potency. The lower mean confidence line is also a one-sided 95% confidence lower bound on the average batch potency. Note this lower line crosses the potency line of 90%LC at month $X_P = 43.6$ months.

8.4.3   Fixed Batch Analysis of Covariance

ICH and FDA guidance permit shelf life assignment with as few as three batches. With stability data from only three batches, it may be unrealistic to attempt to model batch to batch variation. In this case a fixed effects model is employed and inference is limited to the batches in the stability study (Chow and Liu 1995). The shelf life estimation serves as a demonstration that the manufacturing process *is capable of* producing a limited number of batches all of which can meet the estimated shelf life with high (at least 95%) confidence. The selected stability batches may have been manufactured using similar materials in the same facility by the same operators over a limited period of time. When shelf life estimation is based on limited manufacturing experience, it is best to regard the exercise as a preliminary shelf life *assignment* rather than an *estimate* of the shelf life that should apply to all future batches made by the process.

In what follows, we assume that the proper kinetic model is zero order (i.e., slope is constant over time) and that variance is homogeneous. The homogeneity assumption is important. If the error variance cannot be assumed similar among batches, then FDA and ICH guidance recommends that each batch be analyzed separately by the approach described above for a single batch. In other words, the batches cannot be "pooled" as discussed by Ruberg and Stegman (1991). This generally will result in a shorter shelf life assignment because fewer degrees of freedom will be available for estimating uncertainty. If the content uniformity differs among batches, or if there are other differences among batches that affect analytical precision, variance heterogeneity may be present in the data. In what follows, we assume variance homogeneity.

Analysis of covariance (ANCOVA) is a statistical procedure that is a hybrid of analysis of variance (ANOVA) and linear regression in which the model includes both quantitative and qualitative predictor variables. If the underlying assumptions of ANCOVA are met, it may be used to choose a stability model for assignment of shelf life. A good description of ANCOVA including its underlying assumptions is given by Brownlee (1984). At one time the FDA provided SAS macros for stability ANCOVA analysis (Ng 1995), but these macros appear to have been withdrawn by the agency.

The ANCOVA assumes that one of the following stability models is appropriate:

1. CICS = common intercept and common slope for all batches. This is the basic model in Eq. (2.78) of Chap. 2.
2. SICS = separate intercept and common slope for all batches. This model is discussed in Eq. (2.108) of Chap. 2.
3. SISS = separate intercepts and separate slopes for each batch. This model is discussed in Eq. (2.112) of Chap. 2.

The underlying algebraic description of these three models has already been given in Chap. 2. A fourth model (common intercept and separate slopes) is sometimes included if there are scientific reasons for considering it, but it will not be discussed here.

The purpose of a stability ANCOVA is to select the appropriate model (CICS, SICS, or SISS) for shelf life estimation purposes. To do this, three statistical F-tests are conducted as indicated in Table 8.3. Each test is associated with a null hypothesis (less complex model) and an alternate hypothesis (more complex model).

The null hypothesis is accepted unless the p-value associated with the test is less than 0.25. Bancroft (1964) originally proposed 0.25 for such applications as a way of assuring adequate statistical power. This type I error rate is much higher than the usual 0.05 and may lead to adopting an unnecessarily complex model, which in turn may result in an under-estimation of the shelf life. It has been argued that such a liberal decision criterion serves as a disincentive for manufacturers to improve analytical precision or increase the amount of stability data used in shelf life estimation. While the use of 0.25 remains controversial, it has become a regulatory standard.

Table 8.3 Hypotheses associated with each ANCOVA test

Test	Null hypothesis	Alternate hypothesis
A	CICS	SISS
B	CICS	SICS
C	SICS	SISS

Table 8.4 Regression sums of squares for the CICS, SICS, and SISS models

Model	Matrix of independent predictors	Matrix algebra expression for the regression error sum of squares	Degrees of freedom associated with error sum of squares
1 (CICS)	$\mathbf{X}_1$ N rows, 2 columns	$SSE_1 = \mathbf{Y}^T(\mathbf{I} - \mathbf{H}_1)\mathbf{Y}$	$dfe_1 = N - 2$
2 (SICS)	$\mathbf{X}_2$ N rows, k + 1 columns	$SSE_2 = \mathbf{Y}^T(\mathbf{I} - \mathbf{H}_2)\mathbf{Y}$	$dfe_2 = N - k - 1$
3 (SISC)	$\mathbf{X}_3$ N rows, 2k columns	$SSE_3 = \mathbf{Y}^T(\mathbf{I} - \mathbf{H}_3)\mathbf{Y}$	$dfe_3 = N - 2k$

The ANCOVA tests of most interest are tests B and C. Test B is a test for separate intercepts, given that the slope is common among batches. Test C is a test for separate slopes, given that intercepts are separate among batches. Test A is a joint test for separate slopes and intercepts and is not employed for shelf life estimation.

The ANCOVA tests are best described using matrix algebra (see Appendix at the end of the book). Table 8.4 makes use of matrix notation to define the tests for the hypotheses in Table 8.3. Inspection of this table can be insightful. Consider a stability study with k batches and N total test results for a given compound. We place the N test results in a column vector denoted $\mathbf{Y}$. $\mathbf{Y}$ is the response variable vector. The predictor variable $\mathbf{X}$ will then be a matrix with N rows and a column of ones in the first column. The total number of columns in $\mathbf{X}$ will depend on the particular linear model of interest. (You may wish to review the format of the indicator variables and interactions in Sect. 2.12.) ANCOVA is the statistical tool used to perform the F-test that identifies the appropriate stability model.

In the above table, $\mathbf{I}$ is the $N \times N$ identity matrix. For Model 1 (CICS), the $\mathbf{X}_1$ matrix has two columns representing, respectively, the indicator coefficients of the common intercept and common slope for all k batches. For Model 2 (SICS), the $\mathbf{X}_2$ matrix will have k columns representing indicator coefficients for the k separate intercepts and an additional indicator column coefficient for the common slope. For Model 3 (SISS), the $\mathbf{X}_3$ matrix will have 2k columns, represent the indicator coefficients for the intercept and slope for each of the k batches. $\mathbf{H}_i = \mathbf{X}_i(\mathbf{X}_i^T\mathbf{X}_i)\mathbf{X}_i^T$ is the $N \times N$ "hat" matrix for the model i regression. The superscript T indicates the matrix transpose operation. The equations used in the above table are standard regression equations found in most regression textbooks (e.g., see Neter et al. 1996, p. 229, Eq. 6.35). Table 8.5 gives the formulas used in the calculation of each statistic in the ANCOVA table where $MSE = SSE_3/dfe_3$.

Table 8.5 ANCOVA table for model selection

Test from Table 8.3	SS	Df_{Test}	SS/df	F_{Test}	P-value$_{Test}$
A	SSE_1-SSE_3	dfe_1-dfe_3	MS_A	MS_A/MSE	$1 - \Phi(F_A, df_A, dfe_3)$
B	SSE_1-SSE_2	dfe_1-dfe_2	MS_B	MS_B/MSE	$1 - \Phi(F_B, df_B, dfe_3)$
C	SSE_2-SSE_3	dfe_2-dfe_3	MS_C	MS_C/MSE	$1 - \Phi(F_C, df_C, dfe_3)$

In the table, $\Phi(F_i, df_i, dfe_3)$ represents the cumulative F distribution for the observed F ratio for test i with df_i and df_3 degrees of freedom in the numerator and denominator, respectively.

The hypotheses tests associated with tests A and C in the above table are easily justified based on the extra sum of squares principle in linear model building (see Neter et al. (1996), p. 80, Eq. 2.70). In this paradigm, the alternative model represents the "full model" and the null model represents the "reduced model."

The hypotheses test associated with source B is a modification of this principle. The F ratio for test B in the table substitutes MSE in the denominator in place of the traditional SSE_2/dfe_2. This F-test is more clearly understood as a type I test of the batch effect on the intercept in the SISS model.

The ANCOVA test A is usually provided, but not used for model identification. It is a joint test of slopes and intercepts as is easily seen since $(SSE_1 - SSE_2) + (SSE_2 - SSE_3) = (SSE_1 - SSE_3)$. MSE provides a pooled estimate of total analytical plus content non-uniformity variance that may be quite useful. The statistics associated with E are not used for any statistical tests but are traditionally provided for completeness.

From the above P-values associated with tests B and C (P-value$_B$ and P-value$_C$, respectively), the ANCOVA algorithm performs the model selection procedure according to the following rules:

Select Model 1 (CICS) if P-value$_B$ > 0.25 and P-value$_C$ > 0.25
Select Model 2 (SICS) if P-value$_B$ < 0.25 and P-value$_C$ > 0.25
Select Model 3 (SISS) otherwise.

Table 8.6 gives another view of these calculations and statistical tests which relate to these three models which are described in detail in Chap. 2.

In SAS GLM syntax, the ANCOVA could be conducted as an analysis of the SISS Model as follows (order of terms is critical):

```
PROC GLM;
CLASS BATCH;
MODEL LEVEL = TIME BATCH TIME*BATCH/ SS1;RUN;
```

Here, the intercept term is assumed present but not explicitly included in the model statement. The resulting GLM output would automatically print type I sequential tests for each term in the model. In the SAS output, the test associated with the BATCH effect is test B and the test associated with the TIME*BATCH effect is test C.

Table 8.6 Summary of calculations and statistical F-tests for stability model selection by ANCOVA

Model	CICS	SICS	SISS
Model terms	Time	Time	Time
		Batch (categorical coding)	Batch (categorical coding)
			Time*batch (categorical coding)
Model equation	(2.78)	(2.108)	(2.112)
Error df	$df e_1 = N - 2$	$df e_2 = N - k - 1$	$df e_3 = N - 2k$
Error SS	$SSE_1 = \sum_{i=1}^{N} \left(Y_i - \hat{Y}_{CICS,i}\right)^2$	$SSE_2 = \sum_{i=1}^{N} \left(Y_i - \hat{Y}_{SICS,i}\right)^2$	$SSE_3 = \sum_{i=1}^{N} \left(Y_i - \hat{Y}_{SISS,i}\right)^2$
Test B	$F_B = \dfrac{(SSE_1 - SSE_2) \cdot df e_3}{(df e_1 - df e_2) \cdot SSE_3}$		
Test C	$F_C = \dfrac{(SSE_2 - SSE_3) \cdot df e_3}{(df e_2 - df e_3) \cdot SSE_3}$		

Having identified and fit the appropriate regression model, it remains to find the predicted response, $\hat{Y}_h$, and its 95% confidence interval, (L, U), as a function of the predictor, $\mathbf{x_h}$, for all batches (if the CICS model is chosen) or each batch (otherwise). The predictor, $\mathbf{x_h}$, is a row vector with the same structure as a row of the $\mathbf{X}$ matrix except that the desired storage time at which predictions are needed is employed. The predicted response is equal to

$$\hat{Y}_h = \mathbf{x_h} \left(\mathbf{X^T X}\right)^{-1} \mathbf{XY} \tag{8.3}$$

and the confidence interval for the mean predicted response is

$$(L, U) = \hat{Y}_h \pm d \times t_{1-\alpha/2:df e_3} \times \sqrt{\frac{SS_3}{df e_3}} \cdot$$
$$d = \sqrt{\mathbf{x_h} \left(\mathbf{X^T X}\right)^{-1} \mathbf{x_h^T}} \tag{8.4}$$

Note that Eq. (8.4) uses the mean squared error and error degrees of freedom from the SISS model (see Table 8.6) regardless of the final model selected by the ANCOVA algorithm. An alternative approach used by some software packages (such as the Minitab stability platform illustrated below) is to substitute the mean squared error and degrees of freedom for the final model selected into Eq. (8.4). This alternative approach seems to promote an undesirable feature of "using the data twice" (once to choose the model, and again to determine shelf life from the model). Unless $df e_3$ is small (say less than 20), it should make little difference which procedure is used. However, we feel that consistently using the statistics obtained from the full SISS model represents statistical best practice.

If the stability risk is identified as only a decreasing or increasing response (but not both), then a one-sided confidence bound, L or U, respectively, is employed with $\alpha = 0.05$. Otherwise a 2-sided confidence interval is used in which $1-\alpha/2$ is replaced with $1-\alpha$. For models SICS and SISS the predictor vector also includes indicator variables that identify the specific batch for which predictions are desired. Note that Eq. (8.4) performs the same computation as given by Eq. (2.96) of Chap. 2 with $d = \mathbf{x_h}\left(\mathbf{X^T X}\right)^{-1}\mathbf{x_h^T}$.

Following the definition of shelf life given in ICH Q1E, one looks for the longest storage time at which the response remains above L or below U (one-sided) or within the interval (L, U) (two-sided). An analytical expression for the shelf life involves a solution to a quadratic equation. It is generally more convenient and informative to plot $\hat{Y}_h$, L, and/or U as a function of storage time using the above equation and observe the storage times at which the response remains within stability acceptance limits for all batches. Such a relationship is illustrated for three fixed batches in Fig. 8.2.

The R functions ANCOVA, SHELF.LIFE, and slplot at the book website will provide the above ANCOVA model identification, response and confidence interval estimation, and stability profile graphics. If X, Batch, and Y are vectors containing storage times, a batch identification number, and observed responses, then these functions are called using the following R statements:

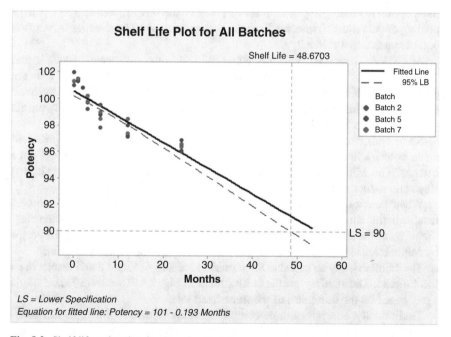

Fig. 8.2 Shelf life estimation for three fixed batches

Table 8.7 Minitab ANCOVA output with potency batches 2, 5, and 7

Source	DF	Seq SS	Seq MS	F-Value	P-Value
Months	1	80.3588	80.3588	117.17	0.000
Batch	2	0.5981	0.2990	0.44	0.651
Months*Batch	2	0.3137	0.1568	0.23	0.797
Error	25	17.1462	0.6858		
Total	30	98.4168			

```
ancova<-ANCOVA(Batch=Batch,X=X,Y=Y)
ancova
predict<-SHELF.LIFE(Batch=Batch,X=X,Y=Y,Model=ancova[[2]],pool=-
TRUE,conf=0.9)
predict
slplot(Batch=Batch,X=X,Y=Y,Model=ancova[[2]],Limit=90,predict=-
predict)
```

Minitab 17 provides a convenient shelf life estimation platform for fixed batches. As an example, Potency batches 2, 5, and 7 (provided on the book website) are analyzed below in Minitab using the menu options Stat/Regression/Stability Study/Stability Study. The data are arranged in three columns: Months, Potency, and Batch (categorical variable). A one-sided lower specification of 90 is used. The fixed option for Batch, a 95% confidence lower bound on the mean and an alpha for pooling of 0.25 must be selected from the options window. The resulting ANCOVA output appears in Table 8.7.

ANCOVA tests B and C correspond to the sources "Batch" and "Months*Batch," respectively, in Table 8.7. Since the P-values for both tests are above 0.25, a CICS model is selected for these batches. Minitab solves the quadratic equation and prints a shelf life, (here 48.67 months) and provides a stability profile plot as shown in Fig. 8.2 that illustrates the estimation process.

Since both intercept and slope are common among batches, a single predicted mean potency line and its lower 95% confidence bound are shown for all three batches. The Minitab stability platform also produces estimates of intercept and slope and their standard errors as well as various regression diagnostics.

When batches 3, 4, and 5 are analyzed similarly in Minitab, a SICS model results. In this situation, each batch has a different shelf life with the minimum shelf life taken as the assigned shelf life. The ANCOVA and shelf life estimates reported by Minitab are shown in Table 8.8 and the stability profile plot is in Fig. 8.3.

The Minitab analysis in Table 8.9 considers batches 4, 5, and 8 and results in an SISS model. The stability profile is illustrated in Fig. 8.4. The overall shelf life is again based on the minimum of the three fixed lots.

Traditionally, analysis of more complex stability studies follows a model building and fitting process similar to the ANCOVA described above. Good reviews are given by Chow and Liu (1995) and Shao and Chow (1995). The statistical aspects of the analysis for a multifactor fixed effects model are discussed by Fairweather et al.

Table 8.8 Minitab ANCOVA output with potency batches 3, 4, and 5

Source	DF	Seq SS	Seq MS	F-Value	P-Value
Months	1	74.489	74.4895	60.01	0.000
Batch	2	53.968	26.9839	21.74	0.000
Months*Batch	2	0.455	0.2273	0.18	0.834
Error	22	27.309	1.2413		
Total	27	156.221			
Batch	Shelf Life				
Batch 3	48.988				
Batch 4	57.316				
Batch 5	43.454				
Overall	43.454				

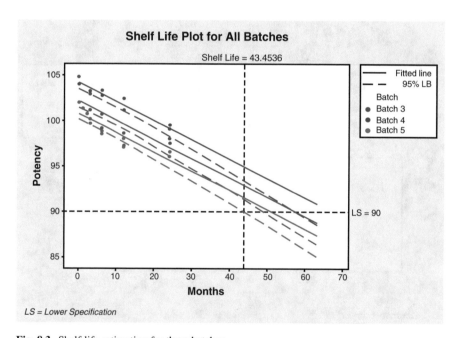

Fig. 8.3 Shelf life estimation for three batches

(1995) and Chen et al. (1997). When a mixed effects model is used (see the discussion in Sect. 8.4.4), the same model identification principles have been employed. Analysis of mixed effects stability models is described by Chen et al. (1995) and Chow and Shao (1991). Such approaches require ad hoc decision rules based on arbitrary criteria (such as a minimum p-value). In practice, such approaches can lead to swings in shelf life projections as data accumulate. It would seem desirable that model selection be based on scientific understanding rather than arbitrary model selection rules.

Table 8.9 Minitab ANCOVA output with potency batches 4, 5, and 8

Source	DF	Seq SS	Seq MS	F-Value	P-Value
Months	1	45.451	45.4513	100.99	0.000
Batch	2	64.918	32.4588	72.12	0.000
Months*Batch	2	1.760	0.8800	1.96	0.170
Error	18	8.101	0.4500		
Total	23	120.230			
Batch	Shelf Life				
Batch 3	59.520				
Batch 4	44.207				
Batch 5	27.166				
Overall	27.166				

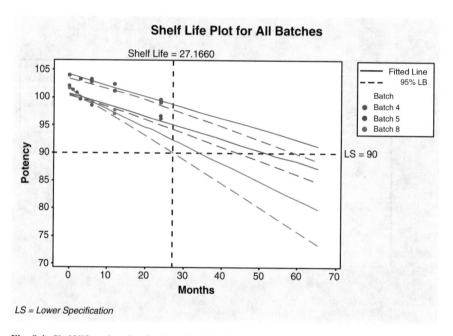

Fig. 8.4 Shelf life estimation for three fixed batches

8.4.4 Random Batch

Because the fixed batch shelf life estimated using the ICH Q1E approach in Sect. 8.4.3 applies only to batches in hand, it cannot be used to make inferences about the shelf life of future batches. To illustrate the problems of using the ICH Q1E approach, consider the related substance data set provided in an Excel sheet on the book website for which 16 batches of stability data are available. The response is "RSTotal" or total weight of undesirable chemicals resulting from the breakdown

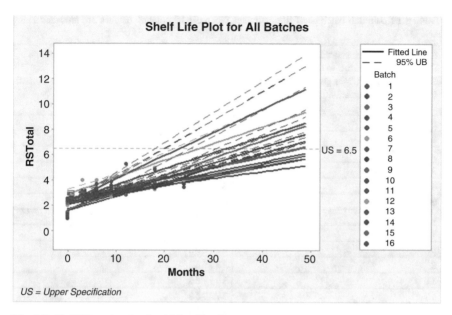

Fig. 8.5 Shelf life estimation for 16 fixed batches

of the active pharmaceutical ingredient expressed as a percent of the label claim or %LC. There is a one-sided upper specification of 6.5%LC. When these data are analyzed using the fixed batch approach in Minitab, an SISS model is selected. The stability profile shown in Fig. 8.5 results. Table 8.10 shows the estimated shelf life for each batch (rounded down to the nearest whole month).

The SISS model results in a separate stability profile and 95% confidence upper bound for each batch with the shelf life being set by the worst-performing batch. The worst-performing batch was batch 13 with a 19.728 month shelf life estimate based on 18 months of data. Batch 16 with only 9 months of stability data was the second-worst batch with an estimated shelf life of 19.998 months. However, 11 of the 15 batches for which shelf life estimates were provided had estimated shelf lives greater than 30 months. The four batches with full 24 month data all exhibited shelf lives above 40 months. In this light, it does not appear that 19.728 months is an appropriate shelf life for future batches. It is rare for all batches to have complete data in a stability data set at the time of analysis because batches are made in sequence (or in limited campaigns) and the initials are therefore staggered. In cases where the SISS model is selected, batches with the least data often exhibit the shortest shelf life simply because their mean levels are most uncertain and have the widest 95% confidence bounds. The estimated slopes for batches with limited data are also highly influenced by random measurement variation in the initial or last data value. As the number of batches being analyzed increases, the probability of underestimating the shelf life for the population of future batches increases.

Table 8.10 Minitab batch shelf life estimates when all 16 potency batches are analyzed using ANCOVA

Shelf Life Estimation
Upper spec limit = 6.5
Shelf life = time period in which you can be 95% confident that at least 50% of response is
below upper spec limit

Batch	Shelf Life
1	44.435
2	30.440
3	25.222
4	40.533
5	43.701
6	43.982
7	38.670
8	42.888
9	33.512
10	38.785
11	37.748
12	21.883
13	19.728
14	32.926
15	*
16	19.998
Overall	*

The mean response slope for batch 15 is not significantly larger than zero. No shelf life estimate for batch 15 is available.

When stability data are available from a sufficient number of batches of a product (say six or more), it is possible to make inferences not only about the batches under study, but also about the process that makes batches itself. Doing so requires hierarchical modeling of the process. As with the fixed batch case, a stability model is assigned to each individual batch. For instance, the following mixed model described in (2.116) may be considered.

$$Y_{ij} = \mu + L_i + (\beta + B_i) \times t_{ij} + E_{ij} \qquad (8.5)$$

where Y_{ij} is the response for lot i at time point j, and the intercept and slope of the ith batch are $\mu + L_i$ and $\beta + B_i$, respectively. The random variables L_i, B_i, and E_{ij} are assumed to be iid normally distributed with zero mean and standard deviations σ_L, σ_B, and σ_E, respectively. Analysis of this model is described in the next section. Section 8.6.1 provides an extension of the model in (8.5) that assumes L_i and E_{ij} are correlated.

8.4.5   *Random Batch Approach Using SAS Proc Mixed*

The predictive model shown above can be thought of as a mixed model and analyzed using the SAS procedure Proc Mixed. This is a very flexible procedure which can be used to analyze a wide variety of linear statistical models that include both fixed and random effects.

Consider the data set of 16 batches analyzed using the fixed effects model in the previous section. The Proc Mixed code below provides a mixed model analysis of this 16 batch degradation data set.

```
data omni;
input batch $ month     RSTotal;
cards;
1    0     1.4
1    3     2.3
...
16    6    3
16    9    3.6
run;

* add missing values for existing batches;
proc sort data=omni;by batch month;run;
data omni;set omni;by batch month;
  output;
  if last.batch then do;
    do month= 0 to 60 by 1;
        RSTotal=.;
        output;
      end;
  end;run;

* add missing values for future batch;
data future;
do month= 0 to 60 by 1;
  batch='f';
  RSTotal=.;
  output;
end;

data both; set omni future;run;

filename myoutput '<file pointer>';
proc printto print=myoutput;run;
```

```
proc print data=both;run;

proc mixed data=both covtest cl;
  class batch;
  model RSTotal = month/solution alphap=0.1 outp=outp;
  random intercept month/type=vc subject=batch;
  prior/out=posterior nsample=60000;
  run;

proc print data=outp;run;

proc print data=posterior;run;
```

The above SAS code implements a Bayesian shelf life estimation. Note that there are alternative non-Bayesian estimation procedures available and these may lead to slightly different shelf life estimates. However, our experiences with Bayesian approaches in stability analysis have been positive and we find them quite informative. More details about Bayesian approaches and the above code can be found in Sect. 8.4.6. At this point we just focus on the key output from the Proc Mixed procedure which is shown in Table 8.11. Proc Mixed provides point and standard error estimates for the five model parameters (fixed effects μ and β, and squares of the standard deviation parameters σ_L, σ_B, and σ_E). The point estimates are shown in Table 8.12.

The Proc Mixed statements above produce an output data set which includes predictions of the average RSTotal levels at months 0, 1, ..., 60 for each of the 16 batches and also for a random future batch whose slope and intercept are unknown, but would be randomly drawn from the respective normal distributions with means and variances estimated above. The output data set includes a one-sided 95% confidence upper bound on this estimated RSTotal mean. The longest whole month at which the RSTotal mean is still below the upper limit of 6.5%LC is given in Table 8.13.

Based on the above list, a future batch, made by the process should be within the acceptance limit for RSTotal for 29 months with a confidence level of 95%. Recall the fixed effects model in the previous section produced a shelf life of only 19.7 months. It should be noted that batches 12 and 13 are projected to be within the acceptance limit for only 24 or 23 months, respectively. Setting the shelf life to only 23 months would be consistent with the worst-case philosophy advocated by ICH Q1E. However, the worst-case philosophy is inappropriate when data from a large number of batches are available. The worst-case philosophy leads to an inverse relationship between estimated shelf life and number of batches which serves as a testing disincentive for sponsors. It seems more appropriate to set a shelf life using the model projection of 29 months for a future batch.

Table 8.11 Proc mixed output from a random batch analysis of the 16 batch data set

Covariance Parameter Estimates

Cov	Parm	Subject	Estimate	Standard Error	Z Value	Pr > Z	Alpha	Lower	Upper
Intercept	σ_L^2	batch	0.07354	0.04439	1.66	0.0488	0.05	0.02961	0.3945
month	σ_B^2	batch	0.000718	0.000458	1.57	0.0586	0.05	0.000278	0.004424
Residual	σ_E^2		0.1571	0.02650	5.93	<.0001	0.05	0.1158	0.2253

Solution for Fixed Effects

Effect	Estimate	Standard Error	DF	t Value	Pr > |t|
Intercept μ	2.3166	0.09443	15	24.53	<.0001
month β	0.09136	0.009655	15	9.46	<.0001

Table 8.12 Parameter point estimates obtained from Table 8.11

Parameters	Description	Computed estimate
μ	Overall average across all lots	2.3166
β	Average slope across all lots	0.09136
σ_L^2	Lot-to-lot variance in intercept	0.07354
σ_B^2	Lot-to-lot variance in slope	0.000718
σ_E^2	Variance of analytical method	0.1571

Table 8.13 Batch shelf lives including a "Future Batch" computed for RSTotal data set

Batch	Shelf Life
1	46
2	36
3	35
4	41
5	41
6	43
7	38
8	40
9	35
10	38
11	40
12	24
13	23
14	40
15	29
16	29
future batch	29

8.4.6   *Considerations with the Traditional Mixed Model Approach*

When using any statistical package, a user should always consult the technical descriptions in the software manual to appreciate the underlying assumptions, approximations, and algorithms used in the analysis. Below we briefly highlight some details that should be considered.

The default estimation method used by Proc Mixed is restricted (or residual) maximum likelihood (REML). REML reduces the bias in estimates of variance parameters (compared to the usual maximum likelihood approach) but includes a "penalty term" that may not be appropriate in certain situations. While strict adherence to the principles of maximum likelihood disallows a negative variance component estimate, REML can sometimes produce negative variance estimates which are (by default) set to zero.

For mixed models, the concept of degrees of freedom can be unclear and as a default Proc Mixed uses a procedure called "containment" to approximate the denominator degrees of freedom. The containment approximation may be adequate unless significant imbalance (correlation among predictors) is present in the experimental design. Sometimes containment can result in an estimate of the degrees of freedom equal to zero.

Often in mixed model analysis, interest focuses on certain parameters and others (such as residual variance) may be regarded as "nuisance parameters." By default, the Proc Mixed procedure will "profile" or "sweep" the model residual variance out of the likelihood expression using a numerical process. In doing so, some information about the parameters of interest may be lost. While this does not usually affect parameter point estimates, it may impact the standard errors or confidence intervals of the parameters of interest.

As a cautionary note, Proc Mixed tends to underestimate sampling variability because no account is taken of the uncertainty in estimating the key covariance matrices. This is a particular issue with small sample sizes (e.g., small numbers of stability batches). To compensate for this, Proc Mixed uses approximate t and F statistics for statistical tests and confidence interval estimation. Other user options include the use of Kenward–Rogers inflation factors and Satterthwaite degrees of freedom approximations. It is important to assess the impact of such approximations on the inferences made.

8.4.7 Random Batch Analysis Using Bayesian Analysis

The preceding technical considerations are an unfortunate, but unavoidable, distraction to the analyst whose goal is to learn about the data at hand rather than the vagaries of the computing methodology. In this regard, Bayesian methods provide some relief as the modeling paradigm is unified under Bayes rule (see Sect. 2.13). Additionally, the associated Markov Chain Monte-Carlo (MCMC) methodology (i.e., Gibbs, Metropolis, or related algorithms) frees the analyst from concern about statistical derivations or mathematical approximations. This is because modern Bayesian software (e.g., BUGS, JAGS, Stan) expresses a statistical model using a syntax that requires only statements of the mathematical and probabilistic relationships that underlie the model. The need for analytical derivations is minimized. Thus many problems that are intractable by traditional approaches are easily implemented—without approximations—in a Bayesian setting.

Background in Bayesian thinking is provided in Sect. 2.13 of this book. For a more comprehensive, and easily readable, introduction to Bayesian approaches, the text by Kruschke (2015) is highly recommended.

SAS Proc Mixed can provide a Bayesian analysis for a class of mixed models called variance component models which ignore covariances between (in this case) random batch slopes and intercepts.

For variance component models, such as the predictive model used here, which assumes zero covariance between random batch slopes and intercepts, Proc Mixed is also capable of generating a sample from the joint posterior of the five model parameters using an independence chain algorithm. This is implemented using the prior statement as shown in the SAS code in Sect. 8.4.5. By default, Proc Mixed uses Jeffreys' prior distributions for the five parameters.

Because this random batch model allows each batch to have its own intercept and slope, we are essentially hypothesizing an SISS model. An important difference between the random batch analysis we describe here and the fixed batch ANCOVA shown in Table 8.10 is that we do not incorporate here any statistical tests for model selection. While such tests are possible, they seem unnecessary and perhaps even unwise. If a simpler model (such as CICS or SICS) better describes the data, this should be evident from the magnitudes of the variance parameters σ_L and σ_B. Rather than make a post hoc model selection based (usually) on limited data, we use the full SISS model and let the estimated parameter values serve to restrict the model as warranted by the data.

The output data set "posterior" provides a sample of 60,000 values from the joint posterior which can be used for shelf life estimation (without the troublesome asymptotic assumptions and approximations described above). However, the PRIOR statement in Proc Mixed requires a variance component model and the default prior selection is limited.

SAS Proc MCMC can provide a more general Bayesian analysis, but we will not discuss this approach here. Instead, we will show below how to conduct Bayesian stability analyses using the well-established freeware package, WinBUGS (Lunn et al. 2013).

8.4.8   Using WinBUGS to Perform a Stability Analysis

A Bayesian analysis of the 16 batch data is performed by assuming diffuse $N(0, 10^6)$ prior distributions on the model parameters μ, and β, and $U(0, 100)$ distributions on the model parameters σ_L, σ_B, and σ_E. These prior distributions contribute little information and give nearly equal weight to parameter estimates that are supported by the data. The WinBUGS file that implements this model is shown below.

```
model
{
    #Likelihood
    for(i in 1:n){
        intercept[i] ~ dnorm(mu,tau.L)
        slope[i] ~ dnorm(beta,tau.B)
    }
```

```
for(k in 1:N){
    Yhat[k] <- intercept[batch[k]] + slope[batch[k]]*t[k]
    Y[k] ~ dnorm(Yhat[k],tau.E)
}

#Priors
mu ~ dnorm(0, 1.0E-6)
beta ~ dnorm(0, 1.0E-6)
sigma.E ~ dunif(0,100);tau.E <- 1/(sigma.E*sigma.E)
sigma.B ~ dunif(0,100); tau.B <- 1 / (sigma.B * sigma.B)
sigma.L ~ dunif(0,100);tau.L <- 1 / (sigma.L * sigma.L)
}
```

This file is called by the R code provided on the website (the library R2WinBUGS is needed). This code generates 60,000 MCMC draws (post burn-in) from the posterior distribution of model parameters. We label each of these draws by the superscript m, where $m = 1, 2, \ldots, 60{,}000$. Draw m from the joint posterior distribution is then a vector of five scalar parameters and two sub-vector values and can be indicated using a bracketed superscript as

$$\left(\mu^{[m]}, \beta^{[m]}, \sigma_L^{[m]}, \sigma_B^{[m]}, \sigma_E^{[m]}, \mathbf{L}^{[m]}, \mathbf{B}^{[m]} \right) \tag{8.6}$$

where $\mathbf{L}^{[m]}$ and $\mathbf{B}^{[m]}$ are sub-vectors of the batch specific random effects L_i and B_i, respectively.

Marginal statistical summaries of each parameter (except $\mathbf{L}^{[m]}$ and $\mathbf{B}^{[m]}$) are output and shown below.

```
parameter   mean    sd   2.5%    25%    50%    75%   97.5%  Rhat  n.eff
mu         2.317 0.104  2.110  2.250  2.318  2.384  2.521 1.001 22000
beta       0.091 0.011  0.071  0.084  0.091  0.098  0.113 1.001 42000
sigma.L    0.294 0.102  0.103  0.227  0.287  0.353  0.516 1.004 13000
sigma.B    0.029 0.011  0.009  0.022  0.028  0.035  0.052 1.001 60000
sigma.E    0.408 0.036  0.344  0.382  0.405  0.430  0.485 1.001 38000
```

For each parameter, n.eff is a crude measure of effective size of the MCMC sample (taking account of the serial correlation in the Markov chain reduces its effective sample size), and Rhat is the potential scale reduction factor (at convergence, Rhat = 1). Each MCMC vector also includes draws for 16 slopes and 16 intercepts. Thus we have 60,000 MCMC posterior draws of the slope and intercept for each of the 16 batches. Samples from the posterior distribution of the ith individual batch intercept and slope can be obtained as follows:

$$Intercept_i^{[m]} = \mu^{[m]} + L_i^{[m]}$$
$$Slope_i^{[m]} = \beta^{[m]} + B_i^{[m]}$$

(8.7)

These intercepts and slopes can be compared conveniently as box plots available from the WinBUGS inference/compare dialogue. Like the boxplots described in Sect. 2.4, the boxes represent inter-quartile ranges. However, in the WinBUGS boxplots, the solid black line near the center of each box is the posterior mean, the arms of each box extend to cover the central 95% of the posterior distribution— their ends correspond, therefore, to the 2.5 and 97.5% quantiles. The WinBUGS boxplots thus differ somewhat from the traditional boxplot description given in Sect. 2.4. These are shown in Figs. 8.6 and 8.7. The default value of the baselines shown in these plots are the global means of the 16 intercept and slope posterior means, respectively.

Samples from the posterior (predictive) distribution of any (random) function of model parameters can easily be obtained from the MCMC posterior samples of model parameters using the obvious equations and (random) functions. For instance, a sample from the posterior distribution of the mean level of the ith batch at time t can be obtained as

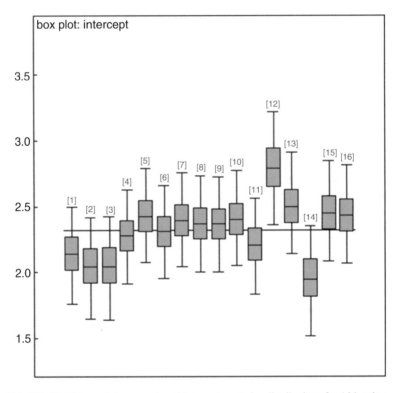

Fig. 8.6 WinBUGS box plot summaries of intercept posterior distributions for 16 batches

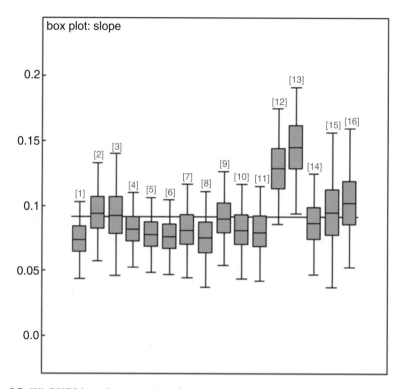

Fig. 8.7 WinBUGS box plot summaries of slope posterior distributions for 16 batches

$$\mu_i^{[m]}(t) = \text{intercept}_i^{[m]} + t \cdot \text{slope}_i^{[m]}. \tag{8.8}$$

Assuming an upper stability acceptance *Limit* on the batch mean level (in this case equal to 6.5%LC), a posterior sample from the shelf life of the ith batch can be obtained as

$$SL_i^{[m]} = \frac{Limit - \text{intercept}_i^{[m]}}{\text{slope}_i^{[m]}}. \tag{8.9}$$

The shelf life samples obtained using Eq. (8.9) are usually left- and right-censored at zero and some large shelf life (such as 200 months), respectively, to avoid negative or infinite values. The censoring will not affect the mode or median point estimates and will have minimal or no effect on interval estimates.

Figure 8.8 shows the posterior distributions (in the form of WinBUGS box plots) for each of the 16 batches in the data set.

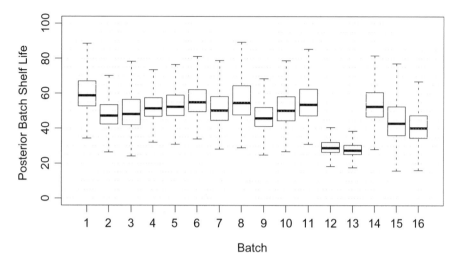

Fig. 8.8 WinBUGS box plots of shelf life posterior distributions for 16 batches

Similarly, the predictive posterior distributions of the slope and intercept of future batches are obtained as follows:

$$\text{intercept}_{fut}^{[m]} \sim N\left(\mu^{[m]}, \left(\sigma_L^{[m]}\right)^2\right) \tag{8.10}$$

$$\text{slope}_{fut}^{[m]} \sim N\left(\beta_1^{[m]}, \left(\sigma_B^{[m]}\right)^2\right) \tag{8.11}$$

where the "$\sim$" indicates the operation of drawing a single pseudo random number from each of the respective m distributions (one for intercept and one for slope). The subscript *fut* indicates that the resulting numbers represent samples from the distribution of the intercept and slope of a random future batch. A sample from the posterior predictive distribution of the mean level of a future batch at time t can be obtained as

$$\mu_{fut}^{[m]}(t) = \text{intercept}_{fut}^{[m]} + \text{slope}_{fut}^{[m]} \cdot t. \tag{8.12}$$

A definition of shelf life consistent with the spirit of ICH Q1E is the earliest storage time at which less than 95% of future batch mean total related substance level is below the acceptance limit of 6.5%LC. The time at which the predictive posterior mean of future batch m intersects 6.5%LC will be

$$SL^{[m]} = \frac{Limit - \text{intercept}_{fut}^{[m]}}{\text{slope}_{fut}^{[m]}} \tag{8.13}$$

As with the batch specific shelf life, to account for the occurrence of negative slopes (which lead to a negative shelf life) and batches with very long (or even infinite shelf lives), the variable SL is truncated between 0 and 200 months as follows:

$$Shelf.Life^{[m]} = \begin{cases} 200 & \text{if } SL^{[m]} \leq 0 \\ SL^{[m]} & \text{if } 0 < SL^{[m]} \leq 200 \\ 200 & \text{if } SL^{[m]} > 200 \end{cases} \quad (8.14)$$

The logic behind the first line in Eq. (8.14) is as follows. For any viable product, it is likely that posterior values for the intercept are well below the *Limit*, so the numerator of Eq. (8.13) should be positive. Therefore, a negative value of $SL^{[m]}$ implies that the slope (denominator of Eq. (8.13)) is non-positive—total related substances are not changing or are decreasing. This suggests a long (i.e., infinite) shelf life. However, common sense argues against an infinite shelf life so we truncate these infinite values at some realistically long shelf life such as 200 months. Similarly in the last line of Eq. (8.14), any values of $SL^{[m]}$ above 200 months are truncated at 200 months.

The posterior distribution of *Shelf.Life* for a future batch is shown in Fig. 8.9.

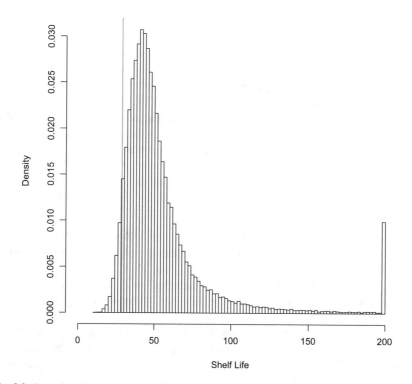

Fig. 8.9 Posterior distribution of shelf life for a future batch

Figure 8.9 is a histogram of the 60,000 values of *Shelf.Life[m]* obtained from application of Eqs. (8.10)–(8.14). *Shelf.Life[m]* is thus a random function of the posterior distributions of underlying model parameters. It is a random function because Eqs. (8.10) and (8.11) require random normal sampling. We refer to the distribution of such random functions as posterior predictive distributions. The estimated shelf life for future batches is taken as the 5th percentile of this posterior predictive distribution of shelf lives of future batches.

The estimated shelf life in this case is 28.39 months (round down to 28 months). The red line in the figure illustrates this estimated shelf life. This result agrees well with 29 month shelf life obtained using the Proc Mixed REML mixed model analysis, but both are considerably different than the 19 months shelf life obtained using the traditional ICH Q1E ANCOVA worst-case batch approach.

It is also of interest to know the probability that future stability test results for batches placed on annual stability monitoring will be greater than the stability acceptance limit of 6.5%LC because this would result in the need for an investigation, regulatory notification, and possibly a recall of the batch. A draw from the posterior predictive distribution of future results at t_{pred} months can be obtained by substituting the parameter values from the mth posterior MCMC draw, Eq. (8.6) into Eq. (8.15).

$$Y_{fut}^{[m]}(t_{pred}) \sim N\left(\text{intercept}_{fut}^{[m]} + \text{slope}_{fut}^{[m]} \times t_{pred}, \left(\sigma_E^{[m]}\right)^2\right) \qquad (8.15)$$

where $\sim N(A,B)$ is shorthand for a random number generated from a normal distribution with mean A and variance B (e.g., using the rnorm function in R). The predicted probability of future test results exceeding 6.5%LC is 0 for samples tested at 0 months and 0.0632 for samples tested at 28.39 months.

The posterior predictive distribution of future test results can also be helpful in setting process controls. The 0.95th quantile of the predictive posterior of future results is equal to the beta expectation 95% tolerance upper bound which, conditional on the observed data, prior distributions, and stability model, should be greater than or equal to 95% of future results. This Bayesian bound is analogous to the traditional beta expectation tolerance bound, but differs from the traditional bound in the following important ways:

1. It represents simply an estimate of the interval within which 95% of future results will fall. It does not require the repeated sampling paradigm for interpretation as the traditional tolerance bound does.
2. It is conditional on the observed data. The traditional interval/bound is conditional on an infinite series of hypothetical repeats of the sampling and estimation process.
3. A "fixed-in-advance" bound is possible. In other words, one can fix the bound and then use the sample from the predictive posterior to estimate the probability of being within or outside the bound by a simple counting exercise. This cannot be done with the traditional interval/bound.

4. Justified historical knowledge could have been included by specifying informative priors. The traditional paradigm provides no mechanism for quantitatively incorporating prior knowledge.

For normal linear models, traditional and Bayesian intervals will often be similar, but their interpretations will differ. The Bayesian 95% upper tolerance bounds for this example are 3.18%LC for samples tested at 0 months and 6.64%LC for samples tested at 28.39 months.

8.5 Stability Risk Assessment

8.5.1 Estimating OOS Probability on Stability

An out of specification (OOS) on stability occurs when an observed stability result exceeds the stability acceptance limit(s). This could be a signal that the true batch mean has exceeded the limit(s) which indicates a batch stability failure. If the stability result came from a batch under distribution the result could be a regulatory notification and possible product recall. Alternatively, the OOS merely reflects analytical uncertainty. At the very least, an OOS requires an investigation to distinguish these possibilities.

Unplanned investigations, notifications, recalls, and batch replacement costs strain a manufacturer's resources, the product supply chain, the burden of regulatory over-sight, and ultimately the cost of pharmaceuticals. So it is in the interest of all stakeholders to anticipate the probability of such adverse events and include the resulting costs into the budget.

Frequentist tools (prediction and tolerance intervals) are commonly employed for such risk assessments. Frequentist probabilities require a repeated sampling interpretation, approximations, and asymptotic assumptions in mixed models. Additionally, they cannot directly provide statements such as "The probability of a future OOS is X." This is because, as noted by Hahn and Meeker (1991), the confidence probability refers to the statistical procedure used to construct the interval and not to any particular interval that is computed. The actual probability that a particular confidence, prediction, or tolerance interval will contain the values it is supposed to contain is unknown because this probability depends on the unknown parameters.

It must be remembered that frequentist intervals (e.g., the estimated L and U in Eq. (2.52)) are random. It is of course possible to invert frequentist equations such as (2.52) to solve for P, given some fixed values for L and U. However, the frequentist paradigm that underlies Eq. (2.52) provides no basis for interpreting this P as a probability associated with the distribution of future predicted values.

Bayesian approaches, on the other hand, can provide such statements directly without any approximations. Bayesian methods also offer the possibility of rigorously incorporating justified prior knowledge into the assessment for more informed decision making. Modern computing tools such as WinBUGS (Lunn et al.), JAGS (Plummer 2012), and Stan (Gelman et al. 2013) make it easy to apply Bayesian methods.

Given an MCMC sample, such as Eq. (8.6), a sample from the posterior predictive distribution of future stability results at time t for a random future batch can be obtained from

$$Y_{fut}^{[m]}(t) \sim N\left(\text{intercept}_{fut}^{[m]} + \text{slope}_{fut}^{[m]} \times t, \left(\sigma_E^{[m]}\right)^2\right). \tag{8.16}$$

It is informative to compare (8.16) with (8.15). As before, the subscript *fut* indicates that the resulting numbers represent samples from the distribution of results from a random future batch. The probability of a future OOS, at any time t, can then be obtained via a simple counting exercise, comparing $Y_{fut}^{[m]}(t)$ to the stability acceptance limit(s). The OOS probability can then be plotted as a function of t as part of an overall risk and resource assessment for the product. The uncertainty in the estimate of the probability is limited only by available data and prior knowledge, appropriateness of the stability model, and the size of the MCMC sample.

8.5.2   Estimating the Probability of a Noncompliant Batch

Occasionally the stability results for a batch will exhibit an apparent "out of trend" (or OOT) condition that causes some concern. Will the trend for this particular batch result in a stability failure? Often there is limited data for the batch and it will be unclear whether the trend is real or the result of analytical uncertainty.

Equation (8.8) provides an MCMC sample from the posterior distribution of the mean level of the ith batch at time t. This level can be compared to the stability acceptance limit as described in Sect. 8.5.1 as a part of risk assessment for a particular batch of interest. A plot of probability of exceeding the shelf life acceptance limit(s) against storage time for this batch can be generated. This plot can be compared to a similar plot for all future batches generated using Eq. (8.12). If the probability profile for the batch of interest is below the probability profile for all future batches, then there is little cause for concern.

For products with a high batch production rate, such comparisons can be conducted pro-actively as part of overall life cycle monitoring and trending.

8.5.3   Release Limit Estimation

8.5.3.1   The Allen et al. (1991) Approach to Release Limit Estimation

The shelf life estimation discussed above depends on the true level at release (i.e., the intercept) being in control. Thus the true release levels of batches used for shelf life determination are assumed to be representative of those of future lots. Often a process control limit is established to assure good control of the release level. If the measured release level exceeds the limit, the product is not released for sale.

In keeping with the definition of shelf life, a one-sided release limit should ensure with at least 95% confidence that the mean level of a released lot will remain above (or below) a given lower (or upper) stability specification at the end of shelf life. Generally, stability data are available and the objective is to calculate a release limit (RL) from the estimated *slope*, it's standard error of estimate ($\hat{\sigma}_{slope}$), an established shelf life acceptance limit (SL), the desired shelf life (D), the estimated total analytical standard deviation ($\hat{\sigma}$), and the number of replicates averaged to obtain the reportable release test result (n). Using this information, the release limit estimate given by Allen et al. (1991) is:

$$RL = SL - \delta \cdot D + t_{p:df}\sqrt{\hat{\sigma}^2/n + D^2 \cdot \hat{\sigma}^2_{slope}}, \qquad (8.17)$$

where $\delta = \min[slope,0]$ or $\max[slope,0]$ for lower or upper RL, respectively, $t_{p:df}$ = pth quantile of the t-distribution with df degrees of freedom, df = degrees of freedom associated with the estimate of both $\hat{\sigma}$ and $\hat{\sigma}_{slope}$, $p = 0.95$ or 0.05 for lower or upper RL, respectively. Allen et al. assume an SISS model with $\hat{\sigma}$ and $\hat{\sigma}_{slope}$ estimated from independent data sets and the value for df obtained using the Satterthwaite approximation (see Eq. (2.120)). This procedure assumes a reasonably large number of batches are available to estimate $\hat{\sigma}_{slope}$. However, at the time of submission there may be as few as three batches of data available. With so few batches the estimate of $\hat{\sigma}_{slope}$ may have considerable uncertainty leading to a small df and consequently an overly wide release limit. In such cases, a Bayesian approach (described in the next section) can be used to provide a more reasonable estimate avoids approximations.

Below we show an example in which both $\hat{\sigma}$ and $\hat{\sigma}_{slope}$ are estimated from the same data set in which a common slope model is assumed so that $\hat{\sigma}_{slope}$ represents only the uncertainty in estimation of the common slope. To illustrate application of Eq. (8.17), the data from Schuirmann, available at the book web site, will be used. A fixed batch analysis of batches 3, 4, and 5 yields a separate intercept, common slope model (SICS) as illustrated in Table 8.8 and Fig. 8.3. The multiple regression output from a SAS analysis is shown in Table 8.14.

With a stability lower acceptance limit of $SL = 90.0\%LC$ and using the methods shown above, shelf lives of 48, 57, and 43 months are obtained for batches 3, 4, and 5, respectively. Thus a shelf life of 43 months could be justified for this product. Thus the development team feels justified in recommending a shelf life for this

Table 8.14 Output from SAS fixed batch analysis

Covariance Parameter

Estimates

Cov Parm	Estimate
Residual	1.1568

Solution for Fixed Effects

Standard

Effect	batch	Estimate	Error	DF	t Value	Pr > |t|
Intercept		100.82	0.3840	24	262.52	<.0001
Month		-0.2131	0.02433	24	-8.76	<.0001
batch	BATCH3	1.3556	0.4849	24	2.80	0.0100
batch	BATCH4	3.4352	0.5032	24	6.83	<.0001
batch	BATCH5	0	.	.	.	.

product of $D = 36$ months. At this point, the development team decided to establish a lower release limit on the mean of $n = 2$ replicate tests to assure that no future batch mean will be below 90%LC through 36 months with 95% confidence. From the above SAS output, the following inputs to Eq. (8.17) are obtained.

$slope = -0.2131\%LC/month$
$\delta = min[slope,0] = -0.2131$ (for lower release limit)
$\hat{\sigma} = 1.1568$ (residual error)
$\hat{\sigma}_{slope} = 0.02433\#\#\#$
$df = 24$
$t_{p;df} = 1.71$
$p = 0.95$ (for lower release limit)

Application of Eq. (8.17) yields a release limit of $RL = 99.7\%LC$ for this product.

For a separate intercept, separate slope model (SISS), a conservative approach is to take the *slope* estimate from the worst-case batch. In cases where the estimated slope implies a (favorable) divergence over time from the limit of interest (e.g., negative slope used to estimate an upper RL), the slope should be set to zero as implied by the δ function above. This provides a worst-case, conservative estimate for the release limit and ensures that lot mean levels will remain in conformance throughout the storage period with at least 95% confidence.

When estimates of $\hat{\sigma}$ and $\hat{\sigma}_{slope}$ are not obtained from the same regression, then the evaluation of df can be problematic. However the Satterthwaite approximation may be applied in this case (see Eq. (2.124)). In principal, a similar approach could be used in the random batch case.

Allen et al. are somewhat vague about the source for an estimate of σ_{slope} and its estimation uncertainty. The number of degrees of freedom associated with this estimate must be known to properly apply the recommended Satterthwaite approximation. So we do not describe it in detail here. A 95% confidence level may not always be appropriate. Indeed it may seem that a prediction (rather than a confidence) bound is more appropriate for release limit estimation. A Bayesian perspective can shed fresh light on the release limit challenge.

8.5.3.2   A Bayesian Approach to Release Limit Estimation

A *stability failure* occurs when the true mean potency of a batch exceeds the stability acceptance range (note that limits are sometimes 2-sided) before the end of shelf life. A *batch release failure* occurs when an observed release result is outside the release limit. A lower release limit is intended not only to reduce the risk of *stability failure* but may also increase the risk of *batch release failures*. Thus, the value of the release test is a compromise that depends on a possible correlation between the release test result and the probability of stability failure.

It is not clear (to this author) how to model this correlation from a traditional perspective to identify an optimal release limit. But it is straightforward using a Bayesian approach. Of course this requires data from a sufficient number of batches to estimate this correlation. We do not explore the sample size requirements of such an approach in any detail here. Clearly three batches seem too few.

Below we illustrate this approach using all eight of the batches from Schuirmann which are illustrated in Fig. 8.10. For illustration, separate simple regression lines are fit to the stability data from each batch.

To evaluate release limit options from a Bayesian point of view, we form samples from

- the posterior distribution of future batch means, and
- the posterior *predictive* distribution of future release test results,

using the parameter values from the mth posterior MCMC draw, Eq. (8.6). Samples from the posterior distribution of future batch means, $\mu_{t_{SL}}^{[m]}$, at the end of shelf life, t_{SL}, are obtained using Eq. (8.18).

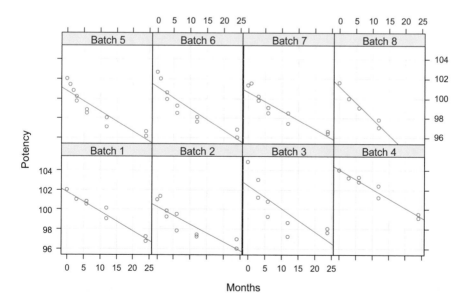

Fig. 8.10 Potency stability results on 8 drug batches

$$\mu_{t_{SL}}^{[m]} = \text{intercept}_{fut}^{[m]} + t_{SL} \cdot \text{slope}_{fut}^{[m]} \qquad (8.18)$$

Samples from the posterior *predictive* distribution of future release test results ($Y_{fut\ rel}^{[m]}$) represent an average of n_{rel} stability test values. These samples are obtained by simulation from a normal distribution described by Eq. (8.19).

$$Y_{fut\ rel}^{[m]} \sim N\left(\text{intercept}_{fut}^{[m]}, \left(\sigma_E^{[m]}\right)^2 / n_{rel}\right). \qquad (8.19)$$

Notice that the modifier *predictive* is included to indicate that the function of parameters in Eq. (8.19) involves random sampling (from a normal distribution).

The correlation between future batch mean potencies at shelf life (36 months) and the release test result for the Schuirmann data is illustrated in Fig. 8.11.

A positive correlation between batch mean potency at the end of shelf life and the corresponding release test result is expected. This positive correlation is necessary if we are to realize any benefit (i.e., a reduction in batch failures on stability) from establishing a release limit. For illustration, a release limit of 98%LC is shown as the vertical line in Fig. 8.11, but interest lies in the consequences of adjusting this value either left or right. Plotted points to the left of this release limit (the "OOS at Release" region in the figure) represent batches that fail to be released. The stability

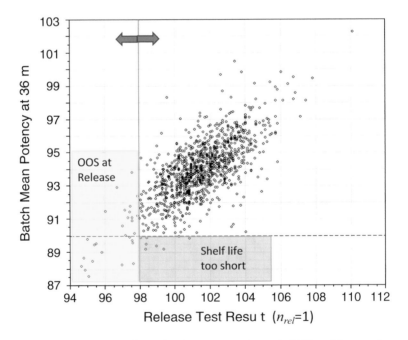

Fig. 8.11 Correlation between future batch mean potencies at shelf life and future release test results

acceptance limit is 90%LC and this is indicated as a fixed horizontal line in Fig. 8.11. Predicted batch mean potencies of released batches below 90%LC at 36 months represent a stability failure. As the release limit increases from left to right it is apparent that the probability of a batch release failure increases, but the probability of released batches with stability failures decreases. By a simple counting exercise of the number of draws that pass or fail the respective limits for $\mu_{tSL}^{[m]}$ and $Y_{fut\ rel}^{[m]}$, we can predict these probabilities as a function of release limit and n_{rel}, which is the number of test results averaged to produce a reportable release value.

The resulting probabilities for this example are plotted in Fig. 8.12. The solid and dashed lines show the predicted probability of shelf life and release limit failure, respectively, as a function of release limit and n_{rel}. Generally as the release limit is increased, the probability of release failure increases and the probability of shelf life failure decreases, as expected. It is somewhat disappointing that the drop in shelf life failures is not more dramatic. We can see also that there is an improvement in the release limit failures as n_{rel} is increased. This graphic can allow decision makers a more informed, and therefore more justifiable, release limit

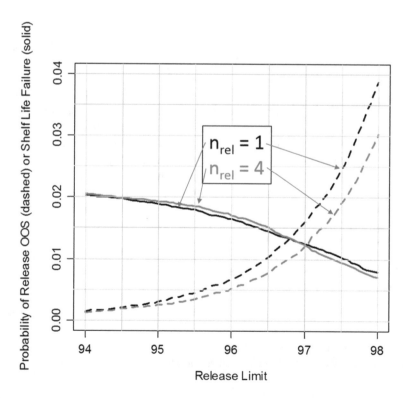

Fig. 8.12 Choosing a release limit based on a compromise between the probability of release failures and stability failures

compromise between these two kinds of failures based on probability. It is note-worthy that as the correlation in Fig. 8.11 increases, the risk of decision errors (i.e., failing to release lots with acceptable shelf life or releasing lots with unacceptable shelf life) will drop. Probability estimates are needed for risk-based decision making and such probabilistic modeling is the hallmark of Bayesian procedures.

8.6   Additional Stability Analysis Issues

8.6.1   Modeling Slope Intercept Correlation

The model of Eq. (8.5) assumes zero correlation between batch slopes and inter-cepts. If the factors that affect batch intercept (e.g., weighing errors, API purity) are independent of those that affect batch instability (e.g., moisture content, packaging integrity), the zero correlation assumption may be appropriate. Otherwise, a model that includes a covariance should be considered. For example, a correlation between L_i and B_i can be added to Eq. (8.5) by modifying the model so that σ_{LB} is a non-zero covariance. That is, assume the model

$$\begin{pmatrix} L_i \\ B_i \end{pmatrix} \sim N\left(\begin{pmatrix} 0 \\ 0 \end{pmatrix}, \begin{pmatrix} \sigma_L^2 & \sigma_{LB} \\ \sigma_{LB} & \sigma_B^2 \end{pmatrix} \right). \tag{8.20}$$

Such models are easily accommodated in SAS Proc Mixed by switching the TYPE = VC option in the RANDOM statement to TYPE = UN. The lmer function in the R library lme4 provides similar functionality. Gelman and Hill (2007) show how to implement such models from both a frequentist and Bayesian perspective.

The Bayesian version of this model is easy to implement in WinBUGS, JAGS, or Stan. As discussed above, the Bayesian version has the advantage of providing a full probability model without analytical approximations or asymptotic approxi-mations. Of course, both traditional and Bayesian approaches depend on distribu-tional assumptions for the likelihood. The Bayesian approach has the additional burden of justifying any prior assumptions.

The MCMC iterations (i.e., Gibbs or Metropolis sampling or the like) with such models can sometimes be speeded up by first centering the time covariate about its grand mean

$$t'_k = t_k - \bar{t}, \tag{8.21}$$

and regressing against t'_k rather than t_k. This leads to MCMC samples from the posterior distribution of a transformed set of parameters

$$\left(\mu'^{[m]}, \beta_1^{[m]}, \sigma_{L'}^{[m]}, \sigma_B^{[m]}, \sigma_E^{[m]}, \sigma'^{[m]}_{L'B}, \mathbf{L}'^{[m]}, \mathbf{B}^{[m]} \right), \tag{8.22}$$

where primed parameters are affected by the time centering transformation.

Inference will most often be desired on the original, un-centered parameters. As described by Paccagnella (2006), the following algebra, performed directly on the MCMC samples, recovers posterior samples from the original, un-centered parameters

$$\mu^{[m]} = \mu'^{[m]} - \beta_1^{[m]} \times \bar{t}$$

$$\mathbf{L}^{[m]} = \mathbf{L}'^{[m]} - B_i^{[m]} \times \bar{t}$$

$$\sigma_{LB}^{[m]} = \sigma_{L'B}^{[m]} - \sigma_B^{[m]2} \times \bar{t} \qquad (8.23)$$

$$\sigma_L^{[m]} = \sqrt{\sigma_{L'}^{[m]2} + \sigma_B^{[m]2} \times \bar{t}^2 - 2\sigma_{L'B}^{[m]} \times \bar{t}}$$

8.6.2   Multifactor Studies

The principles illustrated above also apply when the stability study includes multiple levels of strength, packaging, or other design variables. A good review of model-building aspects in such complex cases is provided by Milliken and Johnson (1992). The model pruning in such studies is done using an alpha level for batch effects of 0.25, while the alpha level for other factors may be at the traditional 0.05 level. Such modeling can, of course, be implemented in a Bayesian framework.

8.6.3   Accelerated Stability Studies

Depending solely on a limited number of pivotal batches for shelf life estimation presents many challenges. It is an entirely empirical process that ignores chemical theory, mechanistic insights and prior knowledge gained during early pharmaceutics studies, preliminary product development, or from similar products. Often, label storage stability studies are incomplete at the time of submission and results from accelerated studies are not used as part of an overall stability model in the shelf life estimation process. While batches are annually placed on stability as part of regulatory commitments, the data from such batches are rarely used to adjust the shelf life after approval.

Waterman and Adami (2005) and Waterman (2011, 2012) have suggested methods for early assessment of pharmaceutical stability, and even early estimation of shelf life, using data from accelerated studies. These methods employ nonlinear predictive stability models that include both thermal and other (e.g., relative humidity) parameters. These methods include both study design and analysis aspects and appear to offer considerable promise in improved decision making with respect to shelf life estimation.

While a thorough discussion of these interesting approaches is beyond the scope of the present text, we note that Bayesian approaches—which do not depend on linear approximations to model nonlinear cases, allow for rigorous incorporation of prior knowledge, and excel at probabilistic inference—may well be used to advantage here.

8.6.4 Managing Storage Excursions

ICH Q1A and USP ⟨1160⟩ advocate the use of mean kinetic temperature (MKT) as a basis for short-dating or destroying pharmaceuticals that experience post manufacturing temperature excursions during storage or transport. The MKT requirement is

$$MKT_{Limit} \geq \frac{\Delta H}{-R \cdot \ln\left(\sum_{i=1}^{k} n_i e^{-\Delta H/RT_i} \bigg/ \sum_{i=1}^{k} n_i\right)} \tag{8.24}$$

where MKT_{Limit} is the mean kinetic temperature upper limit (which, for room temperature storage is 298.15 K, or 25°C), ΔH is the heat of activation for the instability mechanism (generally 42–125 kJ/mol, although USP ⟨1160⟩ recommends 83.144 kJ/mol unless a more accurate experimentally determined value is available), R is the universal gas constant (0.0083144 kJ mol^{-1} K^{-1}), and n_i is the number of hours spent at the storage temperature T_i (°K) for all k temperatures.

It is relatively straightforward to apply Eq. (8.24) to determine if the requirement is met, as long as a value for ΔH can be justified and the storage temperature history is available. It can be challenging to verify this requirement, however, for investigational products—especially for purchased comparators for which the sponsor will generally have little knowledge of physical properties or storage history. Asmussen et al. (2014, 2016) offer a conservative MKT approach based on simple lookup tables that can be easily implemented as part of an enterprise-wide quality management system.

8.7 Stability Study Designs

8.7.1 Full Design

When a stability study includes all combinations of dosage strength, packaging, and storage times, the stability design is referred to as a "full-factorial" (or sometimes as just a "full" or "complete") design. Conducting a full design on a new drug product with multiple strengths and packages can be prohibitively expensive and may stress a company's analytical resources beyond their limits. From a statistical point of view, when the number of study variables is large, a complete factorial may be unnecessary as long as certain assumptions can be made about the stability effects of the variables.

ICH Q1D describes situations in which a "reduced" design can be applied without further justification and some situations in which further justification will need to be provided. If a design deviates markedly from the principles of ICH Q1D, the protocol must be approved by the FDA prior to the initiation of stability studies. Some additional clarification is provided in the literature (see Lin and Chen 2003). In a reduced design, only a specific fraction of the possible combinations of dosage strength, packaging, and storage times are actually tested. ICH Q1D refers to two general approaches to reduced designs: bracketing and matrixing.

8.7.2 Bracketing Designs

In a bracketing approach, the sponsor relies on theory or past experience to identify a small number of variable combinations (say of strength and packaging) that can be assumed to give "worst-case" stability. Often these combinations will be extremes (e.g., of active content, head space, moisture vapor transmission rate) and the sponsor is willing to estimate the product shelf life from a study of these "worst cases" alone. Bracketing assumes that the untested variable combinations will have equal or superior stability and therefore need not be tested at all. Bracketing requires a thorough understanding of the mechanism(s) of instability from theory or from studies on earlier development or clinical batches. Because bracketing makes strong assumptions about the underlying mechanism of instability, it is not applicable when dosage form formulations or package characteristics differ markedly. In a pure bracketing design, the chosen combinations are tested at all time points. Often those combinations not intended to be tested as part of the bracketing design are not even placed on stability.

Bracketing may be applied with no further justification across strength when different strengths have identical or closely related formulations. Examples include capsules of different strength made with different plug sizes from the same powder blend, tablets of different strengths made by compressing varying amounts of the same granulation, and formulations that differ only in minor excipients. Further justification should be considered when amounts of drug substance and excipients change in a formulation. When different excipients are used amongst strengths, bracketing is generally not applied.

Bracketing may be applied with no further justification across packages using the same container closure system where either container size or fill varies while the other remains constant. Further justification should be considered when the closure systems vary for the same container. Justification could include a discussion of relative permeation rates of the bracketed container closure system.

8.7.3   Matrixed Designs

In a matrixing approach, the sponsor takes advantage of traditional principles of experimental design (refer to Box et al. 1978 and Chap. 3 of this book) to reduce study size without sacrificing statistical power, model structure, or parameter estimability. Matrixing depends on choosing a balanced subset of the full factorial set of combinations which supports a predictive model that includes all main effects and critical interactions. Often, all combinations (even those not intended to be tested in the matrix design) are placed on stability. However because of the expense, all are not tested unless there is a need to revert to full testing.

Matrixing with respect to strength may be applied across strengths without further justification when the strength formulations are identical or closely related. Additional justification should be considered when different strengths differ in the drug substance content, excipients change, or different excipients are used. Matrixing across batches of drug product can be applied when batches are made using the same process and equipment, or batches of drug substance. Matrixing across package size and fill is permitted when all packages use the same container closure system. Further justification should be provided when packages use different container closure systems, different contact material, different suppliers, or different orientations. Justification should be based on supportive data (e.g., moisture vapor transmission rates or light protection for different containers).

Matrix designs can be complete (all combination of factors are tested) or incomplete (some combinations are not tested at all). In a complex design, combinations of strength and container size are tested and individual batch of product is not tested in all strength and container size combinations. If the design is broken (some samples are lost or not tested) during the course of the study, testing should revert back to full testing through the proposed retest period or shelf life. Where testing exhibits moderate variability and moderately good stability, the use of a matrix design should provide adequate statistical power and thus be statistically justified.

Matrixing is not without risks. Highly fractional designs, involving factors other than time, generally have less precision in shelf life estimation and may yield shorter shelf life than a full design. With large variability and poor product stability, a matrix should not be applied. Techniques are discussed below for comparing and assessing the statistical power of stability designs.

Before a reduced design is considered, assumptions should be assessed and justified. As a starting point, either the design including all possible combinations, or an appropriately bracketed subset, is taken as the full design. Then reduced designs are obtained by matrixing the full design. The reduced designs may be compared with respect to the following criteria:

1. Probability of justifying the desired shelf life as a function of study duration.
2. Power to detect effects of experimental variables on stability. Any reduced design should retain the ability to adequately detect differences in stability.

3. "Balance" (i.e., each combination of factor levels is tested to the same extent) to assure orthogonality of model parameter estimates.
4. Total number of tests required (or total study cost).
5. Ergonomic spread of testing throughout the study to optimize analytical resources.

ICH Q1D mandates full testing of all studied factor combinations at the beginning and end of the study as well as at the time of submission. It also recommends at least three time points for each studied factor combination be available at the time of submission (nominally at 12 months storage). ICH Q1D provides examples of study designs. These are not the only designs to be considered, but they illustrate many of the principles of balance as well as the practical constraints. These are discussed below.

8.7.3.1 Matrixing on Time Only

A key principal of fractional factorial design is to maximize testing of extremes of continuous variables. In the case of storage time, this means full testing is required at initial, study completion, and submission (typically 12 months). Thus reduced testing for a 36 month study can only be considered at five time points: 3, 6, 9, 18, and 24 months. Consider a study with six combinations: two strengths with three batches per strength. Assume a 1/3 reduction in testing is desired. Which 20 of the 5 time points $\times$ 6 combinations $= 30$ test points should be tested?

One principle of good experimental design is that of balance. A balanced design is one in which the number of replicates in each treatment group is equal. Balanced designs are generally statistically most efficient. The principle of balance requires that

1. Each of the 2 strengths be tested 10 times ($2 \times 10 = 20$).
2. Each of the 3 batches be tested Z times ($Z \times 3 = 20$).
3. Each of the 5 time points be tested 4 times ($5 \times 4 = 20$).
4. Each of the 6 strength*batch combinations be tested Y times ($Y \times 6 = 20$).
5. Each of the 10 strength*time combinations be tested 2 times ($2 \times 10 = 20$).
6. Each of the 15 batch*time combinations be tested W times ($W \times 15 = 20$).

Note that W, Y, and Z cannot be whole numbers. In fact, unless the number of tests is evenly divisible by 2 (strength), 3 (batch), and 5 (time), some loss of balance is inevitable. In this case the lowest number of tests that allows balance is $2 \times 3 \times 5 = 30$, which does not allow for any testing reduction at all.

Similarly, if a 1/2 reduction in testing was desired, which 15 of the 30 test points should be tested? Balance would require that the number 15 is evenly divisible by 2 (strength), 3(batch), and 5(time). Since 15 is not evenly divisible by 2, full balance is not possible in this case either. In the ICH Q1D examples, the compromise made is to allow more testing on some batch, strength*batch, and batch*time combinations than others. However, the continuous time variable is robust to loss of balance because of the assumption of a linear change over time. Thus, the inevitable non-orthogonality in the ICH Q1D design examples is probably negligible.

An example illustrating complete and partial balance is as follows. Assume the sponsor desired an analysis at 24 months (perhaps to justify shelf life extension). Then full testing would be required at 24 months and matrixing would be on the four time points 3, 6, 9, and 18 months only. Thus the full design would require (2 strengths) × (3 batches) × (4 times) = 24 tests. Since 12 is divisible by 2, 3, and 4, a completely balanced 1/2 reduction is possible. A 1/3 reduction would require 18 tests, however, 18 is only evenly divisible by 2 and 3, not by 4. Therefore only a partially balanced design is possible if a 1/3 reduction is desired. To identify which of the 12 or 18 tests to include, the mod arithmetic method of Nordbrock (1994) can be used. The following steps illustrate this approach:

1. Assign a code of $S = 0$ or 1 for each strength.
2. Assign a code of $B = 0$, 1, or 2 for each of the 3 batches within each strength.
3. Assign a code of $T = 0$, 1, 2, or 3 for the time points 3, 6, 9, or 18 months, respectively.
4. For each of the 24 possible strength*batch*time combinations,

   a. for a 1/2-fold reduction, test only combinations where $S + B + T$ mod $2 = 0$.
   b. for a 1/3 reduction, test only combinations where $S + B + T$ mod $3 = 0$ or 1.

Table 8.15 shows the construction of the sum of the S, B, and T indicator variables.

Tables 8.16 and 8.17 illustrate the mod arithmetic for selection of testing schedules providing 1/2 and 1/3 reductions, respectively.

The particular 1/2 or 1/3 fraction selected in Tables 8.16 and 8.17 is of course only one of two or three possible fractions. For instance, one could have decided to test $S + B + T$ mod $2 = 1$ in Table 8.16 or $S + B + T$ mod $3 = 0$ or 2 instead.

The sum $S + B + T$ can be generalized to $n \times S + B + m \times T$, where n and m are constants other than 1, in an attempt to find a design that achieves the desired balance. The same principles can be extended to more complex situations. The example in ICH Q1D involves a product with 3 strengths, 3 packages, 3 batches, and 5 time points. It is typical to first identify a set of balanced (or approximately

Table 8.15 Illustration of use of indicator variable summation to find a matrix design on time points for a product with two strengths

Strength	Batch	$S + B + T$ 3 months (T = 0)	6 months (T = 1)	9 months (T = 2)	18 months (T = 3)
Low (S = 0)	1 (B = 0)	0	1	2	3
	2 (B = 1)	1	2	3	4
	3 (B = 2)	2	3	4	5
High (S = 1)	1 (B = 0)	1	2	3	4
	2 (B = 1)	2	3	4	5
	3 (B = 2)	3	4	5	6

Table 8.16 Example of a balanced one-half reduction matrix design on time points for a product with two strengths

| | | S + B + T mod 2 | | | |
| | | 3 months | 6 months | 9 months | 18 months |
Strength	Batch	(T = 0)	(T = 1)	(T = 2)	(T = 3)
Low (S = 0)	1 (B = 0)	0	1	0	1
	2 (B = 1)	1	0	1	0
	3 (B = 2)	0	1	0	1
High (S = 1)	1 (B = 0)	1	0	1	0
	2 (B = 1)	0	1	0	1
	3 (B = 2)	1	0	1	0

0 = test, 1 = do not test

Table 8.17 Example of a partially balanced one-third reduction matrix design on time points for a product with two strengths

| | | S + B + T mod 3 | | | |
| | | 3 months | 6 months | 9 months | 18 months |
Strength	Batch	(T = 0)	(T = 1)	(T = 2)	(T = 3)
Low (S = 0)	1 (B = 0)	0	1	2	0
	2 (B = 1)	1	2	0	1
	3 (B = 2)	2	0	1	2
High (S = 1)	1 (B = 0)	1	2	0	1
	2 (B = 1)	2	0	1	2
	3 (B = 2)	0	1	2	0

0 or 1 = test, 2 = do not test

balanced) time vectors. Then the various strength × package × batch combinations are then assigned in a balanced (or approximately balanced) way to each of the vectors. The example in ICH Q1D establishes the following testing vectors:

1. $T1 = \{0, 6, 9, 12, 18, 24, 36\}$
2. $T2 = \{0, 3, 9, 12, 24, 36\}$
3. $T3 = \{0, 3, 6, 12, 18, 36\}$

T1 calls for testing at seven total time points, but T2 and T3 only six. However, balance is achieved across the time points 3, 6, and 9. Because of the assumption of linear trend across time, lack of balance does not lead to serious degradation of estimation efficiency. A desirable feature of T1, T2, and T3 is that each of the time points 3, 6, 9, 18, and 24 is represented twice. Thus if the vectors are evenly spread across the combinations, a 1/3 reduction of testing at these time points would be realized. The assignments of the vectors to each of the $3 \times 3 \times 3 = 27$ combinations are illustrated in Table 8.18.

Table 8.18 Using mode arithmetic to matrix on time points only for the example of Table 3a of ICH Q1D

	B + S + P mod 3								
Strength	Low (S = 0)			Med. (S = 1)			High (S = 2)		
Package	A (P = 0)	B (P = 1)	C (P = 2)	A (P = 0)	B (P = 1)	C (P = 2)	A (P = 0)	B (P = 1)	C (P = 2)
Batch 1 (B = 0)	0	1	2	1	2	0	2	0	1
Batch 2 (B = 1)	1	2	0	2	0	1	0	1	2
Batch 3 (B = 2)	2	0	1	0	1	2	1	2	0

0 = use T1, 1 = use T2, 2 = use T3

Table 8.19 Using mod arithmetic to include matrixing against design variables in the example of Table 3b of ICH Q1D

Batch	Do not test if B + S + P mod 3 =
1	2
2	1
3	0

8.7.3.2   Matrixing on Time and Other Variables

Continuing the ICH Q1D matrix example, one can further matrix against batch, strength, and package by omitting the testing based on the value of B + S + P mod 3. In the ICH Q1D Table 3b example, the criterion for elimination depends on batch as described in Table 8.19.

The resulting approximately balanced matrix provides about a 5/9 reduction in the amount of testing at the 3, 6, 9, 12, 18, and 24 month time points. These examples are provided merely for illustration. Other matrixing designs are possible and some may well be superior to those given. Once candidate designs are generated by the methods above—or by other standard experimental design techniques—it is best to perform an assessment of their balance and operation feasibility. Those which pass these assessments may be examined further by the statistical methods described in the next section.

The level of reduction that would be acceptable to a regulatory agency can be addressed by examining the power of the design to detect differences in stability among the different batches, strengths or packages, and/or the precision with which shelf life can be estimated. These characteristics are compared between the candidate matrix design and the full design. If a candidate matrix design sacrifices little relative to the full design, then it should be acceptable from a regulatory point of view.

8.7.4   Accelerated Stability Designs

As noted in Sect. 8.6.3, Waterman and Adami (2005) and Waterman (2011, 2012) have suggested methods for rapid stability assessment and shelf life estimation using accelerated testing studies. Because underlying models of these approaches are nonlinear, traditional factorial arrangements may not be optimal. These authors have discussed some interesting and possibly useful design aspects. A thorough examination of these design considerations is beyond the scope of this text. However, we believe that these designs deserve careful attention and examination by the statistical community.

8.7.5   Comparing Stability Designs

Nordbrock (1992) provided a framework for comparing stability designs with respect to their power to detect slope (stability) differences. This same framework can be applied to a comparison of the probability (power) of meeting a desired shelf life claim among competing stability designs.

8.7.5.1   Power to Detect Slope Differences

A product in which the rate of change (i.e., slope) in potency (or other related substance) is constant across batches is desirable. When batches differ in their slope, it is important that the stability design be capable of estimating slope differences that have safety or efficacy consequences. In this section we illustrate how to determine the power of a stability design to detect such slope differences. First we state some distributional results that will be useful in design evaluation.

Let R and H be mutually independent random variables with distributions $R \sim N(ncp, 1)$ and $H \sim \chi^2(df)$. Then by definition,

$$\frac{R}{\sqrt{H/df}} = W \sim t(df, ncp) \qquad (8.25)$$

where $t(df, ncp)$ represents a random variable distributed as non-central t with degrees of freedom dfe and non-centrality parameter δ. Further, let $pt(Q, df, ncp)$ be the cumulative distribution function of the non-central t random variable W relative to fixed Q such that

$$\text{Prob}(W \leq Q) = pt(Q, df, ncp) \text{ and } \text{Prob}(W > Q) = 1 - pt(Q, df, ncp). \quad (8.26)$$

Non-central t distribution functions are present in software packages such as SAS or R. The following approximations may be useful in implementing these calculations in packages which have a cumulative non-central F (pf()) but not a non-central t function:

$$\mathrm{pt}(t, df, ncp) = \mathrm{pf}(t^2, 1, df, ncp^2) + \mathrm{pt}(-t, df, ncp)$$
$$\approx \mathrm{pf}(t^2, 1, df, ncp^2) \text{ for } ncp \gg 0 \tag{8.27}$$

In some software packages, algorithms for non-central t do not converge when ncp and/or df are large. In those cases, the following approximation may be used:

$$\mathrm{pnorm}(t - ncp) \approx \mathrm{pt}(t, df, ncp) \tag{8.28}$$

with pnorm(z) indicating the cumulative standard normal distribution function for quantile z.

Next we describe the predictive stability model we will use to evaluate stability designs. Consider a linear regression to the fixed model

$$\mathbf{y} = \mathbf{X\beta} + \boldsymbol{\varepsilon} \tag{8.29}$$

with fixed model parameter vector $\boldsymbol{\beta}$ and $\boldsymbol{\varepsilon} \sim MVN(\mathbf{0}, \mathbf{I}\sigma^2)$. In applying the concepts below it is critical to understand the structure of $\boldsymbol{\beta}$ and $\mathbf{X}$. As an example, consider a stability design that includes measurements ($\mathbf{y}$ vector) at a series of time (T) points on three batches (B), each of three strengths (S), with three possible package (P) options. Assume the design will support estimation of all main effects and certain interactions (i.e., others known to be unimportant). In a high level program like SAS, the model might be expressed as

$$Y = \text{intercept T B S P T*B T*S T*B*S.} \tag{8.30}$$

The column vector $\boldsymbol{\beta}$ will be composed of one sub-vector for each model term. The length of each sub-vector is 1 for the continuous variables intercept and T. For categorical main effects B, S, and P, the sub-vector length will be number of categorical levels for that variable minus 1. Thus, the length will be 2 for each main effect. For interactions, the vector length will be the product of the vector lengths of each of the component main effects in the interaction. Working across in the above model order we see the length of $\boldsymbol{\beta}$ will be $1 + (1 + 2 + 2 + 2) + (2 + 2) + (4) = 16$. Thus

$$\boldsymbol{\beta}^T = \left(\beta_{\mathrm{int}}, \beta_T, \beta_{B1}, \beta_{B2}, \beta_{S1}, \beta_{S2}, \beta_{P1}, \beta_{P2}, \beta_{TB1}, \beta_{TB2}, \beta_{TS1}, \beta_{TS2}, \right. \tag{8.31}$$
$$\left. \beta_{TB1S1}, \beta_{TB1S2}, \beta_{TB2S1}, \beta_{TB2S2}\right)$$

Each element of $\boldsymbol{\beta}$ is an effect attributable to the presence of certain levels or combinations of levels of the variables T, B, S, and P.

The $\mathbf{X}$ matrix in this example will have 16 columns corresponding to each of the coefficients in $\boldsymbol{\beta}$, and one row for each result in $\mathbf{y}$. The intercept column in $\mathbf{X}$ will

Table 8.20 Coding of design variable levels in the construction of the X matrix

Sub-vector code	B interpretation	S interpretation	P interpretation
(1 0)	Batch 1	Low strength	Package A
(0 1)	Batch 2	Middle strength	Package B
(0 0)	Batch 3	High strength	Package C

have a "1" in each row and the T column in **X** will have the numeric storage time at which the corresponding test result in **y** was obtained. For categorical main effects, the **X** columns corresponding to B, S, and P will contain codes that specify the level of each corresponding to the respective **y** test result. A convenient 0/1 coding is shown in Table 8.20 in which the presence of a 0 in the first or second element indicates that the corresponding factor level is not acting to produce the corresponding result. If both the first and second elements are 0, then the third level must be active.

Sub-vector codes for interactions consist of the Kronecker product of the sub-vectors of the component main effects. Thus the 2 T*B columns in **X** corresponding to a y result obtained at 3 months on batch 2 would be encoded as $3 \otimes (0\ 1) = (0\ 3)$. The four T*B*S columns in X corresponded to a y result at 6 months on Batch 1, middle strength would be $6 \otimes (1\ 0) \otimes (0\ 1) = (0\ 6\ 0\ 0)$. In this way, the entire **X** matrix is constructed. Consider the **y** result corresponding to the 9 month test on batch 2 of low strength packaged in container type B. The corresponding row of the **X** matrix is then

$$\mathbf{X} = \begin{pmatrix} \cdots \\ 1 & 9 & 0 & 1 & 1 & 0 & 0 & 1 & 0 & 9 & 9 & 0 & 0 & 0 & 9 & 0 \\ \cdots \end{pmatrix} \quad (8.32)$$

Finally we show how the distributional results and predictive model discussed above can be used for design evaluation. The estimates of interest in a stability analysis generally consist of linear functions of the elements of **β**. Functions of interest might be slopes or differences between slopes for specific combinations of design variables or averages across certain design variables. A linear function of **β**, say $\delta = \mathbf{c}^T \boldsymbol{\beta}$, will have an estimate $\hat{\delta} \sim N(\delta, \sigma_{\hat{\delta}}^2)$ with sampling variance

$$\sigma_\delta^2 = \mathbf{c}^T (\mathbf{X'X})^{-1} \mathbf{c} \sigma^2 \quad (8.33)$$

based on degrees of freedom

$$df = \text{number of rows in } \mathbf{X} - \text{number of columns in } \mathbf{X}. \quad (8.34)$$

The dimensions of the contrast column vector **c** will be identical to that of **β**. The elements of **c** will depend on which of the corresponding elements of **β** are active in defining a specific linear function of interest. Finally, note that if $\hat{\sigma}_\delta^2$ is a sample estimate of $\sigma_{\hat{\delta}}^2$ based on df degrees of freedom, then its sampling distribution is given by

$$\frac{\hat{\sigma}_{\hat{\delta}}^2}{\sigma_{\hat{\delta}}^2} \cdot df \sim \chi^2(df). \tag{8.35}$$

Let δ represent some slope difference that is of interest and consider a statistical test of $H_0 : \delta = 0$ against the alternative $H_a : \delta > 0$. H_0 will be rejected at the 0.05 level of significance if

$$\frac{\hat{\delta}}{\hat{\sigma}_{\hat{\delta}}} = \frac{\hat{\delta}/\sigma_{\hat{\delta}}}{\sqrt{\hat{\sigma}_{\hat{\delta}}^2/\sigma_{\hat{\delta}}^2}} > qt(0.95, df) \tag{8.36}$$

where $qt(P,df)$ represents (using R function notation) the quantile of the Student's t distribution with df degrees of freedom having cumulative probability P in its left tail. Now under the assumption that $\delta = \Delta$,

$$\frac{\hat{\delta}}{\sigma_{\hat{\delta}}} \sim N\left(\frac{\Delta}{\sigma_{\hat{\delta}}}, 1\right) \tag{8.37}$$

Combining Eqs. (8.25), (8.26), (8.34)–(8.36) we may state that

$$\Pr\{H_0 \text{ rejected}|\delta = \Delta\} = \Pr\left\{\frac{\hat{\delta}}{\hat{\sigma}_{\hat{\delta}}^2} > qt(0.95, df)|\delta = \Delta\right\}$$
$$= 1 - pnct\left(qt(0.95, df), df, \frac{\Delta}{\sigma_{\hat{\delta}}}\right) \tag{8.38}$$

Equation (8.38) may be evaluated as a function of Δ to provide the operating characteristics of the statistical test. Since df and $\sigma_{\hat{\delta}}^2$ depend, by Eqs. (8.33) and (8.34), only on X which in turn is determined by the stability design, the operating characteristics can be compared for various designs and storage times as an aid in study planning.

As an example, consider three matrix designs presented in ICH Q1D and the model as described above. Let δ be a slope difference between medium and low (medium minus low) strength averaged across all lots. Then the corresponding contrast vector is

$$\mathbf{c}^T = \left(0, 0, 0, 0, 0, 0, 0, 0, 0, 0, -1, +1, -\frac{1}{3}, +\frac{1}{3}, -\frac{1}{3}, +\frac{1}{3}\right) \tag{8.39}$$

Fig. 8.13 Comparison of power to detect slope differences at the 12 month time point for three stability designs in ICH Q1D. See text for legend

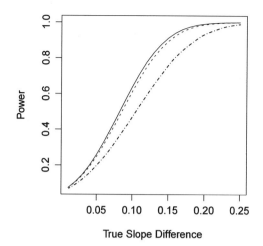

so that

$$\mathbf{c}^T \boldsymbol{\beta} = \beta_{TS2} - \beta_{TS1} + \frac{\beta_{TB1S2} - \beta_{TB1S1} + \beta_{TB2S2} - \beta_{TB2S1}}{3} \tag{8.40}$$

Assume the stability potency analysis for submission will occur at the 12 month time point. Let $\sigma = 1\%LC$ and let the true slope difference, Δ, vary between 0 and 0.25%LC/month. Figure 8.13 shows the power as given by Eq. (8.38) for various true slope differences. Note that very little power is lost in going from the Full design (solid line, 135 tests required) to the design matrixing on time only (dashed line, ICH Q1D Table 3a, 108 tests required). However, considerable power is lost when matrixing on time, strength, batch, and packaging (dash-dotted line, ICH Q1D Table 3b, 72 tests required).

8.7.5.2   Probability of Achieving Desired Shelf Life

Consider a stability determining test, such as potency, with a lower acceptance limit, L. Let D be the desired shelf life for the product. Further assume that the product will have a true release potency (intercept) of η, a true slope of ρ and a true analytical standard deviation of σ.

The regression slope estimate of ρ, $\hat{\rho}$, will have a sampling distribution that can be defined by

$$\frac{\hat{\rho} - \rho}{\sigma_\rho} + ncp \sim N(ncp, 1) \tag{8.41}$$

Combining Eq. (8.41) with Eq. (8.35) and comparing with Eqs. (8.25) and (8.26) we see that

$$\text{Pr}\left\{\frac{\frac{\hat{\rho}-\rho}{\sigma_\rho}+ncp}{\hat{\sigma}_\rho/\sigma_\rho}>Q\right\}=1-\text{pt}(Q,df,ncp) \tag{8.42}$$

Now for a test with a lower limit L, the desired shelf life claim will be achieved whenever

$$\eta+D\cdot\left(\hat{\rho}-\hat{\sigma}_\rho\cdot\text{qt}(0.95,df)\right)>L \tag{8.43}$$

or

$$\frac{\hat{\rho}+(\eta-L)/D}{\hat{\sigma}_\rho}>\text{qt}(0.95,df) \tag{8.44}$$

or

$$\frac{\frac{\hat{\rho}-\rho}{\sigma_\rho}+\frac{(\eta-L)/D+\rho}{\sigma_\rho}}{\hat{\sigma}_\rho/\sigma_\rho}>\text{qt}(0.95,df) \tag{8.45}$$

Comparing this with Eq. (8.42), we see that

$$\text{Pr}\{\text{Achieving desired shelflife claim}\}=1-pnct\left(\text{qt}(0.95,df),df,\frac{(\eta-L)/D+\rho}{\sigma_\rho}\right) \tag{8.46}$$

Note that a similar derivation in the case of an upper acceptance limit would yield

$$\text{Pr}\{\text{Achieving desired shelflife claim}\}=pnct\left(\text{qt}(0.05,df),df,\frac{(\eta-L)/D+\rho}{\sigma_\rho}\right) \tag{8.47}$$

Equation (8.46) or Eq. (8.47) may be evaluated as a function of η, ρ, and σ to provide the operating characteristics of the shelf life estimation goal. Hypothetical η, ρ, and σ may be taken from preliminary stability and analytical studies. Since df and σ_ρ^2 depend, by Eqs. (8.33) and (8.34), only on X which in turn is determined by the stability design, the operating characteristics can be compared for various designs and storage times as an aid in study planning.

As an example, consider three matrix designs presented in ICH Q1D and the model as described above. Let ρ be defined as the slope that represents the stability

Fig. 8.14 Comparison of
probability of achieving a
24-month shelf life claim at
the 12 month time point for
three stability designs in
ICH Q1D

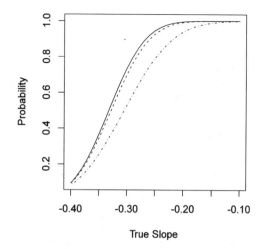

for batch 2 of the low strength formulation (and any package). The corresponding
contrast vector is then

$$\mathbf{c}^T = (0, 1, 0, 0, 0, 0, 0, 0, 0, 1, 1, 0, 0, 0, 1, 0) \qquad (8.48)$$

so that

$$\mathbf{c}^T \boldsymbol{\beta} = \beta_T + \beta_{TB2} + \beta_{TS1} + \beta_{TB2S1} \qquad (8.49)$$

Assume the stability potency analysis for submission will occur at the 12 month
time point. Let $\sigma = 1\%LC$, the lower acceptance limit for potency, L, be 90%LC,
the true initial potency level, η, be 100%LC, and let the true slope, ρ, for the
combination of interest vary between –0.4 and –0.1%LC/month. Figure 8.14 shows
the probability of meeting a shelf life claim of $D = 24$ months as given by Eq. (8.46)
for various true slopes. Note that very little risk is seen in going from the Full design
(solid line, 135 tests required) to the matrix design on time only (dashed line, ICH
Q1D Table 3a, 108 tests required). However, considerable risk is encountered when
matrixing includes both time, strength, batch, and packaging (dash-dotted line, ICH
Q1D Table 3b, 72 tests required).

8.7.5.3 Implementation in R

The R language (R Core Team 2013) provides a good matrix computation and
graphics platform for making stability design comparisons of power and probability
as described above. The design matrix $\mathbf{X}$ (up to the time of proposed analysis) may
be created manually in an Excel spreadsheet and imported as a comma separated
text file to implement the power and probability calculations. R code for estimating
the power of detecting slope differences and estimating the probability of achieving
a desired shelf life is provided in the book web site.

References

Allen PV, Dukes GR, Gerger ME (1991) Determination of release limits: a general methodology. Pharm Res 8(9):1210–1213

Asmussen S, Stroz D, LeBlond D, Stephens D (2014) Use of a two tiered degree-hour measure for the management of investigational medicinal product over-temperature excursions using mean kinetic temperature (MKT). J Validation Technol 20(1), http://www.ivtnetwork.com/article/use-two-tiered-degree-hour-measure-management-investigational-medicinal-product-over-tempera. Accessed 19 Dec 2016

Asmussen S, Stroz D, LeBlond D, Stephens D (2016) Application of a degree-hour approach for the management of over-temperature (OT) excursions for investigational medicinal products (IMPs) and non-investigational medicinal products (NIMPs), submitted for publication

Bancroft TA (1964) Analysis and inference for incompletely specified models involving the use of preliminary test(s) of significance. Biometrics 20:427–442

Box GEP, Hunter WG, Hunter JS (1978) Statistics for experimenters. Wiley, New York

Brownlee KA (1984) Statistical theory and methodology. Robert E. Keiger, Malabar

Carstensen JT (1995) Drug stability: principles and practices, Chapters 2-10. Marcel Decker, Inc., New York

Chen JJ, Hwang JS, Tsong Y (1995) Estimation of the shelf-life of drugs with mixed effects models. J Biopharm Stat 5(1):131–140

Chen JJ, Ahn H, Tsong Y (1997) Shelf life estimation for multifactor stability studies. Drug Inf J 31:573–587

Chow SC, Liu JP (1995) Statistical design and analysis in pharmaceutical science. Marcel Dekker, Inc., New York

Chow SC, Shao J (1991) Estimating drug shelf-life with random batches. Biometrics 47:1071–1079

Fairweather WR, Lin TYD, Kelly R (1995) Regulatory, design, and analysis aspects of complex stability studies. J Pharm Sci 84(11):1322–1326

Gelman A, Hill J (2007) Data analysis using regression and multilevel/hierarchical models. Cambridge University, New York

Gelman A, Carlin JB, Stern HS, Rubin DB (2013) Bayesian data analysis, 3rd edn. Chapman & Hall/CRC Press, Boca Raton

Hahn GJ, Meeker WQ (1991) Statistical intervals: a guide for practitioners. Wiley, New York

International Conference on Harmonization (1995) Q5C Stability testing of biotechnological/biological products

International Conference on Harmonization (1996) Q1B Photostability testing of new drug substances and products

International Conference on Harmonization (1997) Q1C Stability testing for new dosage forms

International Conference on Harmonization (1999a) Q6A Specifications: test procedures and acceptance criteria for new drug substances and new drug products: chemical substances

International Conference on Harmonization (1999b) Q6B Specifications: test procedures and acceptance criteria for biotechnological/biological products

International Conference on Harmonization (2002) Q1D Bracketing and matrixing designs for stability testing of new drug substances and products

International Conference on Harmonization (2003a) Q1A (R2) Stability testing of new drug substances and products

International Conference on Harmonization (2003b) Q1E Evaluation for stability data

International Conference on Harmonization (2003c) Q3A Impurities in new drug substances

International Conference on Harmonization (2003d) Q3B (Revised) Impurities in new drug products

International Conference on Harmonization (2005) Q2 (R1) Validation of analytical procedures: text and methodology

Kiermeier A, Jarrett RG, Vebyla AP (2004) A new approach to estimating shelf-life. Pharm Stat 3:3–11

Kruschke JK (2015) Doing Bayesian data analysis: a tutorial with R JAGS and Stan, 2nd edn. Elsevier, New York

Lin TYD, Chen CW (2003) Overview of stability study design. J Biopharm Stat 13(3):337–354

Lunn D, Jackson C, Best N, Thomas A, Spiegelhalter D (2013) The BUGS book: a practical introduction to Bayesian analysis. CRC Press, Boca Raton. The WinBUGS package was developed jointly by the Cambridge MRC and the Biostatistics Unit at the Imperial College School of Medicine, London. The software and operators manuals may be downloaded from the following website: http://www.mrc-bsu.cam.ac.uk/bugs/welcome.shtml

Milliken GA, Johnson DE (1992) Analysis of messy data, volume III ANCOVA. Chapman & Hall/CRC, London

Neter J, Kutner MH, Nachtsheim CJ, Wasserman W (1996) Applied linear statistical models, 4th edn. Irwin/McGraw-Hill, Inc., Chicago

Ng MJ (1995) STAB stability system for SAS. Division of Biometrics, CDER, FDA. http://www.fda.gov/cder/sas/

Nordbrock E (1992) Statistical comparison of stability study designs. J Biopharm Stat 2(1):91–113

Nordbrock E (1994) Design and analysis of stability studies. ASA Proceedings of the Biopharmaceutical Section, pp 291–294

Paccagnella O (2006) Centering or not centering in multilevel models? The role of the group mean and the assessment of group effects. Eval Rev 30(1):66–85

Plummer M (2012) JAGS version 3.3.0 user manual [computer software manual]

R Core Team (2013) R: A language and environment for statistical computing [computer software manual]. R Foundation for Statistical Computing, Vienna, Austria. http://www.r-project.org/

Ruberg SJ, Stegman JW (1991) Pooling data for stability studies: testing the equality of batch degradation slopes. Biometrics 47:1059–1069

Shao J, Chow SC (1995) Statistical inference in stability analysis. Biometrics 50:753–763

Shuirmann DJ (1989) Current statistical approaches in the Center for Drug Evaluation and Research, FDA. Proceedings of Stability Guidelines, AAPS and FDA Joint Conference, Arlington, VA

Tsong Y (2003) Recent issues in stability study. J Biopharm Stat 13(3):vii–ix

USP 39-NF 34 (2016) General chapter ⟨1160⟩ pharmaceutical calculations in prescription compounding. US Pharmacopeial Convention, Rockville

Waterman KC (2011) The application of the accelerated stability assessment program (ASAP) to quality by design (QbD) for drug product stability. AAPS PharmSciTech 12(3):932–937

Waterman KC (2012) Accelerated stability assessment program (ASAP): using science to set shelf life. Pharmaceutical Outsourcing, posted March 26, 2016 at http://www.pharmoutsourcing.com/Featured-Articles/39128-Accelerated-Stability-Assessment-Program-ASAP-Using-Science-to/. Accessed 10 May 2016

Waterman KC, Adami RC (2005) Accelerated aging: prediction of stability of pharmaceuticals. Int J Pharm 293:101–125

Chapter 9
Analytical Comparability and Similarity

Keywords Accelerated stability • Analytical comparability • Analytical similarity • Biosimilar products • Comparability criterion • Equivalence testing • Non-profile data • Power calculations • Profile data • Scale comparisons • Stability data • Tolerance intervals

9.1 Introduction

In all manufacturing settings, there is an inherent drive to improve product through the reduction in process variation, implementing new technology, increasing efficiency, optimizing resources, and improving customer experience through innovation. In the pharmaceutical industry, these improvements come with added responsibility to the patient such that product made under the post-improvement or post-change condition maintains the safety and efficacy of the pre-change product. As described in FDA comparability guidance (1996) and ICH Q5E (2004), regulatory agencies also recognize the importance in providing manufacturers the flexibility to improve their manufacturing processes. Agencies also acknowledge that some changes may not require additional clinical studies to demonstrate safety and efficacy so that implementation may be more efficient and expeditious to benefit patients. Activities performed when changes are made to the process include demonstration of comparability in product parameters. The actual timing of each activity and the statistical rigor required for the evaluation of pre- and post-change product is linked to the stage of the product development (e.g., clinical versus commercial material) and the scope of the change (e.g., process transfer with similar scale versus a new cell line or formulation).

To set the stage for this chapter, the requirements of comparing pre- and post-change product are reviewed. Comparability is defined by ICH Q5E as a demonstration that the quality attributes of the pre- and post-change product are highly similar and that the existing knowledge is sufficiently predictive to ensure that any differences in quality attributes have no adverse impact on safety or efficacy of the drug product. Guidance provides the manufacturer with flexibility to adjust study rigor based on the stage of development and prior knowledge.

R.K. Burdick et al., *Statistical Applications for Chemistry, Manufacturing and Controls (CMC) in the Pharmaceutical Industry*, Statistics for Biology and Health, DOI 10.1007/978-3-319-50186-4_9

9.2 Statistical Methods for Comparability

The FDA comparability guidance (1996) recognized the need for manufacturers to improve manufacturing processes and analytical methods without performing additional clinical studies to demonstrate product safety and efficacy. This guidance was extended in ICH Q5E to provide additional direction for comparing pre- and post-change manufacturing processes. The direction is related to the scope of the comparability exercise and the type of change under consideration. Major process changes should consider a larger array of testing than those of lesser scope. For example, a change of major scope might reasonably need to consider additional pharmacokinetic (PK) or clinical studies, whereas a change with lesser scope may rely only on analytical comparability for a set of critical quality attributes (e.g., biological activity, purity, and protein structure).

Although product comparability guidance does not cover the comparison of in-process parameters given similar process changes (e.g., site transfers, scale changes, and equipment improvements), these issues are addressed in FDA guidance (2011) and were discussed in Chaps. 3–5 of this book.

Across the regulatory documents, there are only high level recommendations for the design of a comparability study and for setting acceptance criteria to assess the impact of the change. These documents do not contain prescriptive rules for setting acceptance criteria, study design, or statistical methods for analysis. This chapter provides examples of how these issues might be addressed. The study design and statistics associated with clinical, PK, and animal studies are out of scope.

The design and scope of an analytical comparability study will vary depending on the product and process complexity, complexity of the change, and the stage of the clinical/commercial life cycle. The analytical methods used for analytical comparability minimally include lot release. In addition, non-routine methods may be used to further understand the impact of the change on the biochemical, biophysical, and biological properties of the product. Comparison of degradation rates and degradation profiles from select analytical methods may also enhance the understanding of the change on key product degradants. Typically, the conditions considered for evaluating degradation rates are harsher than recommended storage conditions. By design, the degradation observed for a product at recommended storage conditions is small. Given the short period of time typically available for implementing a change, evaluation of degradation rates under recommended storage conditions provides only minimal insight into how a post-change molecule degrades over time. Instead, the pre- and post-change products are held at stressed stability conditions. These conditions may be used to detect potential impurities and structural modifications not otherwise detected by lot release and in-process control testing of non-degraded material. Analytical procedures used during the assessment of drug substance and drug product comparability should be validated or qualified as appropriate for their intended purpose (refer to Chap. 6 for more on this topic).

Chatfield et al. (2011) provide a nice description of statistical techniques that are useful for demonstrating comparability. They differentiate between the statistical

Table 9.1 A summary of the comparability approaches

Comparability approach	Lot release	Stability at recommended storage conditions	Stability at stressed storage conditions	Characterization methods
Comparison of individual values				
Visual comparisons	X	X	X	X
Tolerance intervals	X			X
Specifications	X	X		
Limit evaluations	X	X		X
Comparison of summary measures				
Equivalence testing	X		X	X

equivalence tests described in Chap. 2 and other comparability approaches. Statistical equivalence testing provides a formal statistical approach in which statistical decision errors can be controlled. Equivalence testing is particularly desirable when

1. Summary measures such as means and slopes are relevant to the process change
2. Acceptance criteria that considers scientific importance can be defined a priori, and
3. Data are amenable to the statistical requirements of equivalence testing.

If these conditions are satisfied, then equivalence testing provides the strongest scientific evidence of comparability. Attributes that are not amenable to equivalence testing can be evaluated with alternative comparability approaches including graphical summaries and comparison of individual values to pre-specified ranges.

Table 9.1 presents typical comparability approaches categorized by application. Tolerance intervals and equivalence testing are discussed in Chap. 2 and will be demonstrated with examples in this chapter.

Once an approach has been selected, the comparability study is designed. Since equivalence testing involves a statistical test, concepts typically associated with hypothesis testing such as error rate and statistical power are employed to select an appropriate study design. With the other approaches, the design will be determined by availability of relevant pre-change data and heuristic rules used for defining comparability. The final step is to collect the post-change data and compare it against the pre-stated criteria. If all of the criteria are met, one declares the product manufactured under the post-change process is comparable to the product manufactured under the pre-change process. When one or more acceptance criteria are not met, further investigation is required to determine if the pre- and post-change product is comparable. This could include further characterization, analytical method improvement, or the performance of additional nonclinical or clinical studies.

9.2.1 Lot Release

Minimally, results from the post-change process are compared to the currently approved lot release specification limits regardless of where the molecule is in its development life cycle. As the complexity of the change increases, additional assessment criteria may be required.

It may be desirable to retest pre-change lots at the same time that post-change lots are tested. By recording these measurements in the same analytical run, uncertainty related to precision and changes to analytical methods over time will be mitigated. In order to use such a design, it must be assumed that product stability (freeze-thaw or storage stability) has been demonstrated to be negligible.

Lot release tests are implemented early in clinical drug development to assess safety and efficacy prior to product release. A subset of lot release tests may be selected for comparability assessment. Typically these methods provide a quantitative assessment of critical product quality attributes. As product progresses through clinical development, a data set of analytical test results is accumulated. These data allow an ongoing assessment of patient exposure to levels of product quality attributes that may vary from lot-to-lot. Specifications and comparability assessment criteria may be adjusted using these data during clinical development as patient exposure experience is gained. These limits should factor in the ranges of analytical test data as well as the statistical and operational components that influence the variability of the analytical method (determined during method validation) and the variability of the process (determined during process characterization).

9.2.2 Stability at Recommended and Stressed Storage Conditions

Stability at recommended storage condition is typically assessed by comparing post-change stability results to pre-change stability data. The appropriate stability indicating assays are identified and implemented as part of the normal GMP stability program during clinical and commercial development. As product progresses through clinical development, a data set of stability test results is accumulated. These data allow an ongoing assessment of patient exposure to drug substance/drug product stability profiles that may vary due to manufacturing and formulation variability. Generally speaking, there is typically little to no degradation of product observed under recommended storage conditions. Appropriate recommended storage stability comparability limits can be set using the specification, visual assessments (chromatographic overlays), or limit tests. Because the degradation profile estimated from the post-change data at recommended storage is generally not extrapolated to the established expiry, care must be taken in setting the acceptance criterion if an equivalence test is performed. This is because the pre-change product will have a slope which is estimated across the entire range of shelf life whereas the post-change product will have limited information to estimate the slope. This difference in range causes the variances for the two slope estimates to differ, even if the processes are identical.

Stability at a stressed storage condition is also assessed by comparing post-change stability results to the pre-change stability data. The stressed stability conditions may be conducted under elevated temperatures or other stressed conditions such as chemically induced oxidation. The selection of the stressed condition will be based on the primary degradation pathway. In some cases one or more analytical methods may detect degradation of the product. For stressed stability studies, it is desired to compare the slope of the post-change data to the slope of the pre-change data. Since the comparison of interest concerns the summary measure slope, a test of equivalence is an appropriate choice to assess comparability. Such a test is demonstrated in Sect. 9.4.2. It may also be appropriate to include a visual assessment such as chromatographic overlays at specified time points. Appropriate accelerated storage stability comparability criteria can be established and adjusted during clinical development as patient exposure experience is gained.

9.2.3 Characterization Methods

Biochemical, biological, and biophysical analyses are performed on new process lots as appropriate. These lots are compared side-by-side to representative pre-change lots as well as to the current reference standard. A side-by-side study is conducted when an analytical method is not used routinely. A predetermined number of batches are collected from both the pre-change and post-change process and placed on the assay at the same point in time. This way, any differences associated with the analytical method will not manifest itself as differences between the two processes.

9.3 Comparability Examples for Individual Post-change Values

In this section, several examples are provided where criteria for post-change individual values are represented as ranges based on pre-change expectations. Most typically, comparability is demonstrated if a defined percentage of the post-change individual values fall within these ranges. The range criteria are computed with pre-change data using tolerance intervals. In some cases, specification limits or an LOQ may be appropriate for defining such criteria.

The following examples demonstrate how to compute prediction and tolerance intervals for several types of data structures. Chapter 2 provides the formulas that are demonstrated in this section. The three-step comparability approach is as follows:

1. Plot the data and visually compare the two groups.
2. Compute a tolerance interval using pre-change data.
3. Assess the post-change data by determining the percentage of post-change values that fall in the tolerance interval.

Demonstration of comparability requires a pre-specified proportion of post-change observations falling within the tolerance interval. Therefore, the width of the computed interval is a key component in setting the interval-based acceptance criteria. Confidence levels and proportions contained in a tolerance interval are often based on the amount of both pre- and post-change data. Dong et al. (2015b) offer considerations when using tolerance intervals to define the quality of a pre-change process. In general, if the pre- and post-change data sets are small, confidence intervals and coverage proportions must be reduced to provide meaningful intervals. Use of 99% confidence or 99% coverage with small data sets will result in intervals that are too wide to be useful in assessing comparability. In such cases, specifications or other limit evaluations may be required to serve as criteria. The pre-change data used to compute tolerance intervals must be assessed against the statistical assumption of normality as described in Chap. 2.

9.3.1 Combining Pre-change Data Sets at Different Scales

This example considers a process transfer where pre-change data are available from two manufacturing scales: a clinical scale and a commercial scale from a licensed facility. The process in the licensed facility is to be transferred to a different commercial facility at the same commercial scale (i.e., the post-change facility). The parameter of interest is an in-process control parameter that measures yield in kilograms with a specification of 40.8–75.0 kg. Figure 9.1 presents a plot of the pre-change data. The $n_1 = 5$ lots on the left are from the clinical scale, and the $n_2 = 5$ lots on the right are from the commercial scale. It is clear from the plot that the yields differ between the clinical and commercial scale processes. The spread in the data for the clinical and commercial scale appears similar.

These data are now combined to construct a tolerance interval to provide a comparability criterion. Yields from the post-change process will be expected to fall in this range. Since the spreads of the two scales in Fig. 9.1 are comparable, it is desired to pool (combine) the two data sets for estimating the pre-change variance. Since the commercial scale best represents the expected average of the post-change facility, it is desired to center a tolerance interval on the commercial scale average.

The tolerance interval formula in Eq. (2.23) can be used to compute the desired interval with some slight modifications. In particular, $\bar{Y}$ now represents the sample mean of the commercial scale, and S^2 is replaced with the pooled variance estimate

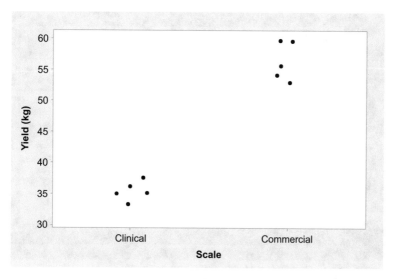

Fig. 9.1 Clinical and commercial scale pre-change data

Table 9.2 Values required to compute tolerance interval

Statistic description	Notation	Example values
Sample mean of commercial scale	$\bar{Y}$	56.36
Pooled variance	S_P^2 in (2.56)	6.21
Error degrees of freedom	$\nu = n_1 + n_2 - 2$	8
Effective sample size	$n_e = n_2$	5
Confidence level	$(1 - \alpha)$	0.95
Proportion contained	P	0.99

of the two scales. This pooled variance is denoted S_P^2 and is defined in Eq. (2.56). The value of K is defined in Eq. (2.25) with $n_e = n_2 = 5$ and $\nu = n_1 + n_2 - 2 = 5 + 5 - 2 = 8$. The required values to compute a two-sided 95% tolerance interval with 99% coverage are shown in Table 9.2.

The computed interval using (2.23) with K defined in (2.25) is

$$L = \bar{Y} - K\sqrt{S^2}$$
$$U = \bar{Y} + K\sqrt{S^2}$$

$$K = \sqrt{\frac{\left(1 + \frac{1}{n_e}\right) Z_{(1+P)/2}^2 \times \nu}{\chi_{\alpha:\nu}^2}} = \sqrt{\frac{\left(1 + \frac{1}{5}\right) 2.58^2 \times 8}{2.73}} = 4.83 \qquad (9.1)$$

$$L = 56.36 - 4.83 \times \sqrt{6.21} = 44.3$$
$$U = 56.36 + 4.83 \times \sqrt{6.21} = 68.4$$

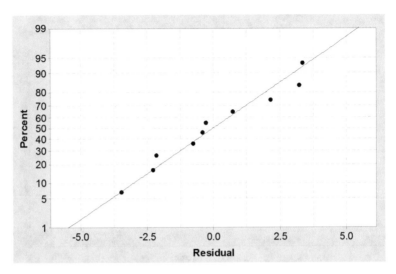

Fig. 9.2 Normal quantile plot of yield residuals

The 95% tolerance interval containing 99% of all future observation is 44.3–68.4 kg which falls within the in-process specification of 40.8–75.0 kg. The residuals formed by subtracting the appropriate scale mean from each observation are plotted in the normal quantile plot shown in Fig. 9.2. The plot suggests the normality assumption is reasonable.

The advantage of combining the two data sets to estimate the variance is seen by comparing the computed interval in (9.1) to an interval based solely on the commercial lots. This calculation is

$$L = \bar{Y} - K\sqrt{S^2}$$
$$U = \bar{Y} + K\sqrt{S^2}$$
$$K = \sqrt{\frac{\left(1 + \frac{1}{n_e}\right)Z^2_{(1+P)/2} \times \nu}{\chi^2_{\alpha:\nu}}} = \sqrt{\frac{\left(1 + \frac{1}{5}\right)2.58^2 \times 4}{0.71}} = 6.69 \qquad (9.2)$$
$$L = 56.36 - 6.69 \times \sqrt{9.90} = 35.3$$
$$U = 56.36 + 6.69 \times \sqrt{9.90} = 77.4$$

This interval is so wide that it exceeds the specification range of 40.8–75.0 kg and has no value as a comparability range.

Figure 9.3 presents the computed tolerance interval (dashed line) and the specifications (solid line) with the pre-change data used in the computations. The yields from the post-change facility are expected to fall in the tolerance interval.

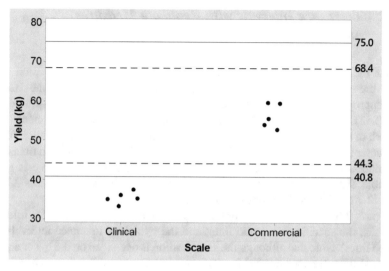

Fig. 9.3 Specifications and tolerance interval for yield data

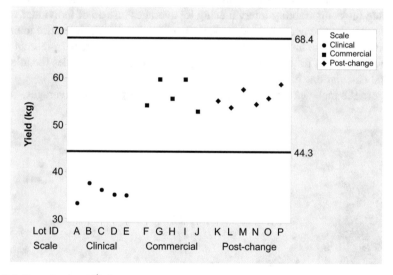

Fig. 9.4 Post-change yield data

Figure 9.4 presents the computed intervals using the pre-change data with the first six post-change yield values. Since all post-change values fall within the tolerance interval, comparability has been demonstrated.

9.3.2 Tolerance Intervals with Replicate Measures on Each Lot

Unlike the data presented in Sect. 9.3.1, this example considers a situation where there are replicate measurements taken on all or a portion of the batches. The appropriate formula for a tolerance interval for these data is provided in Eq. (2.52). Figure 9.5 presents a data set with $r = 5$ purity measurements taken on each of $a = 11$ lots in a pre-change data set. The lot release specification for this parameter is a one-sided specification of $\geq 40.0\%$. The data are plotted in time order. Calculation of the 95% tolerance interval that contains 99% coverage is shown in Table 9.3. The Hoffman and Kringle interval referenced in Sect. 2.7.4 is from 42.7 to 57.8 (calculations on spreadsheet at website).

Figure 9.6 is a plot of the pre-change data used to compute the tolerance intervals, the specification (solid line), and the two-sided tolerance interval (long dashed lines). Note that although the specification is one-sided on the lower end, the comparability tolerance interval is still two-sided. Comparability is a comparison of the two processes, apart from the specification. It is possible that two processes are not comparable, but are both capable of meeting specification.

As an alternative to the interval computed in Table 9.3, one may choose to compute the comparability interval using lot averages instead of individual values. In this case, the averages are independent across lots, and so the tolerance interval is computed using the independent formulas in Sect. 2.6.7. Figure 9.7 presents a plot of the lot averages for the data set in Fig. 9.5, and Table 9.4 provides the tolerance interval calculations based on formulas (2.23) and (2.25).

Figure 9.8 includes the tolerance intervals with the plot of lot averages.

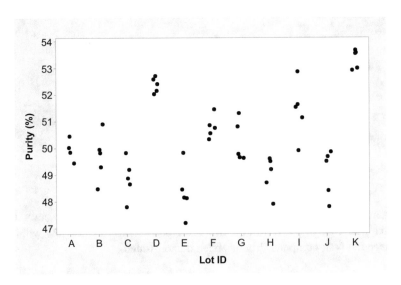

Fig. 9.5 Pre-change batch purity (%) plotted in time order

Table 9.3 Statistics needed to compute tolerance interval

Statistic	Value
Confidence level $100(1 - \alpha)\%$	95%
Proportion covered $100P \%$	99%
a	11
r	5
$\bar{Y}$	50.261
S_A^2	12.260
S_E^2	0.529
S_{Total}^2 from Eq. (2.46)	2.875
m from Eq. (2.50)	13.65
m (rounded)	14
K	3.794
$Z_{\frac{1+P}{2}}$	2.576
L from (2.52)	43.8
U from (2.52)	56.7

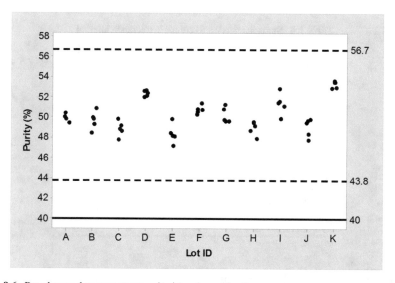

Fig. 9.6 Pre-change data acceptance criteria and specification

As expected, the tolerance interval in Fig. 9.8 is tighter than the tolerance interval in Fig. 9.6. This is because variability in lot means is smaller than variability in individual values. One may construct comparability limits in either manner as long as consistency is maintained between the limits and the post-change values begin compared (i.e., either individual values or lot averages).

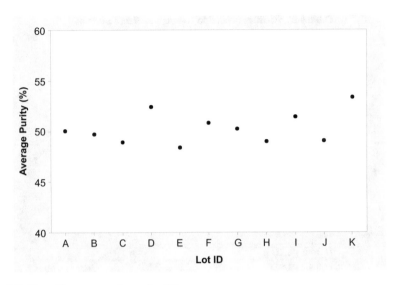

Fig. 9.7 Plot of lot averages for purity ($\%$)

Table 9.4 Statistics needed to compute tolerance interval

Statistic	Value
Confidence level $100(1-\alpha)\%$	95%
Proportion covered $100P\%$	99%
n_e	11
$\bar{Y}$	50.261
S^2	2.452
ν	10
$\chi^2_{\alpha:\nu}$	3.940
$Z_{\frac{1+P}{2}}$	2.576
K from (2.25)	4.286
L from (2.23)	43.6
U from (2.23)	57.0

9.4 Equivalence Testing for Summary Parameters

When summary parameters are informative, equivalence testing provides the strongest statistical evidence of comparability. Equivalence testing is discussed in Sect. 2.11. Data used in an equivalence test can be either profile or non-profile. Non-profile data are collected at a single point in time. Examples of non-profile data include lot release or in-process control measurements. Profile data are collected over time. In the context of comparability, a stability profile is of interest when data are collected in this manner. Typically, non-profile data involve a comparison of averages, and profile data involve a comparison of slopes.

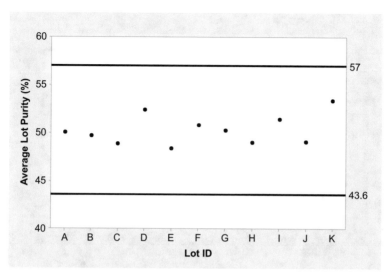

Fig. 9.8 Tolerance interval based on lot averages

As noted in Sect. 2.11, the most challenging part of an equivalence test is often establishment of the equivalence acceptance criterion (EAC). In the next two sections, guidance is offered for both non-profile and profile data.

9.4.1 Equivalence Acceptance Criterion for Non-profile Data

The equivalence hypotheses used to demonstrate comparability with non-profile data are

$$H_0 : |\mu_{Pre} - \mu_{Post}| \geq \text{EAC}$$
$$H_a : |\mu_{Pre} - \mu_{Post}| < \text{EAC}$$
(9.3)

where the subscripts denote the pre- and post-change conditions, respectively. As discussed in Sect. 2.11, equivalence is assessed by constructing a two-sided 100 $(1 - 2\alpha)\%$ confidence interval on the difference $\mu_{Pre} - \mu_{Post}$. The null hypothesis H_0 in Eq. (9.3) is rejected and equivalence is demonstrated if the entire confidence interval falls in the range from $-\text{EAC}$ to $+\text{EAC}$.

When evaluating product comparability, the EAC defines the maximum difference in means that has no practical scientific impact. It is ideal if a subject matter expert (SME) can define the EAC. In the absence of an SME definition, parameters defined by specifications or other decision making limits are used in the decision process.

To provide an example, consider a situation where the lot release specification for protein concentration is $\text{LSL} = 58.5$ mg/mL and $\text{USL} = 71.5$ mg/mL.

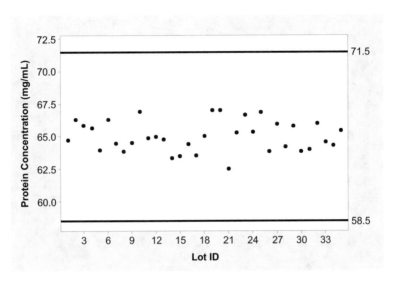

Fig. 9.9   Protein concentration (mg/mL) by lot

Table 9.5   Descriptive statistics for pre-change protein concentration

Statistic	Value (mg/mL)
Mean ($\bar{Y}$)	65.00
Standard deviation (S)	1.18
Minimum	62.47
Maximum	67.02
Lot count	35

The pre-change process data are plotted in Fig. 9.9, and the descriptive statistics are listed in Table 9.5. The data are independent with one value for each lot.

One might ask the question, "Given the pre-change process mean is 65.00 mg/mL and the standard deviation is 1.18 mg/mL, what is the maximum allowable shift in the post-change mean that would not cause an unacceptable probability for an out-of-specification (OOS) observation?" This question can be answered by using a process capability index as presented in Eq. (5.13). This capability measure is

$$\hat{C}_{pk} = \min\left[\frac{USL - \bar{Y}}{3S}, \frac{\bar{Y} - LSL}{3S}\right] \tag{9.4}$$

Using the values in Table 9.5, the computed capability measure is

$$C_{pk} = \min\left[\frac{USL - \mu_Y}{3\sigma_{st}}, \frac{\mu_Y - LSL}{3\sigma_{st}}\right]$$
$$= \min\left[\frac{71.5 - 65.00}{3 \times 1.18}, \frac{65.00 - 58.5}{3 \times 1.18}\right] \qquad (9.5)$$
$$= \min[1.84, 1.84] = 1.84$$

For this example, the two quantities within the parentheses are equal, implying that the process is centered within the specification. For this example, let's assume that a capability of 1.5 is acceptable. This corresponds to a 0.0007% chance of an individual value falling outside of the specification limits (see Montgomery 2013). The largest mean protein concentration for the post-change process that meets this requirement if the process shifts to the right is computed as follows:

$$1.5 = \frac{71.5 - \mu_{Post}}{3 \times 1.18} \qquad (9.6)$$
$$\mu_{Post} = 71.5 - 1.5 \times 3 \times 1.18 = 66.19 \text{ mg/mL}.$$

Thus, the allowable shift from the present position is computed as $66.19\text{-}65.00 = 1.19$ mg/mL, or the equivalence acceptance criterion is EAC $= 1.19$ mg/mL. Because the process is centered, a shift of the post-process change to the left would provide the same EAC.

Figure 9.10 presents a simulated post-change data set with a mean of 66.19 mg/mL and standard deviation of 1.18 mg/mL next to the pre-change data

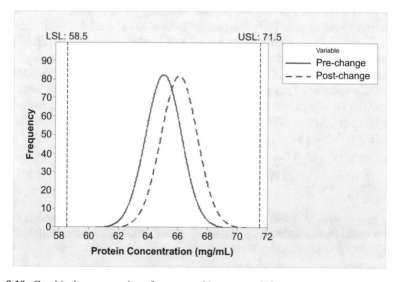

Fig. 9.10 Graphical representation of an acceptable process shift

distribution. This provides an effective representation of a process shift at the EAC. It is clear from Fig. 9.10 that the amount of tolerable shift is small in order to maintain a low percentage of observations exceeding the upper specification limit.

When no specification or SME defined EAC is available, one can define an EAC based on behavior of the pre-change process and a visual assessment as in Fig. 9.10. This approach describes the difference in means based on "expected" behavior of the pre-change process as opposed to "acceptable" behavior in terms of safety or efficacy. The notion of "expected" behavior is proposed by Hauk et al. (2008). Pre-change process behavior is described with a statistical model that incorporates the pre-change process average, lot-to-lot variation, and intermediate precision of the analytical method.

The EAC is defined as the acceptable shift in population means expressed in terms of the standard deviation of the pre-change response variance. This ratio is called the effect size (ES). The ES is defined as

$$ES = \frac{|\mu_{Pre} - \mu_{Post}|}{\sigma_{Pre}} \tag{9.7}$$

and discussed in Sect. 2.8.2. An EAC describing the pre-change process is defined as a function of an acceptable value for ES. In particular,

$$EAC = ES \times \sigma_{Pre} \tag{9.8}$$

where σ_{Pre} is estimated based on a sample of pre-change lots. In some situations, it may be reasonable to replace σ_{Pre} with an upper bound to account for sampling error. This approach has been advocated by Limentani et al. (2005) using a confidence coefficient of 80%.

An acceptable value of ES will depend on the application and rigor required to demonstrate equivalence. For example, demonstration of analytical similarity of a biosimilar may have a smaller ES compared to a demonstration of comparability for a process transfer. Selection of ES in Eq. (9.8) is aided using SMEs and visual representations. By visually representing a variety of ES values, the SME can evaluate the overlap of the pre-change data and simulated post-change data.

Figure 9.11 represents four possible values of ES with the corresponding overlapping coefficient as defined by Inman and Bradley (1989). This overlapping coefficient is defined as

$$OVL = 2 \times \Phi\left(-\frac{|\mu_{Pre} - \mu_{Post}|}{2\sigma}\right) \tag{9.9}$$

where $\sigma_{Pre} = \sigma_{Post} = \sigma$ and $\Phi(\bullet)$ is the cumulative function of a standard normal random variable. For example, if two distributions differ by one standard deviation, then

$$\text{OVL} = 2 \times \Phi\left(-\frac{|\mu_{Pre} - \mu_{Post}|}{2\sigma}\right)$$

$$= 2 \times \Phi\left(-\frac{\sigma}{2\sigma}\right) = 2 \times \Phi\left(-\frac{1}{2}\right) = 2 \times 0.309 = 0.62 \tag{9.10}$$

The pre-change population is represented by a red dashed line and the post-change population is represented by a blue solid line in Fig. 9.11. In the top-left panel, the effect size is zero which corresponds to 100% overlap between the pre- and post-change populations. As the effect size increases, the amount of overlap decreases. The most extreme case presented in Fig. 9.11 is an effect size of three. In this situation there is only 13% overlap between the pre- and post-change populations. There is an important consideration when using these plots to select an acceptable value for ES. When the mean shift is equal to the ES in each panel of Fig. 9.11, there is only a 5% chance that one will pass the statistical test of equivalence. Thus, in order to reasonably pass a test of equivalence, the true mean difference must be much less than the EAC.

Table 9.6 presents a summary table of eight pre-change process lots that are to be used to define a statistically based EAC.

The selected EAC using Eq. (9.8) is

$$\text{EAC} = 2 \times 0.367 = 0.73 \tag{9.11}$$

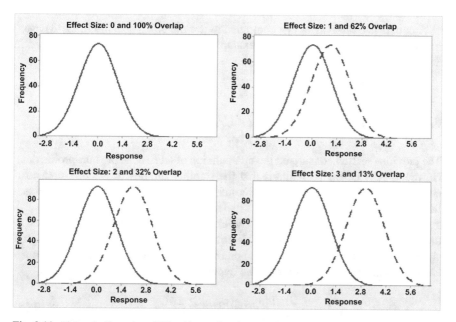

Fig. 9.11 Plots of effect sizes (ES) with overlapping pre-change (*dashed line*) and post-change (*solid line*) populations

Table 9.6 Values required to compute EAC

Description	Value
Number of pre-change lots	$n_{\text{Pre}} = 8$
Pre-change sample standard deviation	0.367
Acceptable effect size	2

Using Eq. (2.10) to incorporate an 80% upper bound on the variance, the EAC is

$$
\begin{aligned}
\text{EAC} &= \text{ES} \sqrt{\frac{(n_{\text{Pre}} - 1)S_{Pre}^2}{\chi_{\alpha:n_{\text{Pre}}-1}^2}} \\
&= 2 \times \sqrt{\frac{(8-1) \times (0.367)^2}{3.82}} = 0.99
\end{aligned}
\tag{9.12}
$$

At this point, the SME can help evaluate the reasonableness of the EAC. If there are repeated measures for each lot, one should consider working with lot averages as discussed in Sect. 9.3.2.

Another option when basing the EAC on effect size is to perform the equivalence test directly on the effect size. That is, change the hypotheses in Eq. (9.3) to

$$
\begin{aligned}
H_0 &: \frac{|\mu_{Pre} - \mu_{Post}|}{\sigma_{\text{Pre}}} \geq \text{EAC} \\
H_a &: \frac{|\mu_{Pre} - \mu_{Post}|}{\sigma_{\text{Pre}}} < \text{EAC}
\end{aligned}
\tag{9.13}
$$

A confidence interval can be computed to test the effect size in Eq. (9.13) using results presented in Sect. 2.8.2. An example of such an application is provided in Sect. 9.7.

9.4.2 Equivalence Acceptance Criterion for Profile Data

The previous section considered the computation of an EAC using non-profile data. However, ICH Q5E also requires that the stability profiles of the pre- and post-change products be highly similar. Burdick and Sidor (2013) provide an approach for defining an EAC with profile data under stressed conditions.

For stability data, the stability profile of the pre-change product is compared to the post-change product in order to determine if the degradation rates are highly similar. The hypothesis test is focused on the difference between the pre- and post-change degradation rates (slopes). The hypotheses are

$$
\begin{aligned}
H_0 &: |\beta_{Pre} - \beta_{Post}| \geq EAC \\
H_a &: |\beta_{Pre} - \beta_{Post}| < EAC
\end{aligned}
\tag{9.14}
$$

where β_{Pre} is the slope of the pre-change profile and β_{Post} is the slope of the post-change profile. As in the non-profile case, the EAC is a pre-selected constant that reflects the maximum allowable difference between two parameters such that they can be deemed equivalent. The challenge with profile data is that degradation at recommended storage conditions may be slow and differences are manifested over a very long time period. Typically, product is exposed to non-recommended storage conditions such as a higher temperature to accelerate degradation. The exposure of product to specific accelerated conditions allows the stability profiles to be compared in a more timely manner. When comparing degradation rates under either recommended storage conditions or accelerated conditions, it is assumed that reaction kinetics driving the stability properties are consistent between the pre-and post-change processes. With this assumption, the slopes can be compared using a statistical test of equivalence.

Under recommended storage conditions, the EAC may be directly linked to product safety and efficacy through the use of the product's specification. However, under non-recommended storage conditions, the linkage to specifications is not meaningful. For small molecules, it might be possible to establish EAC using Arrhenius kinetics to link acceptable degradation rates at accelerated conditions to product specifications at recommended conditions. However, such kinetics are difficult to apply to biological product degradation mechanisms, and thus is not considered a generally useful approach. Instead, with non-recommended storage conditions, the EAC can be expressed as an effect size in much the same manner described for non-profile data.

Assuming that the reaction kinetics driving the stability properties are consistent between the pre- and post-change processes, the random intercept mixed model in Eq. (2.115) is used to define the responses. The assumed model for establishing the preliminary EAC using the pre-change data when all lots are measured at the same time points is

$$Y_{ij} = \mu + L_i + \beta_{Pre} \times t_j + E_{ij}$$
$$i = 1, \ldots, n; \; j = 1, \ldots, T \tag{9.15}$$

where Y_{ij} is a response measured for lot i at time point j, μ is the average y-intercept across all pre-change lots, β_{Pre} is the average slope across all pre-change lots, L_i is a random variable that allows the y-intercept to vary from μ for a given lot, L_i has a normal distribution with mean 0 and variance σ_L^2, t_j is the time point for measurement j of each lot, E_{ij} is a random normal error term created by measurement error and model misspecification with mean 0 and variance σ_E^2, n is the number of sampled lots, T is the number of time points obtained for each pre-change lot, and L_i and E_{ij} are jointly independent.

Once the model has been fit, the EAC is computed. The methodology used to compute the EAC for profile data is similar to the concept presented in Sect. 9.4.1. For the accelerated stability model in (9.15), consider the ordinary least squares estimator of the slope based on the ith lot, $\hat{\beta}_i$. The statistical test of equivalence is now based on comparing the distribution of the $\hat{\beta}_i$ for the pre- and post-change processes. For the pre-change process, the distribution of $\hat{\beta}_i$ is normal with mean β and variance

$$Var(\hat{\beta}) = \frac{\sigma_E^2}{\text{SST}}$$

$$\text{SST} = \sum_{j=1}^{T}(t_j - \bar{t})^2 \tag{9.16}$$

$$\bar{t} = \frac{\sum_{t=1}^{T} t_j}{T}.$$

Treating the pre-change process as the reference distribution, the EAC is

$$\text{EAC} = \text{ES} \times \sqrt{\frac{\sigma_E^2}{\text{SST}}}. \tag{9.17}$$

where an estimate σ_E^2 is based on the pre-change sample. Typically, most of the variability represented by σ_E^2 is due to the analytical method error. If the analytical method is well characterized, the intermediate precision may be used to estimate σ_E^2. Should the intermediate precision not be available, σ_E^2 may be obtained from the pre-change data collected at the storage condition of interest. In some cases, it may be reasonable to use a $100(1 - \alpha)\%$ upper bound on σ_E^2.

The next step in computing (9.17) is to determine an appropriate effect size, ES. Similar to the non-profile case, it is helpful to evaluate the effect size visually. Figure 9.12 displays plots of two processes for four values of ES. The figure presents 15 randomly generated individual slope estimates from each process. The pre-change slope estimates are represented by the solid lines and the post-change process estimates by the dashed lines. All lines are emanating from the same y-intercept in order to better focus on the differences in slopes. One can see from Fig. 9.12 that an ES of three provides essentially two distinct distributions. This suggests that an ES more extreme than three might be too great a separation to declare populations comparable. Overlap of the distributions can be defined as with non-profile data, suggesting a value of ES $= 2$ is reasonable. Recall that for a given EAC, when the true difference in slopes is equal to the EAC, there is only a 5% chance of passing the equivalence test.

To demonstrate, consider a pre-change data set collected for a purity assay over a 3 month time period. Samples are held at the stressed condition of 37°C for the entirety of the study. There are $n = 15$ lots in the pre-change data set and all lots have been evaluated at 0, 1, 2, and 3 months. Figure 9.13 consists of the individual predicted slopes fit through each lot where all regression lines are emanating from the average y-intercept of 86.0% to better visualize the range of slopes for the

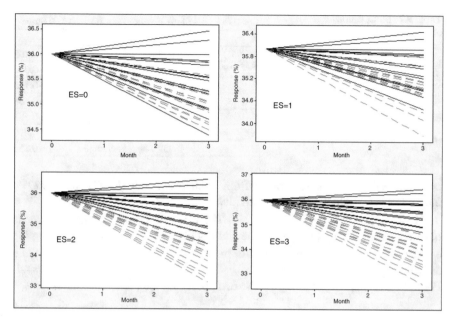

Fig. 9.12 Plots of effect sizes (ES) with stability profiles (*solid lines* are the pre-change process and *dashed lines* the post-change process)

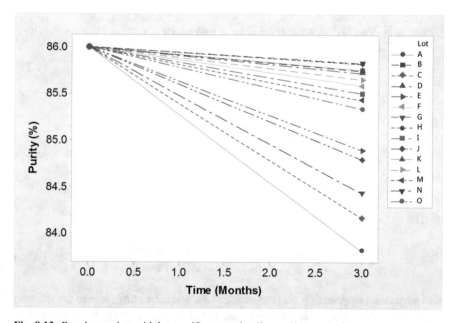

Fig. 9.13 Pre-change data with lot specific regression lines and common intercept

Table 9.7 Slope estimates
of the pre-change lots

Lot ID	Slope	Lot ID	Slope
A	−0.727	I	−0.165
B	−0.090	J	-0.402
C	−0.610	K	−0.082
D	−0.092	L	−0.115
E	−0.368	M	−0.188
F	−0.137	N	−0.521
G	−0.056	O	−0.220
H	−0.059	Average	−0.255

pre-change data. The slopes range from −0.056%/month (lot G) to −0.727%/month (lot A). Table 9.7 lists the individual slope estimates for each lot.

The average slope is $\hat{\beta}_{Pre} = -0.255$. The value for the mean squared error is obtained by regressing Purity on Time and Lot (as a random effect), and using the estimate of σ_E^2. For this example, the mean squared error is $\hat{\sigma}_E^2 = 0.200$. The associated degrees of freedom are $n_{Pre} \times (T - 1) - 1 = 15 \times (4 - 1) - 1 = 44$. Formula (9.17) is used to compute the EAC based on the pre-change data. Here

$$\bar{t} = \frac{\sum_{j=1}^{T} t_j}{T} = \frac{(0 + 1 + 2 + 3)}{4} = 1.5$$

$$SST = \sum_{j=1}^{T}\left(t_j - \bar{t}\right)^2 = (0 - 1.5)^2 + (1 - 1.5)^2 + (2 - 1.5)^2 + (3 - 1.5)^2 = 5$$

$$(9.18)$$

and so

$$\mathrm{EAC} = \mathrm{ES}\sqrt{\frac{\hat{\sigma}_E^2}{SST}}$$

$$= 2 \times \sqrt{\frac{0.200}{5}} = 0.40\% \text{ per month}$$

$$(9.19)$$

The test of the hypotheses

$$H_0 : |\beta_{Pre} - \beta_{Post}| \geq 0.40\% \text{ per month}$$
$$H_a : |\beta_{Pre} - \beta_{Post}| < 0.40\% \text{ per month}$$

$$(9.20)$$

was performed by selecting six post-change lots, and subjecting them to the stressed condition of 37°C at 0, 1, 2, and 3 months. The slopes and average for these six lots are shown in Table 9.8. The mean squared error is 0.183 with $n_{Post} \times (T - 1) - 1 = 17$ degrees of freedom.

Table 9.8 Slope estimates of the post-change lots

Lot ID	Slope
P	−0.547
Q	−0.576
R	−0.613
S	−0.460
T	−0.362
U	−0.196
Average	−0.459

Table 9.9 Summary of slopes and mean squared errors

Group	Slope average	Mean squared error	Error degrees of freedom
Pre-change	−0.255	0.200	44
Post-change	−0.459	0.183	17

By pooling the two mean squared errors in Table 9.9, the estimate of σ_E^2 is

$$\hat{\sigma}_E^2 = \frac{44 \times 0.200 + 17 \times 0.183}{61} = 0.195. \tag{9.21}$$

Because the same time points are used for each data set, $SST = 5$ for both groups, and a 90% two-sided confidence interval on the difference in slopes is

$$\hat{\beta}_{Pre} - \hat{\beta}_{Post} \pm t_{0.95:61} \sqrt{\frac{\hat{\sigma}_E^2}{SST}\left(\frac{1}{n_{Pre}} + \frac{1}{n_{Post}}\right)}$$

$$L = -0.255 - (-0.459) - (1.67)\sqrt{\frac{0.195}{5}\left(\frac{1}{15} + \frac{1}{6}\right)} = 0.045\% \text{ per month}$$

$$U = -0.255 - (-0.459) + (1.67)\sqrt{\frac{0.195}{5}\left(\frac{1}{15} + \frac{1}{6}\right)} = 0.363\% \text{ per month}$$

$$\tag{9.22}$$

As shown in Fig. 9.14, the confidence interval from 0.045 to 0.363% falls between the EAC range of −0.40 to +0.40%/month. This demonstrates equivalence of slopes.

9.5 Design and Power Considerations

Before performing any equivalence test, it is important to plan a design that has a good chance of passing when the groups are indeed equivalent. First, the collected samples for the study should be run in a random order to minimize the impact of bias. The randomization of the samples should be discussed with the laboratory

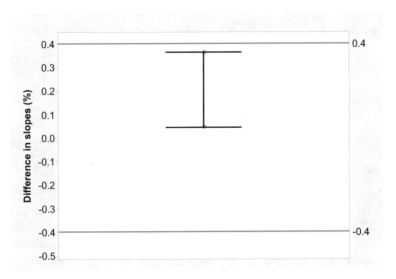

Fig. 9.14 Results of equivalence test on slopes

prior to sample collection so that any logistics with sample freezing and run order can be agreed upon. Second, the study owner should understand the format of the data set. This includes significant figures, labeling format, and administration of the data set. Data precision should align with the maximum precision allowed by the analytical method. If data are overly rounded, the parameter estimates will be over- or underestimated depending on the direction of the rounding. Section 2.3 provides more discussion on this topic.

Power is the probability of rejecting the null hypothesis for a given value of the parameter of interest. It is important to properly power a statistical test in order to ensure that equivalence can be demonstrated when it is present. Recommendations for determining the number of post-change lots in an equivalence study are discussed in the next two sections.

9.5.1   Non-profile Data

Recall the equivalence test used to demonstrate comparability with non-profile data is

$$H_0 : |\mu_{Pre} - \mu_{Post}| \geq \text{EAC}$$
$$H_a : |\mu_{Pre} - \mu_{Post}| < \text{EAC}$$

(9.23)

As discussed in Sect. 2.11, equivalence is assessed by constructing a two-sided $100(1 - 2\alpha)\%$ confidence interval on the difference $\mu_{Pre} - \mu_{Post}$. The null

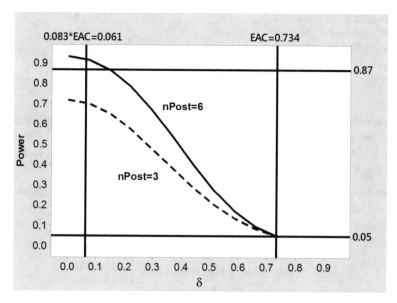

Fig. 9.15 Power curve with $\delta = |\mu_{Pre} - \mu_{Post}|$, $n_{Pre} = 8$, $\sigma = 0.367$, and $EAC = 0.734$

hypothesis H_0 in Eq. (9.23) is rejected and equivalence is demonstrated if the entire confidence interval falls in the range from $-EAC$ to $+EAC$.

Power is defined as the probability of rejecting H_0 and claiming equivalence for a given value of $|\mu_{Pre} - \mu_{Post}|$. By definition, if the value $|\mu_{Pre} - \mu_{Post}| = EAC$, then the power is equal to α. Typically, $\alpha = 0.05$ and one constructs a $100(1 - 2\alpha)\% = 90\%$ confidence interval on the difference $\mu_{Pre} - \mu_{Post}$. For values of $|\mu_{Pre} - \mu_{Post}|$ less than EAC, the power is greater than 0.05, and for values greater than EAC, it is less than 0.05. Figure 9.15 presents a power curve for the $EAC = 0.734$ with a standard deviation of 0.367.

As expected, increasing the number of post-change lots from 3 to 6 increases the power for any given value of $\delta = |\mu_{Pre} - \mu_{Post}|$. In order to determine an appropriate number of post-change lots, we recommend a power somewhere between at least 0.74–0.87 when $|\mu_{Pre} - \mu_{Post}| = 0.083 \times EAC$. This assumes one is using a two-sided 90% confidence interval to conduct the test. Applying this rule to our example, a sample size of three post-change lots is minimally sufficient and six post-change lots provides more than adequate power.

The power curve in Fig. 9.15 was computed using the SAS program PROC POWER. This code is shown below for the case where $n_{Post} = 6$ post-change lots are tested against $n_{Pre} = 8$ pre-change lots assuming $|\mu_{Pre} - \mu_{Post}| = 0.083 \times EAC = 0.061$. The computed power from this code is 0.923.

```
proc power;
twosamplemeans test=equiv_diff
lower=-0.734
```

upper=0.734
meandiff=0.061
stddev=0.367
power=.
groupns=(8 6);
run;

If software is not available to perform this calculation, one can write a simple simulation code to compute the power. Consider the same example shown in Fig. 9.15. A simulation can be constructed using Excel by following these steps and using the pooled confidence interval shown in Eq. (2.56):

1. Select values for $\delta = |\mu_{Pre} - \mu_{Post}|$, σ, n_{Pre}, and n_{Post}. For our example select $\delta = 0.061$, $\sigma = 0.367$, $n_{Pre} = 8$, and $n_{Post} = 6$.
2. Simulate a random value for $\bar{Y}_{Pre} - \bar{Y}_{Post}$ using the formula

$$\bar{Y}_{Pre} - \bar{Y}_{Post} = \mu_{Pre} - \mu_{Post} + Z \times \sqrt{\sigma^2 \left(\frac{1}{n_{Pre}} + \frac{1}{n_{Post}} \right)}$$

$$= 0.061 + Z \times \sqrt{(0.367)^2 \left(\frac{1}{8} + \frac{1}{6} \right)}$$

(9.24)

where Z is a randomly simulated standard normal random variable.

3. Simulate a random value for S_P^2 using the formula

$$S_P^2 = \frac{\sigma^2}{n_{Pre} + n_{Post} - 2} \times W = \frac{(0.367)^2}{8 + 6 - 2} \times W$$

(9.25)

where W is a chi-squared random variable with $n_{Pre} + n_{Post} - 2$ degrees of freedom.

4. Compute L and U using Eq. (2.56) to form a 90% confidence interval on $\mu_{Pre} - \mu_{Post}$.
5. If the confidence interval in step 4 falls between –EAC and +EAC, increase a counter by one, and simulate another iteration of steps 1–5. Repeat 10,000 times.

Figure 9.16 shows the first 25 rows of an Excel spreadsheet with 10,000 iterations of the simulation. (The entire spreadsheet is available at the website for this book.) Note that in Excel, W needs to be determined by first using the random uniform function since a chi-squared generator is not available. The percentage of the simulated 10,000 values that falls within the range –EAC to +EAC is 0.922. This matches to two decimal places the value computed using PROC POWER.

						Power	0.922
Delta: assumed difference in means	0.061						
Assumed SD	0.367						
Sample size for the pre change	8						
Sample size for the post change	6						
Two-sided conf level	0.9						
Two-sided t-value	1.782						
EAC	0.734						

Simulation	Z	W uniform	W chi-square	Diff sample means (9.24)	Pooled Variance (9.25)	L (2.59)	U (2.59)	CI between -EAC EAC
1	-0.357000545	0.42527543	12.2568888	-0.009758455	0.137572341	-0.366773681	0.347256771	1
2	-0.524848929	0.4741966	11.64982271	-0.04302645	0.130758581	-0.391088179	0.305035279	1
3	0.458624072	0.566148869	10.56925263	0.151900508	0.118630172	-0.179626378	0.483427394	1
4	-0.307602477	0.419446394	12.33138051	3.2373E-05	0.138408442	-0.358066096	0.358130842	1
5	-2.053966455	0.436414686	12.11592975	-0.346101601	0.135990205	-0.701057982	0.008854781	1
6	0.280442691	0.261787774	14.63877075	0.116584485	0.164306783	-0.273581064	0.506750035	1
7	0.071702289	0.836085086	7.315417391	0.075211584	0.082108854	-0.20060233	0.351025497	1
8	0.054455995	0.4242378	12.27011139	0.071793323	0.137720753	-0.285414423	0.429001069	1
9	-0.380582605	0.839381085	7.268366943	-0.014432482	0.081580756	-0.289357991	0.260493027	1
10	0.760979901	0.216803491	15.46861979	0.211828235	0.173621078	-0.189243814	0.612900285	1
11	-0.748975708	0.231269265	15.18888123	-0.087448972	0.170481269	-0.48487793	0.309979986	1
12	0.808336154	0.202185125	15.76610812	0.22121437	0.176960111	-0.183695968	0.626124708	1
13	0.223501502	0.382610553	12.81525564	0.105298591	0.143839497	-0.259758049	0.47035523	1
14	0.082598035	0.653004547	9.577125928	0.07737115	0.10749446	-0.238212296	0.392954596	1
15	0.746547357	0.633075961	9.804854988	0.208967667	0.110050509	-0.110345776	0.52828111	1
16	-1.452704055	0.140842921	17.23852588	-0.226929798	0.193486651	-0.650325727	0.196466131	1
17	1.281673576	0.308297983	13.8821385	0.315031103	0.155814279	-0.064917456	0.694979662	1
18	-0.327878524	0.129520554	17.56591448	-0.003986393	0.197161288	-0.431383913	0.423411126	1
19	-0.913512395	0.007263405	27.18644017	-0.12006058	0.30514287	-0.651768081	0.411646921	1
20	-0.400232238	0.913663137	6.047608267	-0.018327091	0.067878859	-0.269104518	0.232450335	1
21	0.827576514	0.383739738	12.80005278	0.225027861	0.143668859	-0.13981218	0.589867901	1
22	1.772223186	0.481612598	11.56022349	0.412259337	0.129752912	0.06553867	0.758980005	0
23	-0.385275598	0.936460463	5.563720062	-0.015362646	0.062447658	-0.255898184	0.225172892	1
24	0.491164656	0.007812738	26.96677776	0.158350138	0.302677361	-0.371204948	0.687905223	1
25	-0.894633558	0.942553647	5.418214126	-0.116318745	0.060814487	-0.353688126	0.121050636	1

Fig. 9.16 Simulated power spreadsheet

9.5.2 Power Considerations with Profile Data

The equivalence test used to demonstrate comparability with profile data as described in (9.14) is

$$H_0 : |\beta_{Pre} - \beta_{Post}| \geq \text{EAC}$$
$$H_a : |\beta_{Pre} - \beta_{Post}| < \text{EAC} \tag{9.26}$$

Equivalence is demonstrated if a two-sided $100(1 - 2\alpha)\%$ confidence interval on $\beta_{Pre} - \beta_{Post}$ falls in the range from $-\text{EAC}$ to $+\text{EAC}$. A simulation to determine power can be constructed using Excel by following these steps and using the pooled confidence interval shown in Eq. (9.22):

1. Select values for EAC, $|\beta_{Pre} - \beta_{Post}|$, σ_E^2, T, SST, n_{Pre}, and n_{Post}. For our example assume we select EAC = 0.40%, $|\beta_{Pre} - \beta_{Post}| = 0.083 \times \text{EAC} = 0.033$, $\sigma_E^2 = 0.20$, $T = 4$, $SST = 5$, $n_{Pre} = 15$, and $n_{Post} = 6$. Assuming we will be pooling the data to estimate error, the error df is $(n_{Pre} + n_{Post}) \times (T - 1) - 2 = 61$.

2. Simulate a random value for $\hat{\beta}_{Pre} - \hat{\beta}_{Post}$ using the formula

$$\hat{\beta}_{Pre} - \hat{\beta}_{Post} = \beta_{Pre} - \beta_{Post} + Z \times \sqrt{\frac{\sigma_E^2}{SST}\left(\frac{1}{n_{Pre}} + \frac{1}{n_{Post}}\right)}$$
$$= 0.033 + Z \times \sqrt{\frac{0.20}{5}\left(\frac{1}{15} + \frac{1}{6}\right)} \tag{9.27}$$

where Z is a randomly simulated standard normal random variable.

3. Simulate a random value for $\hat{\sigma}_E^2$ using the formula

$$\hat{\sigma}_E^2 = \frac{\sigma_E^2}{\text{Error df}} \times W$$
$$= \frac{0.20}{61} \times W$$

(9.28)

where W is a chi-squared random variable with 61 error degrees of freedom for this example.

4. Compute L and U using Eq. (9.22) to form a 90% confidence interval on $\beta_{Pre} - \beta_{Post}$.
5. If the confidence interval in step 4 falls between –EAC and +EAC, increase a counter by one, and simulate another iteration of steps 1–5. Repeat 10,000 times.

Figure 9.17 shows the first 25 rows of an Excel spreadsheet with 10,000 iterations of the simulation. The percentage of the simulated 10,000 values that falls within the range –EAC to +EAC is 0.975, which exceeds the target of 0.87.

It is important to note the impact of *SST* on power. In the previous example, time points were at 0, 1, 2, and 3 months. If the timeframe is increased, the power will increase for the same number of lots. For example, suppose we select the three time points 0, 3, and 6 months. Our range is now 6 months instead of only 3 months. Using Eq. (9.19) we have SST = 18 and given the same EAC, the power is increased to almost 1.0.

Delta: assumed difference in slopes	0.033				
Variance Sigma^2_E	0.2				
Sample size for the pre change	15				
Sample size for the post change	6				
Error df	61				
SST	5				
Two-sided conf level	0.9			Power	0.975
Two-sided t-value	1.729				
EAC	0.4				

Simulation	Z	W uniform	W chi-square	Diff sample means (9.27)	Pooled Variance (9.28)	L (9.22)	U (9.22)	CI between -EAC EAC
1	-0.357000545	0.42527543	62.4261843	-0.001489529	0.204676014	-0.170481168	0.16750211	1
2	-0.524848929	0.4741966	61.04777877	-0.017705224	0.200156652	-0.184820733	0.149410285	1
3	0.458624072	0.566148869	58.52491969	0.077307295	0.191884983	-0.086318676	0.240933265	1
4	-0.307602477	0.419446394	62.59355026	0.003282777	0.205224755	-0.165935245	0.1725008	1
5	-2.053966455	0.436414686	62.10844188	-0.165432011	0.203634236	-0.333993027	0.003129004	1
6	0.280442691	0.261787774	67.60844306	0.060093338	0.221667026	-0.115772826	0.235959502	1
7	0.071702289	0.836085086	50.2238738	0.039927099	0.164668439	-0.111651164	0.191505362	1
8	0.054455995	0.4242378	62.45591999	0.038260949	0.204773508	-0.130770933	0.207292831	1
9	-0.380582605	0.839381085	50.09376796	-0.003767773	0.164241862	-0.155149575	0.14761403	1
10	0.760979901	0.216803491	69.34148331	0.106517643	0.227349126	-0.07158829	0.284623576	1
11	-0.748975708	0.231269265	68.76101338	-0.039357928	0.225445946	-0.216716816	0.13800096	1
12	0.808336154	0.202185125	69.9547871	0.111092692	0.229359958	-0.067799153	0.289984536	1
13	0.223501502	0.382610553	63.67174131	0.054592296	0.208759808	-0.116076917	0.22526151	1
14	0.082598035	0.653004547	56.11914653	0.040979728	0.183997202	-0.119247881	0.201207337	1
15	0.746547357	0.633075961	56.67968794	0.105123327	0.185835042	-0.055902504	0.266149157	1
16	-1.452704055	0.140842921	72.93374974	-0.107344545	0.239127048	-0.290005656	0.075316566	1
17	1.281673576	0.308297983	65.99775278	0.156821431	0.216386075	-0.016937204	0.330580066	1
18	-0.327878524	0.129520554	73.5841801	0.001323925	0.241259607	-0.182149872	0.184797723	1
19	-0.913512395	0.007263405	91.27474884	-0.055253682	0.299261472	-0.25959545	0.149088086	1
20	-0.400232238	0.913663137	46.58077555	-0.005666108	0.152723854	-0.151643358	0.140311143	1
21	0.827576514	0.383739738	63.63809658	0.112951487	0.208649497	-0.057672629	0.283575603	1
22	1.772223186	0.481612598	60.84207369	0.204213026	0.199482209	0.037379308	0.371046743	1
23	-0.385275598	0.936460463	45.10331355	-0.004221159	0.147879717	-0.147864682	0.139422364	1
24	0.491164656	0.007812738	90.89488417	0.080451014	0.298016014	-0.123465099	0.284367126	1
25	-0.894633558	0.942533647	44.64791449	-0.053429813	0.146386605	-0.196346326	0.0894867	1

Fig. 9.17 Simulated power spreadsheet for profile data

9.6   Reporting Analytical Comparability Results

Once the EAC has been defined, it is time to collect the post-change data using the study design outlined in the analytical comparability protocol. Regardless of the approach, plots of the raw data and descriptive statistics should be part of the analysis. For non-profile data, plot the pre-change data and post-change data in time order along with descriptive statistics. For profile data, plot the raw data and the average slope for the pre- and post-change data and report the appropriate descriptive statistics. Additional plots and results are required if an equivalence test is performed. We now provide some examples.

9.6.1   Reports for Individual Post-change Values

When the acceptance criterion is based on pre-change data, the subsequent analysis consists of evaluating each post-change value relative to the acceptance criterion and the specification. It is recommended to plot the post-change and pre-change data by time-ordered batch ID (trend plot). This plot provides a visual assessment of any shift in the post-change mean along with changes in variability. Inclusion of reference lines for a specification or the acceptance criterion is a matter of personal preference. However, in cases where the data are far away from the specification, the excessive white space between the specification and the actual data may not be of value. In addition, the raw data are difficult to see because they are isolated to a narrow range of the y-axis. A rule of thumb for graphing data is to retain approximately one-third of the graph for white space.

In addition to the graphical presentation, descriptive statistics should be presented in the analysis. When the sample size of the post-change data set is at least four lots, the descriptive statistics (mean, standard deviation, minimum, and maximum) should be presented for both the pre- and post-change data sets. If there are fewer than four lots in the post-change data set, providing the minimum and maximum values is adequate. Table 9.10 reports descriptive statistics, types of

Table 9.10 Guidance on presentations for reporting comparability results for individual post-change values

	Item	Sample size 1–3 lots	Sample size ≥4 lots
Descriptive statistics	Mean		✓
	Standard deviation		✓
	Variance		Optional
	Range (minimum and maximum value)	✓	✓
	Confidence interval on the mean		Optional
Plots	Trend plot	✓	✓
	Boxplot		Optional
	Individual value plot		Optional

plots, and guidance on sample size for presentation. Items with a check mark are strongly recommended. Optional plots and descriptive statistics are listed as such. If there is a blank, the use of that plot or statistic is not recommended.

9.6.2   Reports for Equivalence Testing with Non-profile Data

Once the data for the post-change process have been collected, the equivalence test is performed. This test is performed by computing a two-sided 90% confidence interval on the difference in means. Equivalence is demonstrated if both bounds fall within the range from $-EAC$ to $+EAC$. The confidence intervals are computed using one of the intervals shown in Table 9.11.

Table 9.12 summarizes useful plots and descriptive statistics for reporting an equivalence test. Check marks denote recommendations and optional items are defined as such.

To demonstrate, consider a non-profile analysis of lot release data for protein concentration measured in mg/mL. There are $n_{Pre} = 35$ lots of pre-change product and $n_{Post} = 3$ lots of post-change product. The equivalence hypotheses of interest are

Table 9.11 Confidence intervals used with equivalence tests of mean

Data structure	Compute the two-sided confidence interval with equation
Independent measurements with equal variances	(2.56)
Independent measurements with unequal variances	(2.58)
Dependent measurements	(2.71)

Table 9.12 Guidance on presentations for reporting comparability results for equivalence tests with non-profile data

	Item	Guidance
Descriptive statistics	Mean	✓
	Standard deviation	✓
	Variance	Optional
	Range (minimum and maximum value)	Optional
	Confidence interval on the mean	Optional
	Confidence interval on the difference in means	✓
Plots	Trend plot	✓
	Boxplot	Optional
	Individual value plot	✓
	Equivalence plot	✓

$$H_0 : |\mu_{Pre} - \mu_{Post}| \geq 2.0 \text{ mg/mL}$$
$$H_a : |\mu_{Pre} - \mu_{Post}| < 2.0 \text{ mg/mL}. \tag{9.29}$$

Table 9.13 shows the statistics recommended in Table 9.12.

Figure 9.18 presents the trend plot recommended in Table 9.12. Figure 9.19 presents the individual value plot, and Fig. 9.20 the equivalence plot.

It is clear that there are no unexpected trends in the post-change data relative to the pre-change lots. For this example, the specification reference lines are added to the trend plot.

The plus signs in Fig. 9.19 represent the pre- and post-change process means. This plot is valuable as it gives a visual assessment of the two process means relative to each other along with the spread of the data.

Table 9.13 Recommended descriptive statistics in Table 9.12

Statistic	Protein concentration (mg/mL)
Pre-change mean	65.00
Post-change mean	65.29
Difference in means ($\bar{d}$)	−0.29
Pre-change standard deviation	1.18
Post-change standard deviation	0.48
90% margin of error (ME)	0.73
Lower bound of 90% CI on difference from (2.56)	−1.46
Upper bound of 90% CI on difference from (2.56)	0.88
EAC	2.0
Conclusion	Statistically equivalent

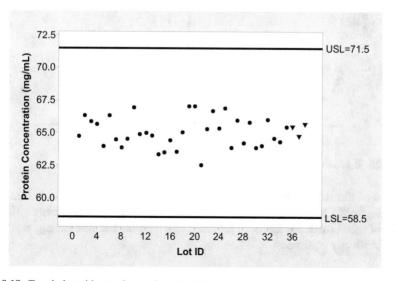

Fig. 9.18 Trend plot with pre-change data (*circle*) and post-change data (*triangle*)

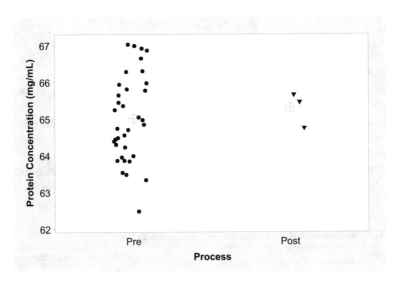

Fig. 9.19 Individual value plot recommended in Table 9.12

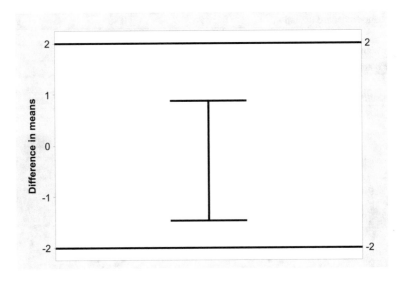

Fig. 9.20 Equivalence plot recommended in Table 9.12

The confidence interval in Fig. 9.20 assumes equal variances and is computed with Eq. (2.56). Since the confidence interval in Fig. 9.20 falls completely inside the EAC of ±2.0, evidence has been provided that the pre- and post-change processes are statistically equivalent.

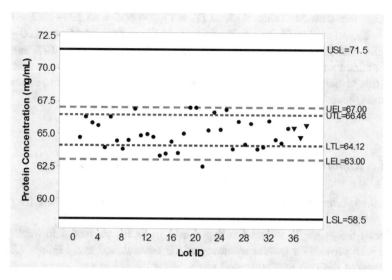

Fig. 9.21 Trend chart with equivalence test

Burdick et al. (2011) developed a trend chart that can be used to visually present the test of equivalence. This chart is shown in Fig. 9.21 for the data described in this section.

The lines in Fig. 9.21 represent algebraic re-expressions of the equivalence inequalities. In the present example, equivalence is demonstrated using Eq. (2.56) if

$$\bar{Y}_{\text{Pre}} - \bar{Y}_{Post} - ME > -EAC \text{ and}$$
$$\bar{Y}_{\text{Pre}} - \bar{Y}_{Post} + ME < EAC \text{ where}$$
$$ME = t_{1-\alpha/2:n_{\text{Pre}}+n_{Post}-2}\sqrt{S^2_{Pooled}\left(\frac{1}{n_{\text{Pre}}} + \frac{1}{n_{Post}}\right)}$$

(9.30)

Equation (9.30) can be rewritten as

$$\bar{Y}_{Post} + ME < \bar{Y}_{\text{Pre}} + EAC, \text{ and}$$
$$\bar{Y}_{Post} - ME > \bar{Y}_{\text{Pre}} - EAC$$

(9.31)

Define the lower test limit $LTL = \bar{Y}_{Post} - ME$, the upper test limit $UTL = \bar{Y}_{Post} + ME$, the lower equivalence limit $LEL = \bar{Y}_{\text{Pre}} - EAC$, and the upper equivalence limit $UEL = \bar{Y}_{\text{Pre}} + EAC$. Average equivalence is demonstrated when $UTL < UEL$ and $LTL > LEL$. Visually this translates into having both the UTL and LTL (short dashed lines) falling within UEL and LEL (long dashed lines).

Using the data in Table 9.13, $LTL = \bar{Y}_{Post} - ME = 65.29 - 1.17 = 64.12$, $UTL = \bar{Y}_{Post} + ME = 65.29 + 1.17 = 66.46$, $LEL = \bar{Y}_{Pre} - EAC = 65.00 - 2.00 = 63.00$, and $UEL = \bar{Y}_{Pre} + EAC = 65.00 + 2.00 = 67.00$. Since LTL and UTL are contained within UEL and LEL, equivalence is demonstrated. Note that when using this visualization technique, some individual values are likely to fall outside the limits, because the limits are based on means.

9.6.3 Reports for Equivalence Testing with Profile Data

Table 9.14 summarizes the recommended descriptive statistics and plots for profile data.

Recall the example to test the hypotheses in (9.20) presented in Sect. 9.4.2. Table 9.15 provides a tabular summary of the equivalence test. Figures 9.22 and 9.23 present the two recommended plots. Since the lower and upper confidence bounds fall within the range from –EAC to +EAC, the two processes are statistically equivalent.

Table 9.14 Guidance on presentations for reporting comparability results for equivalence tests with profile data

	Item	Guidance
Descriptive statistics	Slope for each process	✓
	Confidence interval on the difference in slopes	✓
	Standard deviation for each process slope	Optional
	Range (minimum and maximum slope for each process)	Optional
	Individual slopes for each lot	Optional
	Confidence interval on the slope	Optional
Plots	Regression plot	✓
	Equivalence plot	✓
	Regression plot with normalized y-intercept for each lot	Optional

Table 9.15 Tabular summary of equivalence test results

Statistic	Value
Pre-change slope	−0.255%/month
Post-change slope	−0.459%/month
Difference in slopes	0.204%/month
SST from (9.18)	5
Lower bound of 90% CI on difference from (9.22)	0.045%/month
Upper bound of 90% CI on difference from (9.22)	0.363%/month
EAC	0.40%/month
Conclusion	Statistically equivalent

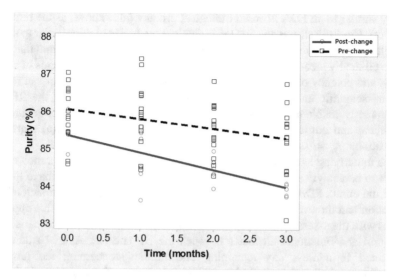

Fig. 9.22 Regression plot

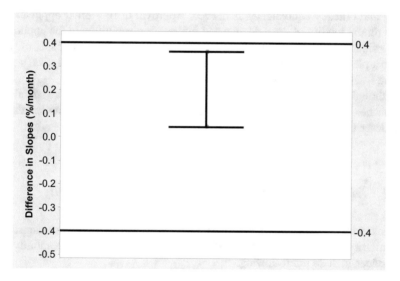

Fig. 9.23 Equivalence plot

9.7 Analytical Similarity for Biosimilar Products

In this section, we provide statistical methods for demonstrating analytical similarity between a proposed biosimilar product (BP) and its associated reference product (RP). The Biologics Price Competition and Innovation Act (BPCIA) of 2009 created an abbreviated licensure pathway for biological products shown to be highly similar to an FDA licensed biological product (also known as the reference product). Biosimilarity means that the biological product is highly similar to the RP notwithstanding minor differences in clinically inactive components and that there are no clinically meaningful differences between the BP and RP in terms of safety, purity, and potency of the product (FDA 2015b). A biosimilar sponsor can rely on existing scientific knowledge about the safety and effectiveness of the RP, and consequently enable a BP to be licensed based on less than a full complement of preclinical and clinical data typically required with a section 351(a) marketing application.

The underlying assumption of this abbreviated pathway is that if a molecule is shown to be analytically and functionally similar to an RP, it will behave like the RP in the clinic. FDA recommends that sponsors use a stepwise approach of data collection and the evaluation of residual uncertainty (FDA 2015b). This approach begins with the assessment of analytical similarity, which includes comparisons of structural and functional attributes between the BP and RP. Animal studies are conducted to address any remaining uncertainties concerning the proposed biosimilar product before initiation of clinical testing of the product in human subjects. The stepwise approach continues with clinical studies including assessment of immunogenicity and pharmacokinetics or pharmacodynamics to establish safety and efficacy equivalency as needed. Approval of biosimilar applications is based on the totality of the evidence and information submitted in the application. FDA guidance on this topic has been published including FDA (2015a, b, c). There is also a planned guidance on statistical methods to demonstrate analytical similarity due in 2016.

Although statistical approaches used to demonstrate similarity are generally the approaches used for comparability, there are some important differences that are now described.

9.7.1 Differences Between Comparability and Similarity

Burdick et al. (2016) have described several features that distinguish demonstration of similarity between an RP and a BP from an assessment of comparability. Key differences include the following:

1. Lack of RP knowledge in a similarity assessment relative to knowledge concerning the pre-change process in a comparability assessment. Lack of RP product knowledge includes such items as

   a. RP process changes which may make pooling of data inappropriate for statistical analysis.
   b. RP process deviations resulting in quality within permitted specifications but outside expected variability. For example, a sampled RP lot may have a measured value that is out of trend with respect to other RP values, even if

the release of the lot was justified based on impact to quality, safety, and efficacy.

   c. Linkage between drug substance (DS) and drug product (DP) lots is not identifiable from sampled RP lots in similarity assessments. If sampled DP lots were manufactured with the same DS, they are correlated, and the assumption of independence required in many statistical calculations is not appropriate.

2. RP target specifications and in-process control (IPC) limits are not known for the majority of the analytical methods in a similarity assessment. This lack of knowledge makes the selection of meaningful acceptance criteria more difficult.
3. The sampling process used to collect the RP lots has an inherent bias that leads to RP lots being generally older than newly manufactured TP lots. This bias is especially problematic for stability indicating methods.

These differences present some important limitations to the statistical methods that have been recommended for demonstration of analytical similarity. We now review a statistical approach suggested by FDA as described in an FDA ODAC briefing document (2015d), Chow (2014, 2015), Dong et al. (2015a), Dong (2015), Shen et al. (2015), Tsong (2015) and Tsong et al. (2015).

9.7.2 Risk Categories for Critical Quality Attributes

Demonstration of analytical similarity begins with the assessment of the relative criticality of quality attributes. Table 9.16 reports the three categories described by the FDA in the previously mentioned references.

Tier 1 attributes require the most statistically rigorous evidence of similarity. This evidence is provided using a statistical test of equivalence. Tier 2 quality attributes require a lesser level of statistical rigor. The recommended approach for Tier 2 attributes are quality ranges. Finally, Tier 3 attributes can be examined using graphical display. Each of these approaches is now described below.

Table 9.16 Risk categories

Risk category	Definition
Tier 1	High impact on activity, PK/PD, safety, or immunogenicity Where practical the attributes measured require a statistical test of equivalence between the proposed biosimilar product and the RP
Tier 2	Moderate impact on activity, PK/PD, safety, or immunogenicity Attributes measured are consistent with a statistical quality range
Tier 3	Low impact on activity, PK/PD, safety, or immunogenicity Descriptive raw data and graphical presentations of similarity

Table 9.17 Summary of adalimumab data

Product	Number of batches	Sample mean (%)	Sample standard deviation (%)	Min (%)	Max (%)
ABP 501(BP)	$n_B = 10$	$\bar{Y}_B = 104$	$S_B = 4.1$	98	110
US-licensed Humira (RP)	$n_R = 21$	$\bar{Y}_R = 105$	$S_R = 5.7$	95	114

9.7.3   Tier 1 Testing

Table 9.17 reports a data set from Table 6 of the FDA (2016) document presented as part of the Arthritis Advisory Committee Meeting for ABP 501, a proposed biosimilar to Humira (adalimumab). The quality attribute comes from an apoptosis inhibition bioassay and is measured in %. This assay measures the primary mechanism of action for the product.

Demonstration of statistical equivalence for a Tier 1 attribute requires testing the following set of hypotheses:

$$H_0 : |\mu_B - \mu_R| \geq 1.5\sigma_R$$
$$H_1 : |\mu_B - \mu_R| < 1.5\sigma_R \tag{9.32}$$

where μ_B is the mean of BP, μ_R is the mean of the RP, and σ_R is the standard deviation for the RP. The value of 1.5 was established by the FDA based on numerous simulation studies and is described in Shen et al. (2015).

Equivalence testing is described in Sect. 2.11, with $\mathrm{EAC} = 1.5\sigma_R$ in this application. A 90% confidence interval on the difference is computed assuming equal variances using the formula in (2.56),

$$S_P^2 = \frac{(n_B - 1)S_B^2 + (n_R - 1)S_R^2}{n_B + n_R - 2} = \frac{(10 - 1)(4.1)^2 + (21 - 1)(5.7)^2}{10 + 21 - 2} = 27.6$$

$$L = \bar{Y}_B - \bar{Y}_R - t_{1-\alpha/2:n_B+n_R-2}\sqrt{S_P^2\left(\frac{1}{n_B} + \frac{1}{n_R}\right)}$$

$$L = 104 - 105 - 1.7\sqrt{27.6\left(\frac{1}{10} + \frac{1}{21}\right)} = -4.1$$

$$U = \bar{Y}_B - \bar{Y}_R + t_{1-\alpha/2:n_B+n_R-2}\sqrt{S_P^2\left(\frac{1}{n_B} + \frac{1}{n_R}\right)}$$

$$U = 104 - 105 + 1.7\sqrt{27.6\left(\frac{1}{10} + \frac{1}{21}\right)} = 2.1$$

$$\tag{9.33}$$

(Note the reported interval in the FDA report was computed to a greater decimal precision.) The EAC is determined by replacing σ_R with S_R to yield EAC $= 1.5 \times 5.7 = 8.6$. Since the interval from -4.1 to 2.1 falls entirely within the range from -8.6 to 8.6, equivalence has been demonstrated.

One problem with this approach discussed by Burdick et al. (2016) concerns the fact that σ_R is estimated using S_R to define EAC. The consequence of estimating EAC is that the confidence interval in Eq. (9.33) does not maintain the desired probability of rejecting H_0 when H_0 is true. This increases the risk of passing the test when the BP is not equivalent to the RP. This problem can be resolved by changing the hypotheses in (9.32) to

$$
\begin{aligned}
H_0 &: \left| \frac{\mu_B - \mu_R}{\sigma_R} \right| \geq 1.5 \\
H_a &: \left| \frac{\mu_B - \mu_R}{\sigma_R} \right| < 1.5
\end{aligned}
\tag{9.34}
$$

Note that (9.34) is equivalent to (9.32). The only difference is that σ_R has moved to the left-hand side of the equation, and EAC $= 1.5$ is now a known constant. The hypotheses in (9.34) are tested by constructing a 90% confidence interval on the effect size, $\frac{\mu_B - \mu_R}{\sigma_R}$, as described in Sect. 2.8.2.

To demonstrate, the confidence interval on the effect size for the data in Table 9.17 using the procedure described in Sect. 2.8.2 yields

$$
\begin{aligned}
t_{calc} &= \frac{\bar{Y}_B - \bar{Y}_R}{\sqrt{S_P^2 \left(\frac{1}{n_B} + \frac{1}{n_R} \right)}} \\
&= \frac{104 - 105}{\sqrt{27.6 \left(\frac{1}{10} + \frac{1}{21} \right)}} = -0.5
\end{aligned}
\tag{9.35}
$$

and the resulting confidence interval is from -0.8 to 0.4. Since the entire confidence interval falls in the range from -1.5 to $+1.5$, equivalence is demonstrated.

Yang et al. (2016) have studied the impact on the recommended Tier 1 test when the RP lots are correlated. They show that when RP lots are correlated, the probability of rejecting H_0 when H_0 is true (i.e., falsely concluding equivalence) will increase, and the probability of passing when the products are equivalent will decrease. As noted in Sect. 9.7.1, linkage between drug substance (DS) and drug product (DP) is often not identifiable from sampled RP lots. If sampled RP lots were manufactured with the same DS, they are correlated, and the Tier 1 equivalence test is impacted. Yang et al. describe approaches to mitigate this problem, but sponsors of biosimilar products are cautioned to avoid correlation by selecting a small number of lots at any given time, and spreading purchase of RP lots over as long a time period as feasible.

9.7.4 Tier 2 Testing

Tier 2 testing uses a quality range approach. The quality range is defined as

$$\mu_R \pm K \times \sigma_R \tag{9.36}$$

where K is appropriately justified. In practice, μ_R is estimated with $\bar{Y}_R$ and σ_R is estimated with S_R. The biosimilar product passes Tier 2 if a predefined proportion (e.g., 90%) of the measured assay responses from the biosimilar lots falls within the quality range. Yang et al. provide results that justify the use of $K = 3$, and demonstrate that correlation among RP lots will cause the quality range to be too tight, because the lot-to-lot variation will not be fully represented.

The quality interval for the data in Table 9.17 is computed as

$$\begin{aligned} \bar{Y}_R &\pm K \times S_R \\ 105 &\pm 3 \times 5.7 \\ L &= 87.9\% \\ U &= 122.1\% \end{aligned} \tag{9.37}$$

To pass the Tier 2 test, nine of the ten biosimilar lots (90%) must fall in the range from 87.9 to 122.1%. Since the range of the biosimilar shown in Table 9.17 is from 98 to 110%, all ten lots fall in the quality range, and Tier 2 similarity is demonstrated.

More information on the strategy demonstrated in this example is presented by Velayudhan et al. (2016).

References

Burdick RK, Sidor L (2013) Establishment of an equivalence acceptance criterion for accelerated stability studies. J Biopharm Stat 23:730–743

Burdick RK, Pferdeort V, Sidor L, Tholudur A (2011) A graphical representation for a statistical test of average equivalence and variance comparison with process data. Qual Reliab Eng Int 27(6):771–780

Burdick R, Coffey T, Gutka H, Gratzl G, Conlon H, Huang C-T, Boyne M, Kuehne H (2016) Statistical approaches to assess biosimilarity from analytical data. AAPS J. doi: 10.1208/s12248-016-9968-0

Chatfield MJ, Borman PJ, Damjanov I (2011) Evaluating change during pharmaceutical product development and manufacture—comparability and equivalence. Qual Reliab Eng Int 27:629–640

Chow S-C (2014) On assessment of analytical similarity in biosimilar studies. Drug Des 3:119. doi:10.4172/2169-0138.1000e124, Accessed 28 Nov 2015

Chow S-C (2015) Challenging issues in assessing analytical similarity in biosimilar studies. Biosimilars 5:33–39

Dong X (2015) Equivalence test for biosimilar analytical assessment. Second Statistical and Data Management Approaches for Biotechnology Drug Development, USP Headquarters, Rockville

Dong X, Shen M, Tsong Y (2015a) EP2006 statistical equivalence testing for bioactivity and content. http://www.fda.gov/downloads/AdvisoryCommittees/CommitteesMeetingMaterials/Drugs/OncologicDrugsAdvisoryCommittee/UCM431118.pdf

Dong X, Shen M, Tsong Y, Zhong J (2015b) Using tolerance intervals for assessment of pharmaceutical quality. J Biopharm Stat 25:317–327

Food and Drug Administration. Center for Drugs Evaluation Research (1996) Demonstration of comparability of human biological products, including therapeutic biotechnology-derived products, guidance for industry

Food and Drug Administration. Center for Drugs Evaluation Research (2011) Process validation: general principles and practices, guidance for industry

Food and Drug Administration. Center for Drugs Evaluation Research (2015a) Quality considerations in demonstrating biosimilarity of a therapeutic protein product to a reference product, guidance for industry

Food and Drug Administration. Center for Drugs Evaluation Research (2015b) Scientific considerations in demonstrating biosimilarity to a reference product, guidance for industry

Food and Drug Administration. Center for Drugs Evaluation Research (2015c) Questions and answers regarding implementation of the biologics price competition and innovation act of 2009, guidance for industry

Food and Drug Administration. Center for Drugs Evaluation Research (2015d) FDA briefing document for the oncologic drugs advisory committee (ODAC) meeting held on January 7, 2015. http://www.fda.gov/downloads/AdvisoryCommittees/CommitteesMeetingMaterials/Drugs/OncologicDrugsAdvisoryCommittee/UCM428781.pdf

Food and Drug Administration. Center for Drugs Evaluation Research (2016) FDA briefing document, arthritis advisory committee meeting, ABP 501, a proposed biosimilar to Humira (adalimumab) by Amgen. http://www.fda.gov/downloads/advisorycommittees/committeesmeetingmaterials/drugs/arthritisadvisorycommittee/ucm510293.pdf

Hauk W, Abernethy D, Williams R (2008) Metrological approaches to setting acceptance criteria: unacceptable and unusual characteristics. J Pharm Biomed Anal 48:1042–1045

International Conference on Harmonization (2004) Q5E Comparability of biotechnological/biological products subject to changes in their manufacturing process

Inman HF, Bradley EL Jr (1989) The overlapping coefficient as a measure of agreement between probability distributions and point estimation of the overlap of two normal densities. Commun Stat Theory Methods 18(10):3851–3874

Limentani G, Ringo M, Ye F, Bergquist M, McSorley E (2005) Beyond the t-test: statistical equivalence testing. Anal Chem 77(11):221A–226A

Montgomery DC (2013) Introduction to statistical quality control, 7th edn. Wiley, New York

Shen M, Dong X, Tsong Y (2015) Equivalence margin determination for analytical biosimilar assessment, 2015 FDA Industry Statistics Workshop, September 16–18, Washington, DC

Tsong Y (2015) Statistical strategies for determining biosimilarities, 2015 Nonclinical Biostatistics Conference, October 13–15, Villanova, PA

Tsong Y, Shen M, Dong X (2015) Development of statistical approaches for analytical biosimilarity evaluation, 2015 DIA/FDA Statistical Forum, April, 2015, Rockville, MD

Velayudhan J, Chen Y-F, Rohrbach A, Pastula C, Maher G, Thomas H, Brown R, Born T (2016) Demonstration of functional similarity of proposed biosimilar ABP 501 to adalimumab. BioDrugs. doi:10.1007/s40259-016-0185-2. Published online July 15, 2016

Yang H, Novick S, Burdick R (2016) On statistical approaches for demonstrating analytical similarity in the presence of correlation. PDA J Pharm Sci Technol 70:6

Appendix: Basic Concepts of Matrix Algebra

Matrix algebra provides a shorthand for describing and implementing some of the statistical procedures described in this book. The objects and operations of matrix algebra can now be executed easily in R and Excel spreadsheet software. This section provides a basic description of these objects and operations.

Vectors and Matrices

The reader is already familiar with the idea of a variable. For instance, the relationship $x = 6$ defines a variable x as a "place-holder" for the number. In this case, the number is equal to 6. We refer to x as a scalar to indicate that the variable x can contain only a single number. There is no reason why we cannot similarly define variables as place-holders for more than one number. For instance,

$$\mathbf{x} = \begin{pmatrix} 4 \\ 5 \\ 6 \end{pmatrix}. \tag{A.1}$$

We refer to the bold-faced lower case $\mathbf{x}$ as a column vector. In this example $\mathbf{x}$ includes three rows. A column vector can contain any number of rows but always contains only a single column. Similarly, a row vector denoted as $\mathbf{x}^T$ is a single row that can have any number of columns.

We can also define place-holders with multiple rows and columns such as

$$\mathbf{X} = \begin{pmatrix} 1 & 2 & 3 \\ 7 & 8 & 9 \end{pmatrix}. \tag{A.2}$$

© Springer International Publishing AG 2017
R.K. Burdick et al., *Statistical Applications for Chemistry, Manufacturing and Controls (CMC) in the Pharmaceutical Industry*, Statistics for Biology and Health, DOI 10.1007/978-3-319-50186-4

We refer to the upper case bold-faced $\mathbf{X}$ as a matrix. A matrix is always described by the number of rows and number of columns. In this example, $\mathbf{X}$ has $r = 2$ rows and $c = 3$ columns. As such, it is described as a 2×3 matrix. A matrix can contain any number of rows and columns. As noted above, to distinguish vectors and matrices from scalar variables, we use a bold font. Vectors are shown in lower case bold and matrices in upper case bold. Note that we can also think of $\mathbf{x}$ as a 3×1 matrix. The scalar values within a matrix are referred to as matrix elements and are identified by their specific row and column positions as numbered from top to bottom and left to right, respectively (i.e., the r,c element).

In Excel, a vector or a matrix can be created simply by entering the elements into adjacent rows and columns of a spreadsheet. The vector or matrix is then referred to by the appropriate array range. For instance, the column vector in (A.1) might be referred to as A1:A3 in an Excel spreadsheet and the 2×3 matrix in (A.2) might be referred to as A1:C2. In R, the column vector in (A.1) can be created by the statement $x < -c(4, 5, 6)$, and the matrix (A.2) can be created by the statement $X < -matrix(c(1, 7, 2, 8, 3, 9), nrow = 2)$. Note that the matrix elements are listed column-wise by default in R.

Matrix Transposition

One convenient matrix (or vector) operation is transposition which is a simple interchange of the rows and columns. A transpose is indicated by placing a $\mathbf{T}$ superscript on the matrix (or vector). For instance, defining $\mathbf{x}$ as in (A.1), one can write

$$\mathbf{x}^{\mathbf{T}} = (4 \quad 5 \quad 6). \tag{A.3}$$

Similarly, defining $\mathbf{X}$ as in (A.2),

$$\mathbf{X}^{\mathbf{T}} = \begin{pmatrix} 1 & 7 \\ 2 & 8 \\ 3 & 9 \end{pmatrix}. \tag{A.4}$$

In Excel, transposition of a vector or matrix is accomplished using the matrix function TRANSPOSE(). To transpose a 2×3 matrix in the array range A1:C2, first select a 3×2 cell range elsewhere in the sheet. Next, type into the selected range the statement "=TRANSPOSE(A1:C2)" (without the quotes), and execute a control-shift-enter. In R, transposition is accomplished using the t() function. For example, the transpose of X is obtained using the statement "t(X)".

Matrix Addition and Multiplication

When two matrices (or vectors) have the same number of rows and columns, then matrix addition is allowed and performed by adding elements in the same position. The resulting matrix has the same number of rows and columns as the summed matrices. For instance, if

$$A = \begin{pmatrix} 1 & 7 \\ 2 & 8 \\ 3 & 9 \end{pmatrix},$$

$$B = \begin{pmatrix} 5 & 8 \\ 2 & 5 \\ 6 & 0 \end{pmatrix}, \text{ then} \tag{A.5}$$

$$A + B = \begin{pmatrix} 1+5 & 7+8 \\ 2+2 & 8+5 \\ 3+6 & 9+0 \end{pmatrix} = \begin{pmatrix} 6 & 15 \\ 4 & 13 \\ 9 & 9 \end{pmatrix}.$$

In Excel, two matrices, say in array ranges A1:B3 and D1:E3, respectively, are added by highlighting a 3×2 empty array range and entering the command "=A1:B3+D1:E3" followed by control-shift-enter. In R, two vectors or matrices are added by simply using the "+" sign. That is, to add the matrices A and B, simply write $A + B$.

Matrix multiplication of two matrices A and B (written as $A \times B$) is defined only if the number of columns in A equals the number of rows in B. For instance, as defined in (A.5), A and B each has three rows and two columns. Thus, $A \times B$ cannot be defined because the number of columns in A (two) is not equal to the number of rows in B (three). However, if we take the transpose of B, B^T, the transpose will have two rows and we can then define the multiplication of $A \times B^T$. In particular,

$$A = \begin{pmatrix} 1 & 7 \\ 2 & 8 \\ 3 & 9 \end{pmatrix},$$

$$B^T = \begin{pmatrix} 5 & 2 & 6 \\ 8 & 5 & 0 \end{pmatrix}$$

$$A \times B^T = \begin{pmatrix} 1 \times 5 + 7 \times 8 & 1 \times 2 + 7 \times 5 & 1 \times 6 + 7 \times 0 \\ 2 \times 5 + 8 \times 8 & 2 \times 2 + 8 \times 5 & 2 \times 6 + 8 \times 0 \\ 3 \times 5 + 9 \times 8 & 3 \times 2 + 9 \times 5 & 3 \times 6 + 9 \times 0 \end{pmatrix} \tag{A.6}$$

$$= \begin{pmatrix} 61 & 37 & 6 \\ 74 & 44 & 12 \\ 87 & 51 & 18 \end{pmatrix}.$$

A close examination of the above shows that the resulting matrix has the same number of rows as $\mathbf{A}$ and the same number of columns as $\mathbf{B}^T$. The element in row r and column c in the product matrix is the sum of the pairwise product of the rth row of $\mathbf{A}$ with the cth column of $\mathbf{B}$. Unlike scalar multiplication, it is not generally true that $\mathbf{C} \times \mathbf{D} = \mathbf{D} \times \mathbf{C}$. Thus, matrix multiplication is not commutative. Order matters!

In Excel, the product of two matrices in array ranges A1:B3 and D1:F2 is obtained by highlighting an empty 3×3 array range, entering "=MMULT(A1:B3,D1:F2)," and typing control-shift-enter. In R, the operator "%*%" is used for matrix multiplication. For example, to multiply the matrices $\mathbf{C}$ and $\mathbf{D}$, type the R code "C%*%D".

A special case of matrix multiplication occurs when the first matrix is a scalar. In this case the resulting matrix is simply the second matrix with each element multiplied by the scalar. For example,

$$\mathbf{A} = \begin{pmatrix} 1 & 7 \\ 2 & 8 \\ 3 & 9 \end{pmatrix},$$

$$3 \times \mathbf{A} = \begin{pmatrix} 3 \times 1 & 3 \times 7 \\ 3 \times 2 & 3 \times 8 \\ 3 \times 3 & 3 \times 9 \end{pmatrix} = \begin{pmatrix} 3 & 21 \\ 6 & 24 \\ 9 & 27 \end{pmatrix}. \tag{A.7}$$

In Excel, let 3 be the value in the cell D1 and let the matrix $\mathbf{A}$ reside in array range A1:B3. Then the product $3 \times \mathbf{A}$ is obtained by highlighting an empty 3×2 array range and entering "=D1*A1:B3" followed by control-shift-enter. In R, the expression "3*A" is used to obtain the product.

Matrix Inverse

A matrix with an equal number of rows and columns is called a square matrix. A square matrix with s rows and columns, say $\mathbf{E}$, sometimes has a partner, also with s rows and columns, which is indicated as $\mathbf{E}^{-1}$ and referred to as the inverse of $\mathbf{E}$. The inverse, if it exists, has the useful property that $\mathbf{E} \times \mathbf{E}^{-1} = \mathbf{E}^{-1} \times \mathbf{E} = \mathbf{I}$, where $\mathbf{I}$ is a special matrix with s rows and s columns called the identity matrix. The identity matrix is a square matrix with the value one in every diagonal position and zeros elsewhere. For example, the identity matrix with $s = 3$ is

$$\mathbf{I} = \begin{pmatrix} 1 & 0 & 0 \\ 0 & 1 & 0 \\ 0 & 0 & 1 \end{pmatrix}. \tag{A.8}$$

If the inverse of a square matrix exists, the matrix is said to be invertible. Otherwise, it is said to be singular.

In Excel, the inverse of a 3×3 square matrix in the array range A1:C3 is obtained by highlighting an empty 3×3 array range, entering "=MINVERSE(A1: C3)" followed by control-shift-enter. In R, the expression "solve(E)" is used to obtain the inverse of a square matrix **E**. Attempting to invert a singular matrix will always result in some type of error message.

Expressions with Multiple Matrix Operations

Derivation of complicated matrix algebra expressions can be challenging. However, using already-derived expressions is straightforward. Such derived expressions are used in this book when needed to provide a convenient way to implement the statistical procedures commonly used in software such as Excel or R.

To illustrate a more complex matrix algebraic expression, consider the simple linear regression model in Eq. (2.78) and rewritten below:

$$Y_i = \beta_0 + \beta_1 X_i + E_i \quad i = 1, \ldots, n. \tag{A.9}$$

Let $n = 4$ and note that the data described by (A.9) can be written as

$$\begin{bmatrix} Y_1 \\ Y_2 \\ Y_3 \\ Y_4 \end{bmatrix} = \begin{bmatrix} 1 & X_1 \\ 1 & X_2 \\ 1 & X_3 \\ 1 & X_4 \end{bmatrix} \times \begin{bmatrix} \beta_0 \\ \beta_1 \end{bmatrix} + \begin{bmatrix} E_1 \\ E_2 \\ E_3 \\ E_4 \end{bmatrix}. \tag{A.10}$$

Now using the previously described notation, this model can be simply written as

$$\mathbf{y} = \mathbf{X} \times \beta + \mathbf{e} \tag{A.11}$$

Here the $n \times 1$ column vectors **y** and **e** are the list of response variable values and unknown normal errors, respectively. The matrix **X** is a $n \times 2$ matrix with first column elements all equal 1 and with second column elements equal to the values of the predictor variable. The 2×1 column vector **β** holds the unknown regression coefficients (i.e., the intercept and slope). The intercept and slope least squares estimates are given by the matrix expression

$$\hat{\beta} = \left(\mathbf{X}^{\mathbf{T}} \times \mathbf{X} \right)^{-1} \times \mathbf{X}^{\mathbf{T}} \times \mathbf{y}. \tag{A.12}$$

Matrix expressions such as this can be evaluated simply by executing the operations much as one would do in evaluating an algebraic expression. One exception is that since multiplication is not commutative, expressions cannot be re-arranged and must be evaluated as written with operations performed hierarchically. That is, those within parentheses are computed first.

In Excel, Eq. (A.12) can be evaluated by selecting an empty 2×1 array range and entering "=MMULT(MMULT(MINVERSE(MMULT(TRANSPOSE(X),X)), TRANSPOSE(X)),y)" followed by a control-shift-enter. In the above Excel formula, the appropriate array range should be substituted for $\mathbf{X}$ and $\mathbf{y}$. In R, the expression "solve(t(X)%*%X)%*%t(X)%*%y" can be used, where $\mathbf{X}$ and $\mathbf{y}$ would be the appropriately defined matrix and vector, respectively.

Index

© Springer International Publishing AG 2017
R.K. Burdick et al., *Statistical Applications for Chemistry, Manufacturing and Controls (CMC) in the Pharmaceutical Industry*, Statistics for Biology and Health, DOI 10.1007/978-3-319-50186-4

Printed in the United States
By Bookmasters